Jean-Pierre Vigier

and the Stochastic Interpretation of Quantum Mechanics

selected and edited by Stanley Jeffers, Bo Lehnert, Nils Abramson & Lev Chebotarev

Apeiron
Montreal

Published by Apeiron
4405, rue St-Dominique
Montreal, Quebec H2W 2B2 Canada
http://redshift.vif.com

First Published 2000

Canadian Cataloguing in Publication Data

Vigier, Jean-Pierre, 1920-
Jean-Pierre Vigier and the stochastic interpretation of quantum mechanics

Includes bibliographical references.
ISBN 0-9683689-5-6

1. Vigier, Jean-Pierre, 1920- 2. Quantum theory.
3. Stochastic processes. I. Jeffers, Stanley II. Title.

QC174.17.S76V54 2000 530'.12'092 C00-900907-8

Table of Contents

Foreword

Professor Jean-Pierre Vigier is a living link to that glorious generation of physicists that included Einstein, De Broglie, Shrödinger, Pauli and others. In fact, Einstein wanted the young Vigier to be his personal assistant. Given Vigier's political positions and the onset of the Cold War, it was not possible for him to obtain a visa to go to Princeton to work with Einstein. Physics and politics have dominated Vigier's life. His philosophical approach has been consistently materialist and accordingly he has sided with Einstein against Bohr in the great disputes over the interpretation of quantum mechanics.

This volume includes a review of the de Broglie-Bohm-Vigier approach to quantum mechanics written by Lev Chebotarev. Many of the papers referenced in this review and which were authored or co-authored by Professor Vigier are reproduced. This volume also includes an extensive listing of Professor Vigier's publications.

This volume is a salute to Professor Vigier on the occasion of his 80th birthday. He has had a long and productive career which continues to this day. The Preface comprises reflections on his life compiled by one of us (S.J.) from a series of interviews with Professor Vigier in Paris during the summer of 1999.

We would like to acknowledge and thank the following for permission to re-print articles: Elsevier Science Publishers, ITPS Ltd., The American Physical Society, the Societa Italiana di Fisica and Kluwer Academic/Plenum Publishers.

Stanley Jeffers, Department of Physics and Astronomy, York University, Toronto

Bo Lehnert, Professor Emeritus, Royal Institute of Technology, Stockholm

Nils Abramson, Professor Emeritus, Royal Institute of Technology, Stockholm

Lev Chebotarev, D.Sc., Ph.D., Professor of Physics

This page is deliberately left blank.

Jean-Pierre Vigier and the Stochastic Interpretation of Quantum Mechanics

A Volume in Honour of the 80th Birthday of Jean-Pierre Vigier

Sponsored by the Royal Swedish Academy

Compiled by:

Stanley Jeffers, Department of Physics and Astronomy, York University, Toronto, Ontario, Canada

Bo Lehnert, Professor Emeritus, Royal Institute of Technology, Stockholm, Sweden

Nils Abramson, Professor Emeritus, Royal Institute of Technology, Stockholm, Sweden

Lev Chebotarev, D.Sc., Ph.D., Montreal, Quebec, Canada.

"Great Physicists Fight Great Battles"

"Great physicists fight great battles"—so wrote Professor Vigier in an essay he penned in tribute to his old friend and mentor Louis de Broglie. However, this phrase could equally well be applied to Vigier himself. He has waged a battle on two fronts—within physics and within politics. Now 80 years of age, he continues to battle.

He was born on January 16, 1920 to Henri and Françoise (née Dupuy) Vigier. He was one of three brothers, Phillipe (deceased) and François, currently Professor of Architecture at Harvard University. His father was Professor of English at the École Normale Supérieure—hence Vigier's mastery of that language. He attended an international school in Geneva at the time of the Spanish Civil War. This event aroused his intense interest in politics, as most of his school friends were both Spanish and Republicans. At the age of 14 he dreamt of going to Spain to help with the Republican cause. While still a teenager, he discovered the works of Marx and Engels and welcomed the victory of the Popular Front in France in 1936. He felt acutely at this time that Europe was heading towards a major conflagration as Hitler developed his plans for European domination. He recalls vividly the treason of Doddier and Chamberlain in the notorious Munich agreement. At the French International Exposition held in Paris in 1936, the German and Russian pavilions were arranged opposite each other, and the sense of impending war was in the air.

Vigier was intensely interested in both physics and mathematics, and was sent by his parents to Paris in 1938 to study both subjects. For Vigier, mathematics is more like an abstract game, his primary interest being in physics as it rests on two legs, the empirical and the theoretical. At the start of the war, it was clear to the young Vigier that large segments of the ruling class in several European countries including England, France and Italy actually sympathized with the Nazi programme. The French army soon fell apart due to a leadership which was not terribly interested in confronting the German army. The French political leadership now comprised open sympathizers to the German cause, such as Marshall Pétain.

All the young soldiers were sent to Les Chantiers de la Jeunesse, and it was here that he joined the Communist Party. The young radicals were involved in acts of sabotage near the Spanish border, such as oiling the highways to impede the progress of the fascists. At this time the French Communist Party was deeply split concerning the level of support to be given to the Resistance. A few leaders went to the Resistance immediately while others, like Thorez, wavered. In the period before the Nazi attack on the Soviet Union, the party equivocated with respect to the Resistance. At this time Vigier was in a part of France controlled by a famous communist leader, Tillion who had participated in the revolt of the sailors in the Black Sea in 1918. Tillion immediately organized groups of resistance fighters called the Organisation Spéciale. Vigier was involved in bombing campaigns against both the Nazis and Vichy collaborators in the Free Zone.

He was able to travel relatively freely within Europe as his parents were now retired and living in Switzerland. This meant that Vigier could travel on Swiss documents and transport material of the Third International from the French Communist Party to the Soviet Union. In Russia he met with a group of German communists dubbed the Red Orchestra, a group including Hadow and Vigier's future mother-in-law Rachel Dubendorfer. He met his future wife, Tamara, in a communist group in Geneva. Now divorced, they had two children, both girls, Maya and Corne. He has since re-married to Andrée Jallon, with whom he has a son, Adrien. In Geneva, Vigier was involved in communicating between the French communist military staff and Russia until he was arrested at the French border in the spring of 1942 and taken to Vichy. Here the French police interrogated him, as he was carrying coded documents. Two policemen took him by train from Vichy to Lyon to be delivered into the hands of the notorious Klaus Barbie. Fortunately, the train was bombed by the English. Vigier managed to jump through the window, escaped to the mountains and resumed his activities with the Resistance until the end of the war. He became an officer in the FTP movement. When De Gaulle returned to France, part of the Resistance forces were converted to regular army units. The famous Communist officer Colonel Fabien, the first man to kill an enemy officer, headed one. He himself was killed by a landmine explosion at the time the French army went over the Rhine. Vigier was part of the French forces which crossed the Rhine near Alsace in the Spring of 1945 almost at the same time as the American forces. Part of the French army comprised former communist Resistance forces, and they faced an army across the Rhine that comprised French Vichy collaborators. During this action Vigier was shot and sent back to Paris for recovery.

The communist forces were very proud of the role they had played during the war and at the time of Liberation. They supported Russia unconditionally, not knowing anything of the Gulags and believing much of the propaganda from Russia. The French government after the war had significant communist representation. The Cold War started almost immediately after the defeat of the Germans. Vigier was still a member of the French General Staff while completing the requirements for a Ph.D. in Mathematics in Geneva. Then the communists were kicked out of the General Staff and Vigier went to work for Joliot-Curie. He in turn lost his job for refusing to build an atomic bomb for the French Government.

Vigier became unemployed for a while, but learned through an accidental meeting with Joliot-Curie that Louis de Broglie was looking for an assistant. When he met de Broglie the only questions asked were "Do you have a Ph.D. in Mathematics?" and "Do you want to do physics?". He was hired in 1948 immediately, with no questions asked about his political views. Although Secretary of the French Academy of Science, de Broglie was marginalized within physics circles given his well-known opposition to the Copenhagen Interpretation of Quantum Mechanics. Notwithstanding his Nobel Prize, de Broglie had difficulty in finding an assistant. Vigier entered the CNRS and worked with de Broglie until his retirement. Vigier's political involvement at this time included responsibility for the French Communist Student movement.

In 1952 a visiting American physicist named Yevick, gave a seminar at the CNRS on the recent ideas of David Bohm. Vigier reports that upon hearing this work de Broglie became radiant and commented that these ideas were first considered by himself a long time ago. Bohm had gone beyond de Broglie's original ideas, however. de Broglie charged Vigier with reading all of Bohm's works in order to prepare a seminar. As a result, de Broglie returned to his old ideas and both he and Vigier started working on the causal interpretation of quantum mechanics. At the 1927 Solvay Congress de Broglie had been shouted down, but now due to the work of Bohm there was renewed interest in his idea that wave and particle could co-exist, eliminating the need for dualism.

Vigier recalls that at this time the Catholic Archbishop of Paris, who exclaimed that everyone knew that Bohr was right, upbraided de Broglie demanding to know how de Broglie could believe otherwise. Although a devout Christian, he was inclined to materialist philosophy in matters of physics. Vigier comments on his days with de Broglie that he was a very timid man who would meticulously prepare his lectures in written form; in fact his books are largely compendia of his lectures. He recalls one particular incident, which illustrates de Broglie's commitment to physics. Vigier was in the habit of meeting with de Broglie weekly to receive instructions as to what papers he should be reading and what calculations he should be focussing on. On one of these occasions he was waiting in an anteroom for his appointment with de Broglie. Also waiting was none other than the French Prime Minister, Edgar Faure, who had come on a courtesy visit in order to discuss his possible membership in the French Academy. When the door finally opened, de Broglie called excitedly for Monsieur Vigier to enter as he had some important calculations for him to do; as for the Prime Minister—he could come back next week! For de Broglie, physics took precedence over politicians, no matter how exalted.

de Broglie sent Vigier to Brazil to spend a year working with David Bohm on the renewed causal interpretation of quantum mechanics. Thereafter, Yukawa got in touch with de Broglie, and Vigier subsequently went to Japan for a year to work with him. Vigier comments that about the only point of disagreement between him and de Broglie was over non-locality. de Broglie never accepted the reality of non-local interactions, whereas Vigier himself accepts the results of experiments such as Aspect's, which clearly imply that such interactions exist.

Looking back on his political commitments, he now regards the October Revolution in Russia as an historical accident. He credits Stalin as a primary instigator of the Cold War along with Truman. He views the former Soviet society as the only third world country that became a world power under communism. The former Soviet regime is now regarded as having some decidedly negative aspects, such as the intervention in Czechoslovakia, but also some worthy aspects such as the support given to Third World countries such as Cuba. Professor Vigier has known personally some of the world leaders such as Fidel Castro and Ho Chi Min.

He still regards himself as a communist, but not a member of any organised group. His response to the question "why do we do science?" is that in part it is to satisfy curiosity about the workings of nature, but also to contribute to the liberation of humanity from the necessity of industrial labour. With characteristic optimism, he regards the new revolution of digital technology as enhancing the prospects for a society based on the principles enunciated by Marx: a society whose members are freed from the necessity of arduous labour as a result of the application of technological advances made possible by science.

Stanley Jeffers

The de Broglie–Bohm–Vigier approach in quantum mechanics

Basic concepts of the Causal Stochastic Interpretation of Quantum Mechanics

L. V. Chebotarev
(Montreal, Canada)

1. Introduction

Quantum mechanics is one of the most beautiful physical theories developed in the 20th century, if not the most elegant. As a mathematical tool, it works impeccably—all of the quantitative results established by quantum mechanics have so far been confirmed by experiment, sometimes with unprecedented accuracy, and not a single experimental fact is available at present that is contrary to the predictions of quantum mechanics.

At the same time, the conceptual situation in quantum mechanics appears to be the most disturbing in modern physics. In the 70 years following the advent of quantum theory, in spite of considerable effort, it has not been possible to achieve a satisfactory understanding of the fundamental physics underlying the mathematical scheme of quantum mechanics. There is still no clear idea as to what its mathematics is actually telling us. Nor is there a satisfactory answer to the question of the physical nature of the wave function; in fact, we are still far from a clear understanding of what it actually describes.

The conventionally accepted (yet rather formal and incomplete) interpretation of quantum mechanics, as formulated by the Copenhagen and Göttingen schools led by Bohr, Heisenberg, Born, and Pauli was strongly opposed by Einstein, Planck, Schrödinger, and de Broglie. The disagreement over the fundamental concepts of quantum mechanics, often referred to as the Bohr-Einstein controversy, appeared as early as the beginnings of quantum physics, at Solvay conferences in 1927, and it is still not resolved. Moreover, as Heisenberg noted, "the paradoxes of quantum theory did not disappear during this process of clarification; on the contrary, they became even more marked and more exciting."

The basic views shared by Planck, Einstein, Schrödinger, and de Broglie rest upon "the idea of an objective real world whose smallest parts exist objectively in the same sense as stones or trees exist, independently of whether or not we observe them" (according to Heisenberg). The development of this idea gave rise to the Stochastic Interpretation of Quantum Mechanics, also referred to as the *de Broglie–Bohm–Vigier approach*, which, to an appreciable extent, owes its conceptual coherence to the works of Prof. J. P. Vigier and his co-workers.

2. Origins of the idea and its development

2.1 The basic idea of the de Broglie–Bohm–Vigier approach

The central idea of the Stochastic Interpretation of Quantum Mechanics consists in treating a microscopic object exhibiting a dual wave-particle nature as composed of a particle in the proper sense of the word (a small region in space with a high concentration of energy), and of an associated wave that guides the particle's motion. Both the particle and the wave are considered to be real, physically observable, and objectively existing entities.

The following analysis will trace the origins of the idea and its development.

2.2 A. Einstein

The above-mentioned interpretation of quantum mechanics goes back to Einstein's discovery of the wave-particle duality for photons:

> "When a ray of light expands starting from a point, the energy does not distribute on ever increasing volumes, but remains constituted of a finite number of energy quanta localized in space and moving without subdividing themselves, and unable to be absorbed or emitted partially" [1].

In other words, Einstein regarded the radiation field as constituted of indivisible particles (carrying energy and momentum) along with the field of an accompanying electromagnetic wave. The latter was considered by Einstein to be an 'empty wave', that is, a wave propagating in space and time but devoid of energy-momentum. In 1909, Einstein proposed to treat light quanta as singularities surrounded and guided in their motion by a continuous wave phenomenon.

The difficulties with the concept of empty waves (if the wave is empty, then it cannot produce any physical changes, so there is no way to observe it—then how should we speak about its existence?) so annoyed Einstein (he called them *Gespensterfelder* - "ghost fields") that he wrote:

> "I must look like an ostrich hiding always his head in the relativistic sand for not having to face the ugly quanta." [Quoted by Louis de Broglie [2].]

2.3 L. de Broglie

Einstein's idea of duality was extended by de Broglie to electrons and other particles with non-zero rest mass [3].

In contrast to Einstein, de Broglie considered the particle and its associated wave to be both real, *i.e.*, both existing objectively in space and time:

> "For me a particle is a very small object which is constantly localized in space and a wave is a *physical process* [italics by *L.Ch.*] which propagates in space" [4].

Later he wrote:

> "The particle is a very small region of high concentration of energy which is embodied in the wave in which it constitutes some kind of singularity generally in motion."

The evolution of physics was such that nearly 30 years later, de Broglie's representation of quantum particles was incorporated, along with some additional new ideas, into Bohm's theory of hidden variables (1952), thus giving rise to the de Broglie-Bohm-Vigier interpretation of quantum mechanics (1954).

2.4 D. Bohm

The probabilistic (Copenhagen) interpretation of quantum mechanics was criticized by Einstein, who believed that even at the quantum level there should exist precisely definable dynamic variables determining, as in classical physics, the *actual* behavior of each individual particle, not merely its *probable* behavior. Following this course of thinking, D. Bohm developed an alternative interpretation, conceiving of each individual quantum system as being in a precisely definable state, whose changes with time are determined by definite (causal) laws analogous to the classical equations of motion [5].

Similar proposals for an alternative interpretation of the quantum theory had been put forward earlier by de Broglie [6]. As de Broglie himself related,

> "For nearly twenty five years, I remained loyal to the Bohr-Heisenberg view, which has been adopted almost unanimously by theorists, and I have adhered to it in my teaching, my lectures, and my books. In the summer of 1951, I was sent the preprint of a paper by a young American physicist David Bohm, which was subsequently published in the January 15, 1952 issue of the Physical Review. In this paper, Mr. Bohm takes up the ideas I had put forward in 1927, at least in one of the forms I had proposed, and extends them in an interesting way on some points. Later, J.-P. Vigier called my attention to the resemblance between a demonstration given by Einstein regarding the motion of particles in General Relativity and a completely independent demonstration I had given in 1927 in an exercise I called the 'theory of the double solution"' [7, 8].

Within this interpretation, called the Causal Interpretation of Quantum Mechanics, quantum-mechanical probabilities were regarded much like their counterparts in classical statistical mechanics, that is, as merely a practical necessity, not a manifestation of an inherent lack of complete knowledge at the quantum level. The physical results obtained with this alternative interpretation, suggested by Bohm, were shown to be precisely the same as those obtained with the usual interpretation.

This approach laid emphasis on the possibility of interpreting quantum mechanics in terms of a hidden-variable theory. However, at the time Bohm's paper appeared, any kind of "hidden-variable" theory seemed to be excluded by von Neumann's theorem, which was unanimously accepted. No wonder that Bohm's attempt at reviving a causal approach to quantum mechanics was generally met with scepticism. Pauli, who had been among the most strident critics of de Broglie's views at the 5th Solvay Congress in 1927, raised cutting arguments against Bohm's approach as well, which he regarded as a direct development of de Broglie's theory.

An essential feature of Bohm's approach, in addition to representing an electron as a particle following a continuous and causally defined trajectory with a well-defined position, $\xi(t)$, and accompanied by a physically real wave field $\psi(x,t)$, [cf. the views of de Broglie given above] was the assumption that the probability distribution in an ensemble of electrons having the same wave function, ψ, is $P = |\psi|^2$.

Pauli regarded this assumption as the most significant flaw in Bohm's theory on the grounds that such a hypothesis is not appropriate in a theory that seeks to provide a causal explanation of quantum mechanics. Instead, one should be free in choosing an arbitrary probability distribution that is (at least in principle) independent of the ψ field (as $P = \delta(x - x_0)$, for instance) and dependent only on the incompleteness of our information concerning the location of the particle.

In order to solve this problem, in his next paper D. Bohm introduced the idea of a random collision process that is responsible for establishing the relation $P = |\psi|^2$ [9]. In this paper, a simple specific model was proposed to show that a statistical ensemble of quantum mechanical systems with an arbitrary initial probability distribution decays in time to an ensemble with $P = |\psi|^2$ (which is equivalent to a proof of Boltzmann's H-theorem in classical statistics). However, due to mathematical difficulties, an extension of these results to an arbitrary system was found to be very difficult. Besides, in its original formulation, the theory contained nothing to describe the actual location $\xi(t)$ of the particle, which is indispensable if one wishes to have a consistent causal theory.

Hence the next step in developing Bohm's approach was to supplement it by introducing two new concepts, as formulated in 1954 by D. Bohm and J. P. Vigier [10], namely,

1. the idea of irregular fluctuations affecting the motion of the particle due to its interaction with an underlying stochastic medium, and

2. the concept of an extended particle core in the form of a highly localized inhomogeneity that moves with an average local velocity $\mathbf{v}(x, t)$.

The first of these, that is, the idea of "subquantal medium" as a source of randomness in quantum motion, was considered by de Broglie to be of capital importance, comparable in significance to Boltzmann's hypothesis of "molecular chaos" [11]. Moreover, as de Broglie noted, he himself had followed the same line of thinking:

> "Soon, it appeared to me with growing evidence that this concept [the guidance principle in de Broglie's theory of the double solution - *L. Ch.*] is not sufficient, and that the regular, in a sense, "average" movement of the particle, as defined by the formula of the guidance, should be superposed by a kind of a "Brownian" movement of random nature" [11].

In fact, both these concepts had their precursors in quantum mechanics.

2.5 The Schrödinger equation and classical mechanics

The appearance of Schrödinger's paper (1926), where he proposed his famous equation, immediately triggered a discussion about the meaning and physical content of the equation. It was noted that in some respects, the Schrödinger equation showed a remarkable affinity to classical mechanics.

2.5.1 The Madelung fluid

In 1926, E. Madelung [12], called attention to the fact that Schrödinger's equation allowed a hydrodynamical interpretation of quantum mechanics. Introducing a special representation for the wave function $\psi = R\exp(iS/\hbar)$ (which would be reproduced later in Bohm's paper), Madelung was able to interpret Schrödinger's equations as describing a 'fluid' having a density $\rho = |\psi|^2$ and composed of identical particles of mass m, each moving with a velocity $\mathbf{v} = \nabla S/m$. This was an attempt to give the quantum motion a classical interpretation. However, following Schrödinger, who viewed the wave function of an electron as a real field that represented the electron's charge spread continuously over space, Madelung regarded the fluid as a dynamical model of this spatially distributed charge. It was difficult, however, for him to explain the local nature of 'quantum forces' (the term *"quantum potential"* would be introduced by de Broglie later on), which depended only on the local density ρ rather than on the properties of the distribution as a whole.

2.5.2 The Brown-Markov interpretation of quantum mechanics

Another remarkable feature of the Schrödinger equation resides in its close affinity to the Fokker-Planck equation, which describes diffusion processes in classical stochastic mechanics, such as Brownian motion. This fact was noticed shortly after the equation was suggested; Schrödinger himself was aware of the similarity. Later on, in 1933, R. Fürth [13] in his paper "Über einige Beziehungen zwischen klassischer Statistik und Quantenmechanik"[1] investigated this resemblance in more detail. He showed that the effective diffusion coefficient D relative to the quantum motion of a particle with a mass m, should be written as $D = \hbar/(2m)$. Fürth observed that the stochastic nature of a Brownian-like process imposed restrictions upon the accuracy with which positions x and velocities v of particles could be measured. Namely, the respective uncertainties Δx and Δv were shown to be subjected to an inequality $\Delta x \cdot \Delta v \geq D$, that is, exactly the same as Heisenberg's uncertainty relation.

This approach was given a more rigorous form by I. Fényes [14] in the paper "Eine wahrscheinlichkeitstheoretische Begründung und Interpretation der Quanten Mechanik"[2] published in 1952. Fényes was able to derive the Schrödinger equation within the framework of the general mathematical theory of stochastic Brown-Markovian processes, thus confirming Fürth's conclusions. In particular, the expression $D = \hbar/(2m)$ for the diffusion coefficient D and the Brown-Markovian uncertainty relation $\Delta x \cdot \Delta v \geq D$ were rederived by Fényes on a rigorous mathematical basis.

Still later, the Brownian interpretation of quantum mechanics was developed in papers by E. Nelson "Derivation of the Schrödinger equation from Newtonian mechanics" [15] and L. de La Peña-Auerbach "New formulation of stochastic theory and quantum mechanics" [16].

2.6 De Broglie's "theory of the double solution"

As mentioned above in Sec. 2.3, an essential feature of de Broglie's picture of the wave-particle duality lay in regarding the particle and its associated wave as simultaneously existing, physically real entities. Much like Einstein's representation of photons as singularities within an electromagnetic wave field, de Broglie's idea was to represent electrons as mathematical singularities in the field of the wave function ψ which moved under the "guiding" action of this wave. According to the "principle of the double solution", as formulated by de Broglie in 1927, to every linear solution $\psi = R\exp(iS/\hbar)$ of the wave equation there should correspond a singular solution $\phi = U\exp(iS'/\hbar)$ representing the motion of the singularity associated with the particle's core. Later on [8], de Broglie came to the conclusion that this singular solution ϕ should be governed by a *non-linear* equation which would

[1] "On some Relationships between Classical Statistics and Quantum Mechanics" - (*L.Ch.*)

[2] "A Probability-Theory Justification and Interpretation of Quantum Mechanics" - (*L.Ch.*)

support a *non-dispersive* wave packet as an adequate representation for the particle's core. However, in his papers on the theory of the double solution de Broglie did not specify the non-linear equation for singular waves, while restricting himself to analyzing the physical properties of these waves.

Assuming that the amplitude U of the singular wave decreased with distance r as $U \sim r^{-1}$, and that the particle's core moved as a whole with the velocity $\mathbf{v}' = \nabla S'/m$, de Broglie concluded that $\mathbf{v}'$ should coincide with the velocity $\mathbf{v}$ associated with the phase S of the linear wave ψ via $\mathbf{v} = \nabla S/m$, $\mathbf{v}' = \mathbf{v}$, thus requiring the phases of the two waves to coincide. The relation $S' = S$ is an expression of de Broglie's "guiding" principle, meaning that the particle beats in phase and coherently with its pilot wave. This coherence ensures that the energy exchange (and thus coupling) between the particle and its pilot wave is most efficient. As a result, the singularity of type r^{-1} moves along the lines of flow determined by the linear wave ψ.

2.7 The Bohm-Vigier model

In order to explain the fact that the equilibrium relation $P = |\psi|^2$ is established for arbitrary quantum motion, Bohm and Vigier [10] proposed a hydrodynamic model supplemented with a special kind of irregular fluctuations. They observed that there are always random perturbations of any quantum mechanical system which arise outside that system. Moreover, one can "also assume that the equations governing the ψ field have nonlinearities, unimportant at the level where the theory has thus far been successfully applied, but perhaps important in connection with processes involving very short distances. Such nonlinearities could produce, in addition to many other qualitatively new effects, the possibility of irregular turbulent motion." Furthermore one "may conceive of a granular substructure of matter ... analogous to (but not necessarily of exactly the same kind as) the molecular structure underlying ordinary fluids."

The authors assumed that for any or all of these reasons, or perhaps for still other reasons, the fluid undergoes a more or less random type of fluctuation about its mean (potential) flow. As a result, Bohm and Vigier were able to prove that an arbitrary probability density ultimately decays into $|\psi|^2$. The proof was extended to the Dirac equation and to the many-particle problem.

The difference between the de Broglie-Bohm-Vigier approach, on the one hand, and the usual (Copenhagen) interpretation of quantum mechanics, on the other hand, is this:

> "In the usual interpretation, the irregular statistical fluctuations in the observed results obtained... when we make very precise measurements on individual atomic systems are assumed, so to speak, to be fundamental elements of reality, since it is supposed that they cannot be analyzed in more detail, and that they cannot be traced to anything else. In the model that we have proposed..., the statistical fluctuation in the results of such measurements are shown to be ascribable consistently to an assumed deeper level of irregular motion in the ψ field" [10].

It is worth noting that in the first paper by Bohm and Vigier, the idea of a nonlinear equation governing the ψ field was introduced, and the importance of its possible soliton-like solutions was pointed out:

> "Such nonlinear equations can lead to many qualitatively new results. For example, it is known that they have a spectrum of stable solutions having local-

ized pulse-like concentrations of field, which could describe inhomogeneities such as we have been assuming in this paper" [10].

Hence, in this first paper by Bohm and Vigier we already find the ideas and concepts that, when further developed in subsequent works by Prof. J.-P. Vigier and his co-workers, would result in the formulation of the Stochastic Interpretation of Quantum Mechanics.

2.8 Picture of quantum motion in the Bohm–Vigier representation

The paper by Bohm and Vigier (1954) forms a conceptual basis for all further evolution of the de Broglie-Bohm-Vigier approach. In fact, the physical associations and images incorporated in this approach, as they were introduced in this first paper (although modified in the course of subsequent development), have not undergone substantial changes so far as their basic sense is concerned. Only specific mechanisms have been changed or their nature specified—not, however, their main effects. Therefore, in order to elucidate subsequent modifications to the approach, it is important to analyze in more detail the physical picture of quantum motion as proposed by Bohm and Vigier.

As mentioned above, in the first paper by D. Bohm [5], an electron was represented as a *point-like* particle following a continuous and causally defined trajectory with a well-defined position, $\boldsymbol{\xi}(t)$. Moreover, this particle was accompanied by a physically real wave field, $\psi(\mathbf{r}, t)$. The agreement with the usual (Copenhagen) interpretation of quantum mechanics was achieved by making the following supplementary assumptions:

1. $\psi(\mathbf{r}, t)$ satisfies the Schrödinger equation.

2. $d\boldsymbol{\xi}(t)/dt = \nabla S/m$, where S is related to the phase φ of $\psi = Re^{i\varphi}$ by $S(\mathbf{r}, t) = \hbar\varphi(\mathbf{r}, t)$.

3. The probability distribution in an ensemble of electrons described by the wave function ψ, is $P = |\psi|^2$.

Hydrodynamic picture. A hydrodynamic model with just a *potential*, or *Madelung*, flow is obtained from these assumptions on writing the Schrödinger equation for ψ in terms of the variables R and S, where $\psi = R\,\exp(iS/\hbar)$:

$$\frac{\partial R^2}{\partial t} + \mathrm{div}\left(\frac{R^2}{m}\nabla S\right) = 0, \tag{1}$$

$$\frac{\partial S}{\partial t} + \frac{1}{2m}(\nabla S)^2 - \frac{\hbar^2}{2m}\frac{\nabla^2 R}{R} + V = 0. \tag{2}$$

Indeed, if we interpret $\rho(\mathbf{r}) = R^2$ as the density of a continuous fluid that has the stream velocity $\mathbf{v} = \nabla S/m$, then Eq. (1) will express the conservation of fluid, while Eq. (2) will determine the evolution of the "velocity potential" S under the combined effect of the classical potential, V, and the "quantum potential", Q,

$$Q = -\frac{\hbar^2}{2m}\frac{\nabla^2 R}{R} = -\frac{\hbar^2}{4m}\left[\frac{\nabla^2\rho}{\rho} - \frac{1}{2}\left(\frac{\nabla\rho}{\rho}\right)^2\right]. \tag{3}$$

Equations (1) and (2) do not contain the actual location of the particle, $\boldsymbol{\xi}(t)$, which is required if one wants to arrive at a causal interpretation of the quantum theory.

Idea of a particle's extended core. For this reason, the model was completed by Bohm and Vigier [10] "by postulating a particle, which takes the form of a highly localized inhomogeneity that moves with the local fluid velocity, $\mathbf{v}(\mathbf{r}, t)$." The precise nature of this inhomogeneity was not specified. "It could be, for example, a foreign body, of a density close to that of the fluid, which was simply being carried along with the local velocity of the fluid as a small floating body is carried along the surface of the water at the local stream velocity of the water. Or else it could be a stable dynamic structure existing in the fluid; for example, a small stable vortex or some other stable localized structure, such as a small pulse-like inhomogeneity. Such structures might be stabilized by some nonlinearity that would be present in a more accurate approximation to the equations governing the fluid motions than is given by (1) and (2)." [Bohm and Vigier, 1954]

Idea of fluctuations. Fluctuations were introduced in the model by assuming that the "fluid undergoes a more or less random type of fluctuation about the Madelung motion as a mean. Thus, the velocity will not be exactly equal to $\nabla S/m$, nor will the density, ρ, be exactly equal to $|\psi|^2$." It was required, however, that "the relations $\rho = |\psi|^2$ and $\mathbf{v} = \nabla S/m$ be valid as averages." As a result, the exact velocity is not derivable from a potential, and so the flow is no longer a potential one. Instead, one has $d\boldsymbol{\xi}(t)/dt = \nabla S'/m + \nabla \times \mathbf{A}$, with $\langle \nabla \times \mathbf{A} \rangle_{\mathrm{Av}} = 0$ and $\langle \nabla S' \rangle_{\mathrm{Av}} = \langle \nabla S \rangle_{\mathrm{Av}}$ where $\langle \ldots \rangle_{\mathrm{Av}}$ means averaging over fluctuations. Consequently, the Schrödinger equation does not apply to fluctuations. However, the conservation equation $\partial\rho/\partial t + \mathrm{div}(\rho\mathbf{v}) = 0$ is assumed to hold even with fluctuations present.

Random walks of particle-like inhomogeneities between different lines of flow (trajectories). It was further assumed that *even in a fluctuation*, the particle-like inhomogeneity "follows the fluid velocity $\mathbf{v}(\mathbf{r}, t)$." The reasons for this assumption were clarified by observing that "such a behavior would result if the inhomogeneity were a very small dynamic structure in the fluid (e.g., a vortex, or a pulse-like inhomogeneity) or if it were a foreign body of about the same density as the fluid, provided that the wavelength associated with the fluctuations were appreciably larger than the size of the particle. For in this case, he inhomogeneity would have to do more or less as the fluid did, since t would act, for all practical purposes, like a small element of fluid."

Consequently, "if we followed a given fluid element, we would discover that it undergoes an exceedingly irregular motion, which is able in time *to carry it from any specified trajectory of the mean Madelung motion to practically any other trajectory* [italics by *L.Ch.*]. Such a random motion of the fluid elements would, if it were the only factor operating, lead eventually to a uniform mean density of the fluid. For it would on the average carry away more fluid from a region of high density than it carried back. The fact that the mean density remains equal to $|\psi|^2$, despite the effects of the random fluctuations, implies then that a systematic tendency must exist for fluid elements to move toward regions of high mean fluid density, in such a way as to maintain the stability of the mean density, $\bar{\rho} = |\psi|^2$." Bohm and Vigier did not discuss the origins of this tendency, but mentioned, among possible mechanisms, "the internal stresses in the fluid" such that "whenever ρ deviates from $|\psi|^2$, a kind of pressure arises that tends to correct the deviation automatically. Such a behavior is analogous to what would happen, for example, to a gas in irregular turbulent motion in a gravitational field, in which the pressures automatically adjust themselves in such a way as to maintain a local mean density close to $\rho = \rho_0\, e^{-mgz/kT}$ if the temperature T is constant."

As a consequence of the above assumptions, "it is evident that the inhomogeneity will undergo an irregular motion, analogous to the Brownian motion."

3. The Vigier model

For the subsequent development of the de Broglie-Bohm-Vigier approach, two concepts introduced by Dirac were of special importance, namely, "Dirac's aether" and "Dirac's extended electron".

3.1 Dirac's aether

In 1952, in a short note "Is there an Æther?" published in *Nature* [17], P. A. M. Dirac pointed out that, in the light of the present-day knowledge, not only the notion of an *æther* is no longer ruled out by relativity[3], but moreover "good reasons can now be advanced for postulating an æther." Assuming the 4-vector v_μ of the velocity of the æther to be distributed uniformly over the hyperboloid $v_0^2 - v_1^2 - v_2^2 - v_3^2 = 1$ (with $v_0 > 0$), Dirac observed that the wave function representing a state where all æther velocities are equally probable should be independent of v's, and so it must be a constant over the hyperboloid. As a result, "we may very well have an æther, subject to quantum mechanics and conforming to relativity, provided we are willing to consider the perfect vacuum as an idealized state, not attainable in practice. From the experimental point of view, there does not seem to be any objection to this." However, "we must make some profound alterations in our theoretical ideas of the vacuum. It is no longer a trivial state, but needs elaborate mathematics for its description" [17].

The idea was developed further by C. Petroni and J. P. Vigier (1983). In their paper "Dirac's Æther in Relativistic Quantum Mechanics" [18] they pointed out that one should distinguish between the notions of Dirac's *vacuum* and Dirac's *æther*. The notion of *vacuum* was originally introduced by Dirac for the spin-$1/2$ particles in order to resolve the problem with negative energies. It implied that all single-particle states with negative energies are filled, while those with positive energies are empty. To comply with the general requirements of special relativity (invariance under Lorentz's transformations), the notion of Dirac's vacuum would have to be supplemented with an additional hypothesis. Namely, one must assume that the 4-momenta $p = (p_0, \mathbf{p})$ of particles with negative energies are distributed *uniformly* over the lower (filled) mass shell defined by the equation $p_0^2 - \mathbf{p}^2 = m_0^2 c^2$, with m_0 the rest mass of the particles. If the Dirac vacuum satisfies the latter condition, it is referred to as Dirac's æther. Later on, J. P. Vigier [19] formulated an advanced model of Dirac's æther, treating the latter as built up of superfluid states of particle-antiparticle pairs (see also Refs. [20]).

3.2 Dirac's extended electron

In the paper "Classical theory of radiating electrons" [21], Dirac proposed to set up (within the framework of the classical theory) a self-consistent scheme of equations describing the interaction of electrons with radiation. Initially, the electron was treated as a point charge, which led to the difficulties of the infinite Coulomb energy. To avoid these, Dirac used a procedure of subtracting divergent terms similar to what was used in the theory of the positron. The result was remarkable. The equations so obtained were found to have the

[3]It is interesting to note that, contrary to the often expressed opinion, Einstein himself also did not deny the existence of the ether. In his lecture given at the University of Leyden (1920), Einstein stressed that "the negation of ether is not necessarily required by the principle of special relativity. We can admit the existence of ether, but we have to give up attributing it to a particular motion ... The hypothesis of the ether as such does not contradict the theory of special relativity." In fact, what Einstein rejected completely was only the existence of the *absolute frame of reference*.

same form as those currently in use, but in their physical interpretation "the final size of the electron reappeared in a new sense." Namely, the interior of the electron appeared as a region of space through which signals could be transmitted *faster than light*. More precisely, Dirac's conclusion was that "the interior of the electron" was "a region of failure, not of the field equations of electromagnetic theory, but of some of the elementary properties of space-time" [21].

3.3 Bell's inequalities and quantum nonlocality

It is not rare in physics that certain assumptions appear to be so obvious and natural that their very plausibility seems to rule out any need for further discussion or justification. The hypothesis (sometimes even called the principle) of the *spatial suppression of correlation* belongs to such "natural" assumptions. According to this hypothesis, *any* kind of physical connection between two (or more) spatially separated parts of a physical system should vanish as soon as the distances separating these parts become large enough. In statistical physics, this principle was formulated and successfully applied in the theory of many-body systems by N. N. Bogolyubov and his co-workers.

In quantum mechanics, however, the situation was found to be more complicated. The main physical reason for this lies in the fact that in quantum mechanics the phases of the wave functions, while deeply involved in the formation of quantum correlations in physical systems, impose qualitatively new features upon the nature of such correlations [4].

It turned out that the nature of phase correlations, as reproduced by *local* hidden-variable theories, is different from that encoded in the wave functions. This remarkable fact was pointed out in 1965 by J. S. Bell [22] in his studies on the problem of quantum measurement. Bell came to the conclusion that any local hidden-variable theory, if applied to subsystems of one and the same quantum system in a well-defined quantum state, should lead to certain restrictions upon the relationships between the series of parallel measurements, each one of the latter pertaining to a particular part of the system. Mathematically, these restrictions were put by J. S. Bell into the form of special inequalities.

The physical meaning of Bell's inequalities resides in the fact that they establish upper bounds to possible correlation rates between spatially separated (and quite remote) parts of a quantum system (such as an EPR pair, for example) attainable from the viewpoint of any *local* hidden-variable theory. Hence, Bell's inequalities are experimentally testable. As a consequence, if experiments on some quantum systems showed higher correlation rates than those allowed by Bell's inequalities (thus violating the latter), then this would provide experimental evidence for quantum mechanics being a non-local theory. All *local* hidden-variable theories would thus have to be excluded.

By the end of 1980s, many different experiments were carried out in order to test Bell's inequalities, such as those by Freedman and Clauser [23], Clauser [24, 25], Fry and Thompson [26], Aspect *et al.* [27, 28], Perrie *et al.* [29], Hassan *et al.* [30]. All these experiments clearly showed that

- Bell's inequalities *are violated*, and
- the *quantitative* predictions of quantum mechanics *are confirmed.*

[4] Long-distance quantum correlations due to the specific nature of phase relations carried by the wave functions are sometimes referred to as "quantum entanglement".

Similar conclusions were drawn about 20 years earlier by Bohm and Aharonov [31] by analyzing the experiments of Wu and Shaknov [32] on double scattering of two photons produced by the annihilation of an electron-positron pair.

Consequently, there now seems to be strong experimental evidence (with a probability approaching 90% if one takes into account some details of the experimental setups which might leave room for differing interpretations) in favor of the fact that quantum mechanics is a *non-local* theory. Hence, in quantum mechanics we currently face a very disturbing problem, i.e., how to understand the nature of this non-locality. Various solutions have been suggested, including some that might appear rather unusual, such as, for instance, admitting signals that propagate backwards in time, or various modifications of quantum theory, including negative probabilities. In particular, the possibility of superluminal connections between remote parts of a quantum system has been suggested and investigated. The Bohm theory of "unbroken wholeness" furnishes an example of such approach. An altenative way of thinking is represented by the Vigier model.

3.4 The Vigier model

The Vigier model [33] is an advanced implementation of the Bohm-Vigier approach which suggests a solution to the problem of quantum nonlocality. This model is essentially relativistic.

In Vigier's representation, the irregular fluctuations of the Bohm-Vigier model (1954) are interpreted as being due to a "random subquantal level of matter", in the sense of Dirac's "æther" or de Broglie's "hidden thermostat" [34]. This idea reflects Einstein's viewpoint according to which quantum statistics should be due to a real subquantal physical vacuum alive with fluctuations and randomness.

The notion of an extended particle, as introduced by Bohm and Vigier in 1954 (see also Ref. [35]) has been developed further by Vigier. If Dirac's picture of an extended electron is accepted, then the motion of the core of the electron should be represented in 4-spacetime not by a line, but by a time-like "hypertube" lying inside the light cone. Accordingly, in the Vigier model particles are regarded as "extended time-like hypertubes" that "move along time-like paths and can only transmit superluminal information localized within their internal structure" (see Refs. [19, 36]).

In Vigier's model, the stochastic jumps introduced by Bohm and Vigier (1954) as a mechanism to carry particles from one line of flow to another, are interpreted as "stochastic jumps on the light cone", meaning that "the stochastic fluctuations occur at the velocity of light" [33]. Here, the relativistic extension of the continuity equation (1), namely, $\partial_\mu j^\mu = 0$, is shown to be equivalent to the set of two (forward and backward) Fokker-Planck equations

$$\frac{\partial \rho}{\partial \tau} + \nabla(v_\pm \rho) \pm D\,\Box\rho = 0\,, \qquad (\rho = R^2) \tag{4}$$

where the diffusion coefficient D is obtained in exactly the same form, $D = \hbar/(2m)$, as in Fürth's paper [13].

Lastly, the notion of "superluminal propagation of the quantum potential" was introduced in the Vigier model [33]. Specifically, for a particle of rest mass m, the quantum potential Q, as defined by $Q = \log M$ with

$$M = \left[m^2 + \frac{\hbar^2}{c^2}\frac{\Box\rho^{1/2}}{\rho^{1/2}}\right]^{1/2},$$

is a function of the density $\rho = (\psi^* \psi)^{1/2}$ alone, and propagates with superluminal velocities within the drift current. The quantum potential is interpreted "as a real interaction among the particles and the subquantal fluid polarized by the presence of the particles" [37]. It is considered to be a "true stochastic potential" [38].

It is important to note that the quantum potential is essentially *non-local*, so that the Vigier model, like Bohm's theory, appears as a particular implementation of non-local hidden-variable theories. Therefore, this model does not conflict with Bell's inequalities.

An essential feature of the Vigier model is that it preserves Einstein's causality in experiments of the Einstein-Podolsky-Rosen type, while at the same time explaining the quantum-mechanical nonlocality through a "nonlocal superluminal information" transfer. The latter is not brought about by individual particles, but rather is due to the propagation of collective excitations (considered to be real and physical) on top of the "material vacuum" as described above in this section.

4. Stochastic interpretation of quantum mechanics

The ideas and concepts described above, make up the content of the so called (causal) "stochastic interpretation of quantum mechanics" [36]. The differences between this approach and the conventional (Copenhagen) interpretation of quantum mechanics are summarized below.

4.1 Stochastic interpretation of quantum mechanics versus the Copenhagen interpretation

There are three principal (conceptual) differences between the two alternative interpretations of quantum mechanics [39].

1. In the Copenhagen interpretation, quantum waves are associated with individual particles and represent an ultimate, statistical knowledge. Micro-objects manifest themselves either as particles or as waves, but never both simultaneously.

   In the Stochastic interpretation, quantum waves are considered to be real, physical fields associated with individual particles as well as with ensembles of identically prepared systems. Micro-objects are thus viewed as complexes involving both particles and waves that co-exist simultaneously.

2. In the Copenhagen interpretation, a measurement on a system implies a discontinuous, instantaneous collapse of its quantum state, known as the *reduction of the wave function*. No microphenomenon is a phenomenon until it is an observed phenomenon [40].

   In the Stochastic interpretation, quantum states, represented by particles along with their associated real waves, evolve causally and continuously in time. No wave-packet reduction occurs. Waves do not collapse, but rather are modified (or split) when interacting with the measuring apparatus. As a result of the measurement, the particle enters one of the apparatus's measurable eigenstates.

3. In the Copenhagen interpretation, the uncertainty principle imposes restrictions upon the simultaneous measurability of complementary observables.

   In the Stochastic interpretation, the Heisenberg uncertainty principle does not restrict the measurability of complementary observables, but represents dispersion relations resulting from (i) the dual nature (wave plus particle, both existing at the same time) of micro-objects, and (ii) the associated subquantal stochastic motions.

In spite of these differences, it would be inappropriate to think that the Stochastic interpretation seeks to re-establish classical views about the physical world. Indeed, even though within this approach particles are considered to have definite values of all their dynamical variables at every given moment, quantum forces due to the quantum potential bring about the dependence between these variables and the quantum state as a whole, thereby mediating the influence of the environment on quantum dynamics.

4.2 Structure of the quantum particle in the de Broglie-Bohm-Vigier approach

In the Stochastic interpretation, a quantum particle with a rest mass m_0 is represented as consisting of the sum $\Phi + U$ of two different waves Φ and U such that

- the *pilot* wave Φ, or P-wave, is a *linear* wave described by $\Phi = R \exp(iS/\hbar)$, with real-valued R and S, satisfying linear equations of quantum theory; for a particle with zero spin, this is the Klein-Gordon equation

$$(\Box - m_0^2 c^2/\hbar^2)\,\Phi = 0\,, \tag{5}$$

whereas

- the *soliton* wave U, or S-wave, is *nonlinear*, highly localized, and nondispersive. This wave is described by $U = H \exp(iS/\hbar)$ with the same phase S as the P-wave. The S-wave describes the core of a quantum particle, the latter being regarded as an extended entity.

The conventional wave function of quantum mechanics, ψ, which is an associated probability wave, is proportional to the P-wave, that is, $\psi = C\Phi$ with C the normalizing constant. The basic properties of the P- and S-waves are specified in such a way that de Broglie's assumption about the "guiding" action which the "pilot" wave exerts on the quantum particle is implemented in the theory.

Specifically, the P-waves have the following properties.

1. The P-wave is a superposition of de Broglie's plane waves

$$B(x,t) = a \exp[2\pi i\nu(t - x/V)] \tag{6}$$

where ν is the observed frequency, $V = c^2/v$ is the phase velocity, and v is the velocity of the particle. The original idea of de Broglie was that, since any particle at rest has the mass m_0 and hence the energy $E = m_0c^2$, there should be a definite frequency ν_0 connected with the particle through $h\nu_0 = m_0c^2$. He then associated with the particle a monochromatic wave (6), which de Broglie considered to be real. If the particle is at rest, then its internal frequency ν_0 coincides with the frequency ν of the wave $B(x,t)$ (6). De Broglie always believed that the frequency ν_0 corresponds to real physical oscillations occurring inside the particle, thus admitting an extended structure of all micro-objects and, inevitably, some *local* hidden variables.

2. The P-wave defines the particle's drift motion along the lines of flow in the 4-spacetime. The direction of the drift is determined by the 4-vector $u_\mu = \partial_\mu S/M_0^2$ where

$$M_0^2 = m_0^2 - \frac{\hbar^2}{c^2}\frac{\Box R}{R} \tag{7}$$

represents de Broglie's and Bohm's quantum potential that appears in the relativistic Jacobi equation $\partial_\mu S \partial^\mu S + M_0^2 c^2 = 0$, the latter obtained as the real part of Eq. (5).

3. The P-wave carries a density $\rho = \sqrt{-g}(M_0/m)R^2$ that is conserved in the drift motion according to $\partial\rho/\partial\tau = 0$ (derivation along the line of flow).

4. The P-wave, having the dispersion relation $\omega = (c^2 k_B^2 + m_0^2 c^4/\hbar^2)^{1/2}$, where k_B is the wave vector, necessarily disperses with time.

5. The P-wave can be regarded as a Brown-Markov stochastic wave propagating on a random covariant thermostat.

On the other hand, the soliton S-wave is governed by a non-linear covariant equation (of the Klein-Gordon type for zero-spin particles). Representing the core of the particle, the S-wave carries an energy $E = \hbar\omega$ and momentum $p_B = \hbar k_B$. It can be shown [41] that in its motion this soliton wave follows the lines of flow as determined by the P-wave, provided that the phases of the S- and P-waves coincide. Under the effect of random subquantal fluctuations, the center of the soliton S-wave moves randomly from one line of flow to another, which establishes the quantum mechanical probability distribution $\psi^*\psi$. As was shown by Mackinnon [42], for a particle traveling in the x-direction with a constant velocity v, the S-wave takes the form

$$U(x, y, z, t) = \frac{\sin kr}{kr} e^{i(\omega t - k_0 x)} \tag{8}$$

with $k = m_0 c/\hbar$, $k_0 = mv/\hbar$ and

$$m = \frac{m_0}{\sqrt{1-\beta^2}}, \quad \beta = \frac{v}{c}, \quad r = \left[\frac{(x-vt)^2}{1-\beta^2} + y^2 + z^2\right]^{1/2}. \tag{9}$$

The spatial distance (in the x direction) between the first two zeros of the S-wave (the effective diameter of the particle) is thus equal to $h/(mc) = [h/(m_0 c)]\sqrt{1-\beta^2}$, *i.e.*, to the Compton wavelength contracted due to the usual relativistic (Lorentz) mechanism. Ph. Gueret and J.-P. Vigier [43] showed that this soliton-like wave function is a solution of a non-linear equation written (for zero-spin particles) as

$$\Box U - \frac{m_0^2 c^2}{\hbar^2} U = \frac{[\Box(U^* U)]^{1/2}}{(U^* U)^{1/2}} U \tag{10}$$

where the non-linearity has the form of a quantum potential $Q \sim \Box|U|/|U|$. The same equation was obtained in a quite different way by F. Guerra and M. Pusterla [44] and A. Smolin [45] They derived the non-linear Schrödinger equation and the Klein-Gordon equation containing a non-linearity of the quantum-potential type by following Nelson's stochastic approach (see Sec. 2.5.2).

From Eqs. (8) and (9) it follows that in the non-relativistic approximation, the S-wave disappears. At the same time, it can be shown that, as $c \to \infty$, equation (5) for the P-wave goes over into the corresponding Schrödinger equation.

4.3 Theoretical consequences and possible experimental tests

A comprehensive review of the Stochastic interpretation of quantum mechanics, along with many of its noteworthy applications, was given by J. P. Vigier and co-authors in *Essays in Honour of David Bohm* [46]. They demonstrated that the de Broglie-Bohm-Vigier approach is able to explain, among others, the basic double-slit experiments, the EPR paradox [37], as well as various experiments on neutron interferometry [47] (and references therein). Moreover, this approach was shown to be successful in suggesting plausible solutions to such difficult problems of quantum theory as the negative probabilities associated with relativistic Klein-Gordon equation [46, 48]. The approach was extended to include particles having a non-zero spin s, namely, particles with spin $s = 1$ [33, 49, 50, 51] as well as particles with spin $s = 1/2$ [52].

In its essence, the de Broglie-Bohm-Vigier approach represents an advanced model of de Broglie's wave-particle duality, and so invokes explicit experiments that are likely to suggest a choice between this approach and the Copenhagen interpretation. The basic idea underlying such experiments is to prove the reality of de Broglie's pilot waves. One possible test of this reality (in the double-slit experiments on neutrons) was proposed by J. P. Vigier [53]. Other experimental setups suitable for detecting the real existence of the P-waves have been proposed and discussed (but not yet performed) for photons [39, 54, 55], neutrons [56, 57, 58], and intersecting laser beams [59]. Further discussion can be found in Ref. [60].

5. Conclusions

As we have seen, the de Broglie-Bohm-Vigier approach originated from the basic ideas of Einstein and de Broglie concerning the relationship between two fundamental properties of a quantum object—its wave and particle aspects. Moreover, the de Broglie-Bohm-Vigier approach rests upon the same philosophical foundations as those to which Planck, Schrödinger, Einstein, and de Broglie adhered—all of them sharing the belief in the objective reality of the physical world as well as man's ability to understand it correctly in its most subtle details. "Physical theories try to form a *picture of reality* [italicized by *L.Ch.*] and to establish its connection with the wide world of sense impressions" [61]. For this reason, the Stochastic interpretation of quantum mechanics is sometimes also referred to as the Einstein-de Broglie-Bohm interpretation [53]. However, in the opinion of the author, referring to it as the *de Broglie-Bohm-Vigier theory* would be a better reflection of the actual state of things.

What can be expected in terms of further developments in the de Broglie-Bohm-Vigier approach in quantum mechanics? It is hardly possible to summarize the outlook better than Prof. J. P. Vigier did himself. "In my opinion the most important development to be expected in the near future concerning the foundations of quantum physics is a revival, in modern covariant form, of the ether concept of the founding fathers of the theory of light (Maxwell, Lorentz, Einstein, etc.). This is a crucial question, and it now appears that the vacuum is a real physical medium which presents surprising properties (superfluid, *i.e.* negligible resistance to inertial motions), so that the observed material manifestations correspond to the propagation of different types of phase waves and different types of internal motions within the extended particles themselves. The transformation of particles into each other would correspond to reciprocal transformations of such motions. The propagation of phase waves on the top of such a complex medium, first suggested by Dirac in his famous 1951 paper in *Nature*, yields the possibility to bring together relativity theory and quantum mechanics as different aspects of motions at different scales. This ether, itself being built from spin one-half ground-state extended elements undergoing covariant stochastic motions, is reminiscent of old ideas at the origin of classical physics proposed by Descartes and in ancient times by Heraclitus himself. The statistics of quantum mechanics thus reflects the basic chaotic nature of ground state motions in the Universe.

Of course, such a model also implies the existence of non-zero mass photons as proposed by Einstein, Schrödinger, and de Broglie. If confirmed by experiment, it would necessitate a complete revision of present cosmological views. The associated tired-light models could possibly replace the so-called expanding Universe models. Non-velocity redshifts could explain anomalous quasar-galaxy associations, etc., and the Universe would possibly be infinite in time. It could be described in an absolute spacetime frame corresponding to the observed 2.7 K microwave background Planck distribution. Absolute 4-momentum and angular momentum conservation would be valid at all times and at every point in the Universe" [62].

References

[1] A. Einstein, *Ann. Phys. (Leipzig)* **17**, 132 (1905).

[2] L. de Broglie, *Ann. Fond. L. de Broglie* **4**, 13 (1979).

[3] L. de Broglie, *Ann. Phys. (Paris)* **3**, 22 (1925).

[4] L. de Broglie, *Ann. Fond. L. de Broglie* **2**, 1 (1977).

[5] D. Bohm, *Phys. Rev.* **85**, 166, 180 (1952).

[6] L. de Broglie, *Compt. Rend.* **183**, 447 (1926); **184**, 273 (1927).

[7] L. de Broglie, *La physique quantique restera-t-elle indéterministe* (Paris, 1953).

[8] L. de Broglie, *Une tentative d'interprétation causale et non-linéaire de la Mécanique Ondulatoire* (Gauthier-Villars, Paris, 1956).

[9] D. Bohm, *Phys. Rev.* **89**, 1458 (1953).

[10] D. Bohm and J. P. Vigier, *Phys. Rev.* **96**, 208 (1954).

[11] L. de Broglie, La thermodynamique "cachée" des particules, *Ann. Inst. Henri Poincaré*, **I**, 1, 1964.

[12] E. Madelung, *Z. Physik* **40**, 332 (1926).

[13] R. Fürth, *Z. Physik* **81**, 143 (1933).

[14] I. Fényes, *Z. Physik* **132**, 81 (1952).

[15] E. Nelson, *Phys. Rev.* **150**, 1079 (1969).

[16] L. de La Peña-Auerbach, *J. Math. Phys.* **10**, 1620 (1969).

[17] P. A. M. Dirac, *Nature (London)* **169**, 702 (1952).

[18] C. Petroni and J. P. Vigier, *Found. Phys.* **13**, 253 (1983).

[19] J. P. Vigier, *Lett. Nuovo Cim.* **29**, 467 (1980).

[20] K. P. Sinha, C. Sivaram, and E. C. G. Sudarshan, *Found. Phys.* **6**, 65 (1976); **6**, 717 (1976); **8**, 823 (1978).

[21] P. A. M. Dirac, *Proc. Roy. Soc.* **167A**, 148 (1938).

[22] J. S. Bell, *Physics* **1**, 195 (1965).

[23] S. J. Freedman and J. F. Clauser, *Phys. Rev. Lett.* **28**, 938 (1972).

[24] J. F. Clauser, *Phys. Rev. Lett.* **37**, 1223 (1976).

[25] J. F. Clauser, *Nuovo Cimento* **33B**, 740 (1976).

[26] E. S. Fry and R. C. Thompson, *Phys. Rev. Lett.* **37**, 465 (1976).

[27] A. Aspect, P. Grangier, and G. Rogier, *Phys. Rev. Lett.* **47**, 460 (1981); **49**, 91 (1982); **49**, 180 (1982).
[28] A. Aspect and P. Granier, *Lett. Nuovo Cimento* **43**, 345 (1985).
[29] W. Perrie, A. J. Duncan, H. J. Beyer, and H. Kleinpoppen, *Phys. Rev. Lett.* **54**, 1790 (1985).
[30] A. Duncan, in: *Quantum Mechanics versus Local Realism: The Einstein, Podolsky, and Rosen Paradox*, Ed. F. Selleri, (Plenum, New York, 1988).
[31] D. Bohm and Y. Aharonov, *Phys. Rev.* **108**, 1070 (1957).
[32] C. S. Wu and I. Shaknov, *Phys. Rev.* **77**, 136 (1950).
[33] J. P. Vigier, *Lett. Nuovo Cimento* **24**, 258, 265 (1979).
[34] L. de Broglie, *La thermodynamique de la particule isolée* (Paris, 1964).
[35] D. Bohm and J. P. Vigier, *Phys. Rev.* **109**, 882 (1958).
[36] F. Selleri and J. P. Vigier, *Lett. Nuovo Cimento* **29**, 1 (1980).
[37] N. Cufaro Petroni and J. P. Vigier, *Lett. Nuovo Cimento* **26**, 149 (1979).
[38] N. Cufaro Petroni, Ph. Droz-Vincent, and J. P. Vigier, *Lett. Nuovo Cimento* **31**, 415 (1981).
[39] Ph. Gueret and J. P. Vigier, *Found. Phys.* **12**, 1057 (1982).
[40] N. Bohr, *Phys. Rev.* 48, 696 (1935).
[41] F. Halbwachs, *Théorie Relativiste des Fluides à Spin* (Gauthier-Villars, Paris, 1960).
[42] L. Mackinnon, *Lett. Nuovo Cimento* **32**, 10 (1981).
[43] Ph. Gueret and J.-P. Vigier, *Lett. Nuovo Cimento* **35**, 256 (1982); **38**, 125 (1983).
[44] F. Guerra and M. Pusterla, *Lett. Nuovo Cimento* **34**, 351 (1982).
[45] L. Smolin, *Phys. Lett. A* **113**, 408 (1986).
[46] J.-P. Vigier, C. Dewdney, P. R. Holland, and A. Kyprianidis, in: *Essays in Honour of David Bohm*, Eds. B. J. Hiley and F. D. Peat (Routledge, London, 1987).
[47] J.-P. Vigier, *Physica B* **151**, 386 (1988).
[48] P. R. Holland and J. P. Vigier, *Nuovo Cimento* **88B**, 20 (1985).
[49] W. Lehr and J. Park, *J. Math. Phys.* **18**, 1235 (1977).
[50] N. Cufaro Petroni and J. P. Vigier, *Int. J. Theor. Phys.* **18**, 807 (1979).
[51] A. Garuccio and J. P. Vigier, *Lett. Nuovo Cim.* **30**, 57 (1981).
[52] N. Cufaro Petroni and J. P. Vigier, *Phys. Lett.* **81A**, 12 (1981).
[53] J. P. Vigier, *Found. Phys.* **24**, 61 (1994).
[54] J. A. Silva, F. Selleri, and J. P. Vigier, *Lett. Nuovo Cim.* **36**, 503 (1983).
[55] A. Garuccio, A. Kyprianidis, D. Sardelis, and J. P. Vigier, *Lett. Nuovo Cim.* **39**, 225 (1984).
[56] C. Dewdney, Ph. Gueret, A. Kyprianidis, and J. P. Vigier, *Phys. Lett.* **102A**, 291 (1984).
[57] C. Dewdney, A. Garuccio, A. Kyprianidis, and J. P. Vigier, *Phys. Lett.* **104A**, 325 (1984).
[58] C. Dewdney, A. Kyprianidis, J. P. Vigier, A. Garuccio, and Ph. Gueret, *Lett. Nuovo Cim.* **40**, 481 (1984).
[59] C. Dewdney, A. Kyprianidis, J. P. Vigier, and M. A. Dubois, *Lett. Nuovo Cim.* **41**, 177 (1984).
[60] G. Tarozzi and A. van der Merwe, *Open Questions in Quantum Physics* (Reidel, Dordrecht, 1985).
[61] A. Einstein and L. Infeld, *The Evolution of Physics* (Simon and Schuster, New York, 1938).
[62] J. P. Vigier, *Apeiron 2*, 114 (1995).

This page is deliberately left blank.

Reprinted from *Physical Review*, Vol. 96, No. 1, pp. 208-17, Copyright (1954) with permission from the American Physical Society

Model of the Causal Interpretation of Quantum Theory in Terms of a Fluid with Irregular Fluctuations

D. Bohm, *Faculdade de Filosofia, Ciências e Letras, Universidade de São Paulo, São Paulo, Brazil*

AND

J. P. Vigier, *Institut Henri Poincaré, Paris, France*

(Received June 14, 1954)

In this paper, we propose a physical model leading to the causal interpretation of the quantum theory. In this model, a set of fields which are equivalent in many ways to a conserved fluid, with density $|\psi|^2$, and local stream velocity, $d\xi/dt=\nabla S/m$, act on a particle-like inhomogeneity which moves with the local stream velocity of the equivalent fluid. By introducing the hypothesis of a very irregular and effectively random fluctuation in the motions of the fluid, we are able to prove that an arbitrary probability density ultimately decays into $|\psi|^2$. Thus, we answer an important objection to the causal interpretation, made by Pauli and others. This result is extended to the Dirac equation and to the many-particle problem.

1. INTRODUCTION

A CAUSAL interpretation of the quantum theory has been proposed,[1,2] involving the assumption that an electron is a particle following a continuous and causally defined trajectory with a well-defined position, $\xi(t)$, accompanied by a physically real wave field, $\psi(\mathbf{x},t)$. To obtain all of the results of the usual interpretation, the following supplementary assumptions had to be made:

1. $\psi(\mathbf{x},t)$ satisfied Schrödinger's equation.
2. $d\xi/dt=\nabla S/m$, where $\psi=R\exp(iS/\hbar)$.
3. The probability distribution in an ensemble of electrons having the same wave function, is $P=|\psi|^2$.

These assumptions were shown to be consistent.

Assumption (3), however, has been criticized by Pauli[3] and others[4] on the ground that such a hypothesis is not appropriate in a theory aimed at giving a causal explanation of the quantum mechanics. Instead, they argue it should be possible to have an arbitrary probability distribution [a special case of which is the function $P=\delta(\mathbf{x}-\mathbf{x}_0)$, representing a particle in a well-defined location], that is at least in principle independent of the ψ field and dependent only on our degree of information concerning the location of the particle.

In a more recent paper,[5] one of us has proposed a means of dealing with this problem by explaining the relation, $P=|\psi|^2$ in terms of random collision processes. It was shown in a simplified case that a statistical ensemble of quantum-mechanical system with an arbitrary initial probability distribution decays in time to an ensemble with $P=|\psi|^2$. This is equivalent to a proof of Boltzmann's H theorem in classical mechanics. Thus, we can answer the objection of Pauli, for no matter what the initial probability distribution may have been (for example, a delta function), it will eventually be given by $P=|\psi|^2$.

In the work cited above, however, certain mathematical difficulties make a generalization of the results to an arbitrary system very difficult. (The difficulties are rather analogous to these appearing in classical statistical mechanics when one tries rigorously to treat the approach of a distribution to equilibrium, by means of demonstrating a quasi-ergodic character of the motion). In the present paper, we shall avoid these difficulties by taking advantage of the fact that the causal interpretation of the quantum theory permits an unlimited number of new physical models, of types not consistent with the usual interpretation, which lead to the usual theory only as an approximation, and which may lead to appreciably different results at new levels (e.g., 10^{-13} cm). The model that we shall propose here furnishes the basis for a simple deduction of the relation, $P=|\psi|^2$; and in addition, gives a possible physical interpretation of the relation $d\xi/dt=\nabla S/m$ (postulate 2), which follows rather naturally from the model. This model is an extension of the causal interpretation of the quantum theory already proposed, which provides a more concrete physical image of the meaning of our postulates than has been available before, and which suggests new properties of matter that may exist at deeper levels.

2. THE HYDRODYNAMIC MODEL

The model that we shall adopt in this paper is an extension of a hydrodynamic model, originally proposed by Madelung[6] and later developed further by Takabayasi[7] and by Schenberg.[8] To obtain this model, we first write down Schrödinger's equation in terms of the

[1] L. de Broglie, Compt. rend. 183, 447 (1926); 184, 273 (1927); 185, 380 (1927).
[2] D. Bohm, Phys. Rev. 85, 166, 180 (1952).
[3] *Les Savants et le Monde*, Collection dirigée par André George, *Louis de Broglie, Physicien et Penseur* (Editions Albin Michel, Paris, 1953).
[4] J. B. Keller, Phys. Rev. 89, 1040 (1953).
[5] D. Bohm, Phys. Rev. 89, 458 (1953).
[6] E. Madelung, Z. Physik 40, 332 (1926).
[7] T. Takabayasi, Progr. Theoret. Phys. (Japan) 8, 143 (1952); 9, 187 (1953).
[8] M. Schenberg, Nuovo cimento (to be published).

variables, R and S, where $\psi=R\exp(iS/\hbar)$:

$$\partial R^2/\partial t+\mathrm{div}(R^2\nabla S/m)=0, \tag{1}$$

$$\frac{\partial S}{\partial t}+\frac{(\nabla S)^2}{2m}-\frac{\hbar^2}{2m}\frac{\nabla^2 R}{R}+V=0. \tag{2}$$

Now Madelung originally proposed that R^2 be interpreted as the density $\rho(\mathbf{x})$ of a continuous fluid, which had the stream velocity $\mathbf{v}=\nabla S/m$. Thus, the fluid is assumed to undergo only potential flow. Equation (1) then expresses the conservation of fluid, while Eq. (2) determines the changes of the velocity potential S in terms of the classical potential V, and the "quantum potential":

$$\frac{-\hbar^2}{2m}\frac{\nabla^2 R}{R}=\frac{-\hbar^2}{4m}\left[\frac{\nabla^2\rho}{\rho}-\frac{1}{2}\left(\frac{\nabla\rho}{\rho}\right)^2\right].$$

As shown by Takabayasi[7] and by Schenberg,[8] the quantum potential may be thought of as arising in the effects of an internal stress in the fluid. This stress depends, however, on derivatives of the fluid density, and therefore is not completely analogous to the usual stresses, such as pressures, which are found in macroscopic fluids.

The above model is, however, not adequate by itself; for it contains nothing to describe the actual location, $\xi(t)$, of the particle, which makes possible, as we have seen in previous papers,[2,5] a consistent causal interpretation of the quantum theory. At this point, we therefore complete the model by postulating a particle, which takes the form of a highly localized inhomogeneity that moves with the local fluid velocity, $\mathbf{v}(\mathbf{x},t)$. The precise nature of this inhomogeneity is irrelevant for our purposes. It could be, for example, a foreign body, of a density close to that of the fluid, which was simply being carried along with the local velocity of the fluid as a small floating body is carried along the surface of the water at the local stream velocity of the water. Or else it could be a stable dynamic structure existing in the fluid; for example, a small stable vortex or some other stable localized structure, such as a small pulse-like inhomogeneity. Such structures might be stabilized by some nonlinearity that would be present in a more accurate approximation to the equations governing the fluid motions than is given by (1) and (2).

3. FLUCTUATIONS OF THE MADELUNG FLUID

Thus far we have been assuming that the Madelung fluid undergoes some regular motion, which can in principle be calculated by solving Schrödinger's equation with appropriate boundary conditions. We know, however, that in all real fluids ever met with thus far (and indeed, in all physically real fields also) the motions never take precisely the forms obtained by solving the appropriate equations with the correct boundary conditions. For there always exist random fluctuations. These fluctuations may have many origins. For example, real fluids may be subject to irregular disturbance originating outside the fluid and transmitted to it at the boundaries. Moreover, because the equations of motion flow of the fluid are, in general, nonlinear, the fluid motion may be unstable, so that irregular turbulent motion may arise within the fluid itself. And finally, because of the underlying constitution of the fluid in terms of molecules in random thermal motion, there may exist a residual Brownian movement in the fluid, even for fluid elements that are large enough to contain a great many molecules. Thus, in a real fluid, there are ample reasons why the usual hydrodynamical equations will, in general, describe only some mean or average aspect of the motion, while the actual motion has an addition some very irregular fluctuating components, which are effectively random.

Since the Madelung fluid is being assumed to be some kind of physically real fluid, it is therefore quite natural to suppose that it too undergoes more or less random fluctuations in its motions. Such random fluctuations are evidently consistent within the framework of the causal interpretation of the quantum theory. Thus, there are always random perturbations of any quantum mechanical system which arise outside that system. (Indeed, as we have already shown in a previous paper,[5] the effects of such perturbations are by themselves capable of explaining the probability distribution, $P=|\psi|^2$, at least for certain simple systems.) We may also assume that the equations governing the ψ field have nonlinearities, unimportant at the level where the theory has thus far been successfully applied, but perhaps important in connection with processes involving very short distances. Such nonlinearities could produce, in addition to many other qualitatively new effects, the possibility of irregular turbulent motion. Moreover, we may conceive of a granular substructure of matter underlying the Madelung fluid, analogous to (but not necessarily of exactly the same kind as) the molecular structure underlying ordinary fluids.

We may therefore assume that for any or all of these reasons, or perhaps for still other reasons not mentioned here, our fluid undergoes a more or less random type of fluctuation about the Madelung motion as a mean. Thus, the velocity will not be exactly equal to $\nabla S/m$, nor will the density, ρ, be exactly equal to $|\psi|^2$. All that we require is that the relations $\rho=|\psi|^2$ and $\mathbf{v}=\nabla S/m$ be valid as averages. Indeed, it is not even necessary that the exact velocity be derivable from a potential. Thus, we would have $d\xi/dt=\nabla S'/m+\nabla\times\mathbf{A}$, more generally,[7–9] where $\langle\nabla\times\mathbf{A}\rangle_{\mathrm{Av}}=0$ and $\langle\nabla S'\rangle_{\mathrm{Av}}=\langle\nabla S\rangle_{\mathrm{Av}}$. Hence Schrödinger's equation will not apply to the fluctua-

[9] Such vortex components of the velocity may also explain the appearance of "spin" provided that they could have a regular component as well as a random component. Indeed, in another paper, the Pauli equation will be treated from this point of view. But here we concern ourselves only with a level of precision in which the spin can be neglected, so that Schrödinger's equation is a good approximation for the mean behavior of the fluid.

tions. However, the conservation equation $\partial\rho/\partial t + \text{div}(\rho\mathbf{v})=0$ will be assumed to hold even during a fluctuation. Such an equation is implied almost by the very concept of a fluid; for if there were no conservation, then the model of a fluid would lose practically all of its content.

From the above assumptions, it is clear that if we followed a given fluid element, we would discover that it undergoes an exceedingly irregular motion, which is able in time to carry it from any specified trajectory of the mean Madelung motion to practically any other trajectory. Such a random motion of the fluid elements would, if it were the only factor operating, lead eventually to a uniform mean density of the fluid. For it would on the average carry away more fluid from a region of high density than it carried back. The fact that the mean density remains equal to $|\psi|^2$, despite the effects of the random fluctuations, implies then that a systematic tendency must exist for fluid elements to move toward regions of high mean fluid density, in such a way as to maintain the stability of the mean density, $\bar{\rho}=|\psi|^2$. As for the origin of such a tendency, the question is, of course, not important for the problem that we are treating in this paper. We may, however, suggest by way of a possible explanation that the internal stresses in the fluid are such that whenever ρ deviates from $|\psi|^2$, a kind of pressure arises that tends to correct the deviation automatically. Such a behavior is analogous to what would happen, for example, to a gas in irregular turbulent motion in a gravitational field, in which the pressures automatically adjust themselves in such a way as to maintain a local mean density close to $\rho=\rho_0 e^{-mgz/KT}$ if the temperature T is constant. (In this connection, note that as shown in theoretical treatments of turbulence, the irregular turbulent motions themselves raise the effective "pressure" in the fluid, so that the effective "temperature" T is equal to the sum of the mean kinetic energy of random molecular motion and that of irregular turbulent motion.)

We must now make some assumptions concerning the behavior of the particle-like inhomogeneity. We assume that *even in a fluctuation*, it follows the fluid velocity $\mathbf{v}(\mathbf{x},t)$. Such a behavior would result if the inhomogeneity were a very small dynamic structure in the fluid (e.g., a vortex, or a pulse-like inhomogeneity) or if it were a foreign body of about the same density as the fluid, provided that the wavelengths associated with the fluctuations were appreciably larger than the size of the particle. For in this case, the inhomogeneity would have to do more or less as the fluid did, since it would act, for all practical purposes, like a small element of fluid.

The presence of fluctuations with wavelengths smaller than the size of the body could complicate the problem, especially if we were considering inhomogeneities, such as vortices and pulses, which were dynamically maintained structures in the fluid itself. For, such fluctuations would treat different parts of the inhomogeneity differently, and thus, in general, would tend to lead to a dispersal of the inhomogeneity. Let us recall, however, that we are by hypothesis considering only equations having such nonlinearities in them as to lead to *stable* inhomogeneities. It is true that the equations of ordinary hydrodynamics do not do this. But it is not necessary that the sub-quantum-mechanical Madelung fluid should have exactly the same kinds of properties as are possessed by ordinary fluids. Indeed, we have already seen that instead of the usual classical pressure term, it has a quantum-mechanical internal stress, which depends on the derivatives of the fluid density, rather than on the density itself. Thus, we may reasonably postulate that it also has some characteristically new kind of nonlinear term which leads to stable inhomogeneities. Hence, small fluctuations of wave length much less than the size of the body will merely cause irregular oscillations in the inhomogeneities, the effects of which will, for practical purposes, cancel out. Large fluctuations may destroy the inhomogeneity or transform it into new kinds of inhomogeneity. This could, however, represent certain aspects of the "creation," "destruction," and transformation of "elementary" particles, which is characteristic of phenomena connected with very high energies and very short distances. But in the low-energy domain, which we are treating now, where Schrödinger's equation is a good enough approximation, such processes will not occur.

We see then that if there are fluctuations of wavelength a great deal shorter than the size of the body, they will have a negligible effect on the over-all motions of the body (whether it be a foreign body or a stable dynamic structure in the fluid). In this case, the body will follow the mean velocity of the fluid in a small region surrounding it. To take into account the possibility that such fluctuations may exist, we shall therefore hereafter let $\mathbf{v}(\mathbf{x},t)$ and $\rho(\mathbf{x},t)$ represent respectively the mean velocity and mean density in a small neighborhood surrounding the body, while $\nabla S(\mathbf{x},t)$ and $\rho(\mathbf{x},t)$ represent the means of these quantities in a region that is much larger than the size of the body, but still small enough so that $\psi(\mathbf{x},t)$ does not change appreciably within this region. The consistency of these assumptions evidently requires that the body be very small; but with a choice, for example, of something of the order of 10^{-13} cm for its size, one obtains ample opportunity to satisfy the above assumptions in a consistent way.

It is clear, of course, that fluctuations having a wavelength close to the size of the body will neither cancel out completely, nor will they necessarily cause the body to move exactly with the mean of the fluid velocity in a small neighborhood surrounding it. We may assume, however, that the magnitude of the longer-wavelength fluctuations is so great that we can neglect the effects of fluctuations of these intermediate wavelengths. Thus, a rather wide range exists of kinds of fluctuations that could lead to the type of motion that we are assuming for the inhomogeneity.

On the basis of the above assumptions, it is evident that the inhomogeneity will undergo an irregular motion, analogous to the Brownian motion.[10] Let us now consider a statistical ensemble of fluids, each having in it an inhomogeneity, and let us denote the probability density of such inhomogeneities in the ensemble by $P(\mathbf{x},t)$. Let us further assume that the fluid motion is so irregular that in time a fluid element initially in an arbitrary region $d\mathbf{x}'$ in the domain in which the mean fluid density $|\psi(\mathbf{x},t)|^2$ is appreciable, has a nonzero probability of reaching any other region $d\mathbf{x}$ in this domain. We can then quite easily see in qualitative terms that the probability density $P(\mathbf{x},t)$ must approach $|\psi(\mathbf{x},t)|^2$ as an equilibrium value.

First of all, it is clear that if, for any reason whatever, the distribution $P=|\psi|^2$, is once established, then it will be maintained for all time, despite the random fluctuations in the fluid motion. For the inhomogeneities simply follow the fluid velocity in a small neighborhood surrounding the body. Now by hypothesis the fluid fluctuations are just such as to preserve the equilibrium mean density of $P=|\psi|^2$. Therefore, they must also preserve the equilibrium probability density of particles in the same way.

Let us now consider what happens when P is not equal to $|\psi|^2$. Suppose, for example, that there were a larger number of particles in a specified element of volume than is given by $P=|\psi|^2$. Now, the random motions carry particles away from such an element at a rate proportional to their density in this element. The systematic tendency for particles to come back to the element, which results from their following the fluid, as it drifts back at a rate sufficient to maintain the mean equilibrium density of $\bar{\rho}=|\psi|^2$, will however be just large enough to cancel the loss that would have taken place if the probability density of particles had been $P=|\psi|^2$. Since the density was actually greater than this, more particles are lost than are compensated by the drift back and the density therefore approaches $P=|\psi|^2$. If the probability density of particles in this element had been less than $P=|\psi|^2$, the element would, of course, have tended to gain particles until it had a density of $|\psi|^2$.

In the next section, we shall give a mathematical demonstration of the above result, the correctness of which should however, already be evident from the qualitative considerations cited above.

Finally, we may mention that the picture of a fluid undergoing random motion about a regular mean is only one out of an infinite number of possible models leading to the same general type of theory. Indeed, all the properties that we have assumed for our fluid could equally well belong to some 4-vector field $(\rho,\mathbf{j})$ which was conserved, and which underwent random fluctuations about a mean given (in the nonrelativistic limit) by $\rho=|\psi|^2$ and $\mathbf{j}=(\hbar/2mi)(\psi^*\nabla\psi-\psi\nabla\psi^*)=R^2\nabla S/m$, where ψ is a solution of Schrödinger's equation. And if ρ and $\mathbf{j}$ were assumed to satisfy sufficiently nonlinear equations, there could also exist pulse-like solutions[11] for ρ and $\mathbf{j}$ that moved with a 4-velocity parallel to $(\rho,\mathbf{j})$.

Although it is important to keep in mind these more general possibilities when one is actually trying to formulate a more detailed theory, we have found it convenient in this paper to express our assumptions and results in terms of a hydrodynamical model, because this model not only provides a very natural and vivid physical image of the behavior of the ψ field, but also a simple explanation of the formula, $d\xi/dt=\nabla S/m$, (postulate 2) expressing the velocity of an inhomogeneity in terms of the local mean stream velocity.

4. PROOF THAT PROBABILITY DENSITY APPROACHES FLUID DENSITY IN RANDOM FLUCTUATIONS OF A FLUID

We shall now prove the following theorem. Suppose that we have a conserved fluid that undergoes random fluctuations of the velocity, $\mathbf{v}(\mathbf{x},t)$, and of the density, $\rho(\mathbf{x},t)$, about respective mean values $\mathbf{v}_0(\mathbf{x},t)$ and $\rho_0(\mathbf{x},t)$ [so that $\partial\rho/\partial t+\text{div}(\rho\mathbf{v})=0$ and $\partial\rho_0/\partial t+\text{div}(\rho_0\mathbf{v}_0)=0$]. Suppose in addition that there is an inhomogeneity that follows the fluid motions, with the local stream velocity, $\mathbf{v}(\mathbf{x},t)$. Then if the fluctuations are such that a fluid element starting in an arbitrary element of volume, $d\mathbf{x}'$, in the region where the fluid density is appreciable has a nonzero probability of reaching any other element of volume $d\mathbf{x}$ in this region, it follows that an arbitrary initial probability density of inhomogeneities will in time approach $P=\rho_0(\mathbf{x},t)$.

This theorem is seen to apply to our problems as a special case, in which we set $\rho_0=|\psi(\mathbf{x},t)|^2$ and $\mathbf{v}_0(\mathbf{x},t)=\nabla S(\mathbf{x},t)/m$, where $\psi(\mathbf{x},t)$ satisfies Schrödinger's equation, provided that we regard $\rho(\mathbf{x},t)$ and $\mathbf{v}(\mathbf{x},t)$ as the mean fluid density and velocity in a small region surrounding the inhomogeneity. This theorem is a generalization of a well-known theorem concerning the approach to equilibrium in a Markow process.[12] Essentially, we have generalized the theorem to treat the time-dependent probabilities of transition and time-dependent limiting distributions with which we have to deal in our problem.

To prove this theorem, we note that, as shown in the previous section, a given fluid element follows an extremely irregular trajectory, in which its density $\rho(\mathbf{x},t)$ fluctuates near the mean density $\rho_0(\mathbf{x},t)$. Now because the volume of a given fluid element is always

[10] Brownian motion models of the quantum theory have already been proposed elsewhere, but on a very different basis. See, I. Fenyes, Z. Physik **132**, 81 (1952); W. Weizel, Z. Physik **134**, 264 (1953); **135**, 270 (1953).

[11] See L. de Broglie, *La Physique Quantique, Restera-t-elle Indeterministe* (Gauthier-Villars, Paris, 1953), where the idea of L. de Broglie and J. P. Vigier on this subject are discussed.

[12] W. Feller, *Probability Theory and Its Applications* (John Wiley and Sons, Inc., New York, 1950).

changing in accordance with the changing mean fluid density in the new regions that it enters, it is rather difficult in rectangular coordinates to keep track of how much fluid is transferred on the average from one element of volume to another. To facilitate the treatment of the problem, we shall therefore take the preliminary step of introducing a new set of coordinates, $\xi_1(\mathbf{x})$, $\xi_2(\mathbf{x})$, $\xi_3(\mathbf{x})$, which are so defined that an elementary cell in the space of ξ_1, ξ_2, ξ_3 always contains a mean quantity of fluid proportional to its volume.

Such a set of coordinates is easily defined. For the mean quantity of fluid in a given volume element is

$$dQ=\rho_0(\mathbf{x},t)d\mathbf{x}=\rho_0(\mathbf{x},t)J(\partial x_\mu/\partial\xi_\nu)d\xi_1d\xi_2d\xi_3$$
$$=\rho_0 d\xi_1d\xi_2d\xi_3/J_0(\partial\xi_\nu/\partial x_\mu),$$

where $J(\partial\xi_\nu/\partial x_\mu)$ is the Jacobian of the transformation. Now we want to have $J(\partial\xi_\nu/\partial x_\mu)=c\rho_0(\mathbf{x},t)$ (where we shall choose c to be unity for convenience).

Since there is only one equation, it is clear that only one of the ξ_ν can be defined in this way, so that the other two can be chosen according to what is convenient. Thus, if we fix the forms of ξ_2 and ξ_3, we see that the above equation becomes a linear differential equation defining ξ_1, in terms of ξ_2, ξ_3, and ρ_0. Such an equation always has solutions wherever ξ_2, ξ_3, and ρ_0 are regular. There may exist singular points or curves, but we shall later show how these are to be dealt with.

As an example, consider a cylindrically symmetric density function $\rho(R)=e^{-R}/R$. We first express the volume element in cylindrical polar coordinates (with $R^2=X^2+Y^2$):

$$\rho(R)RdRd\phi dZ=e^{-R}dRd\phi dZ.$$

Now we want $e^{-R}dR=d\xi_1$, or $\xi_1=e^{-R}$. As for ξ_2 and ξ_3, we can in this case leave them equal to ϕ and Z respectively.

Here we see that when R goes from 0 to ∞, ξ_1 goes from unity to zero. This is an example of a characteristic property of the ξ_ν space to be limited in volume when the function $\rho_0(R)$ is appreciable only in a limited domain. Such a property is to be expected, because we are mapping the x_μ on the ξ_ν in just such a way that each region maps into a new volume proportional to the amount of fluid originally in that region. Thus even infinite regions of x_μ space may map onto negligible regions of ξ_ν space, if they contain negligible quantities of fluid.

The solution of the differential equation for ξ_1, will lead in general to multiple-valued functions. This, however, causes no trouble, as we need merely establish a convenient cut somewhere which defines which branch of the function that we are using. Thus the transition to cylindrical polar coordinates, $R^2=X^2+Y^2$; $\phi=\tan^{-1}\times(Y/X)$, leads to a multiple valued function for ϕ, but we deal with this problem by establishing a cut, say at $\phi=0$, and then defining the range of variation of ϕ as being from zero to 2π. In order to cover the entire XY plane only once, a similar definition can be made with any multiple-valued function.

If $\rho_0(\mathbf{x},t)$ vanishes at certain points, then at those points we cannot solve for all the ξ_ν in terms of the x_μ (as, for example, in cylindrical polar coordinates we cannot solve for ϕ at $R=0$). As long as $\rho_0(\mathbf{x},t)$ vanishes only at a set of isolated points, or at most, on a set of one-dimensional curves, where will be no real difficulty. For the vanishing of $\rho_0(\mathbf{x},t)$ means only (as in the case of cylindrical polar coordinates) that some of the ξ_ν are not defined along these curves. To avoid any ambiguities arising from the lack of definition, we may surround each of these curves with a tube, as small in radius as we please, and thus exclude them from the region under consideration without excluding any significant physical effects.

If, however, there are two-dimensional surfaces where $\rho_0(\mathbf{x},t)=0$, this creates more serious mathematical difficulties. Since such surfaces do not, in fact, arise in any real problem of interest to us,[13] we shall assume that $\rho_0(\mathbf{x},t)$ vanishes at most on a set of one-dimensional curves.

Finally, let us note that since ρ_0 changes with time, our ξ_ν will change with time correspondingly. Thus, we are adopting a moving set of coordinates (but not in general one that moves with the mean motion of the fluid elements).

In the space of the ξ_ν, the mean fluid density will be a constant which also does not change with time. As a result, the problem of describing the fluctuations will be greatly simplified. For in the ξ_ν space there is no tendency for the fluctuation to favor any special region since the equilibrium density, which was $\rho_0(\mathbf{x},t)$ in rectangular coordinate, is now a constant. Thus, in the ξ_ν space, the fluctuations have a truly random character, independent of the fluid density at any particular point.

We are now ready to set up the equations governing

[13] In the case of interest to us, $\rho_0=|\psi(\mathbf{x},t)|^2$. At first sight, it may seem that we shall have to be concerned with surfaces on which ρ_0 vanished, because in a perfectly stationary state, $\psi(\mathbf{x},t)$ can be zero on certain nodal surfaces. In the case of a perfectly stationary state, ψ can be real [or more generally, writing $\psi=U(\mathbf{x},t)+iV(\mathbf{x},t)$, we may have a functional relationship between $U(\mathbf{x},t)$ and $V(\mathbf{x},t)$ permitting both to vanish on some two-dimensional surface]. However, for the general complex function ψ, which we obtain in a nonstationary state, it may be shown that there is no such functional relation between U and V, so that ψ can vanish at most on a set of one-dimensional curves.

Now a *perfectly* stationary state is an abstraction that never really exists. For all systems that have ever been dealt with are perturbed to some extent by interactions with other systems. Thus, in a gas, a hydrogen atom suffers 10^{12} collisions per second. In a metal, the electrons suffer a correspondingly large number of collisions with each other and with the cores. In the nucleus, there is a continual process of perturbation due to the fluctuating electronic and ionic fields acting on the spin and quadripole moments of the nuclei. Even in interstellar space, atoms undergo at least one collision with electrons in 10^7 seconds. Thus, all states are slightly nonstationary, and no perfectly nodal planes of the ψ function ever really appear in nature.

A set of perfectly nodal surfaces could interfere with our proof that $P\to|\psi|^2$; for they would represent surfaces that would never be crossed so that the regions on different sides of these surfaces could be completely isolated from each other.

the changes of the probability density $P(\mathbf{x},t)$ for the inhomogeneities. We first transform to the ξ_ν space, writing

$$P(\mathbf{x},t)d\mathbf{x}=\frac{P(\mathbf{x},t)d\xi}{J(\partial\xi_\nu/\partial x_\mu)}=\frac{P(\mathbf{x},t)}{\rho_0(\mathbf{x},t)}d\xi=F(\xi,t)d\xi, \quad (3)$$

where we have defined the vector $\xi=(\xi_1,\xi_2,\xi_3)$ in the ξ_ν space, with the volume element, $d\xi=d\xi_1 d\xi_2 d\xi_3$. The probability density for the space of the ξ_ν is clearly $F=P/\rho_0$. To prove that $P\rightarrow\rho_0$, we then merely have to show that in ξ_ν space, $F(\xi,t)$ approaches a constant.

We now define the probability that fluid in an element $\delta\xi$, centered at the point ξ at the time t, has in the process of fluctuation come from an element $\delta\xi'$ at an earlier time t' with its center ξ' lying in a region $d\xi'$. (Note that $\delta\xi'$ is the magnitude of the volume element,[14] whereas $d\xi'$ is the size of the cell in which the center of the volume element was located at the time t'). This probability is

$$dP=K(\xi,\xi',t,t')d\xi'. \quad (4)$$

Clearly, by definition,

$$\int K(\xi,\xi',t,t')d\xi'=1. \quad (5)$$

Now the exact form of $K(\xi,\xi',t,t')$ will depend on the precise nature of the fluctuations that are taking place in the fluid. We shall see, however, that in order to prove that $F(\xi,t)\rightarrow 1$, it is sufficient to assume that $K(\xi,\xi',t,t')$ fails to be zero over the part of ξ space corresponding to the region of $\mathbf{x}$ space in which $\rho_0(\mathbf{x},t)$ is appreciable. This is clearly just a mathematical expression of the assumption appearing in the first part of this section that there is a nonzero probability that an element starting at any point $\mathbf{x}$ in this region has a nonzero probability of arriving at any other point x' in the region.

Note, however, that the region of $\mathbf{x}$ space in which ρ_0 is appreciable will include, for practical purposes, the whole of the ξ space (except for a region of negligible dimensions). Thus, we may postulate that $K(\xi,\xi',t,t')$ fails to be zero in the whole of ξ space (except possibly along some one-dimensional curves where $\rho_0(\mathbf{x},t)$ may be zero, which we can exclude by means of tubes of negligible dimensions).

As for other properties of K, they are irrelevant for our purposes here, although we shall discuss some of them in Sec. 6, in another connection.[15]

Let us now discuss the motions of the inhomogeneities. Since these latter follow the fluid in its fluctuations, it is easily seen that the probability density of inhomogeneities, $F(\xi,t)$, is just the average of $F(\xi',t')$ weighted with the probability $K(\xi,\xi',t,t')$. Thus,

$$F(\xi,t)=\int K(\xi,\xi',t,t')F(\xi',t')d\xi'. \quad (6)$$

Now, let $\xi_M(t)$ represent the value of ξ for which $F(\xi,t)$ is a maximum, $\xi_m(t)$ the value for which it is a minimum. (If there is more than one pair of such points, let us consider any single pair.) We also let $F_{\max}(\xi,t)=M(t)$, and $F_{\min}(\xi,t)=m(t)$. Setting $\xi=\xi_M(t)$ in Eq. (6), and using (5), we obtain

$$M(t)=\int K(\xi_M(t),\xi',t,t')F(\xi',t')d\xi'$$
$$\overline{\overline{<}}\int K(\xi_M(t),\xi',t,t')M(t')d\xi'=M(t'); \quad (7)$$

and with $\xi=\xi_m(t)$ in Eq. (6), we get similarly

$$m(t)=\int K(\xi_m(t),\xi',t,t')F(\xi',t')d\xi'$$
$$\overline{\overline{>}}\int K(\xi_m(t),\xi',t,t')m(t')d\xi'=m(t'). \quad (8)$$

Thus,

$$M(t)\overline{\overline{<}}M(t'), \quad (9a)$$
$$m(t)\overline{\overline{>}}m(t'). \quad (9b)$$

In order for the equal sign to hold in Eq. (9a), it is necessary that $F(\xi',t')$ be a constant. For by hypothesis, $K(\xi,\xi',t,t')$ fails to vanish anywhere in the ξ space; and if $F(\xi',t')$ is *not* a constant, then the integral (7) must obtain contributions from regions in which $F(\xi',t')<M$. Similarly, we can show that the equal sign can hold in (9b) only if $F(\xi',t')$ is a constant. But if $F(\xi',t')$ is a constant in ξ space, then by (6) we have

$$F(\xi,t)=F(\xi',t')\int K(\xi,\xi',t,t')d\xi'=F(\xi',t').$$

Thus, $F(\xi',t')=$constant is also an equilibrium solution, since it does not change with the passage of time. The result, of course, is more or less to be expected from the physical argument given at the beginning of this section showing that $P=\rho_0(\mathbf{x},t)$ is an equilibrium solution, so that $F=P/\rho_0=$constant must likewise be one. We conclude then that if $F(\xi',t')$ is not a constant, Eqs. (9a) and (9b) must be written as

$$M(t)<M(t'), \quad (10a)$$
$$m(t)>m(t'). \quad (10b)$$

[14] On the average, $\delta\xi$ will not change as the fluid element moves because the fluid density fluctuates near a constant volume in space.

[15] It may be noted at this point that the kernel $K(\xi,\xi',t,t')$ already contains implicit within it a description of the mean fluid velocity $\nabla S/m$. To show this, consider $t-t'=\delta t$ to be a small interval of time. Then $K(x,x',t,t'-\delta t)$ will be large in only a small region of ξ space corresponding in $\mathbf{x}$ space to a region centered around $(x-x'-\nabla S\delta t/m)=0$. The motion of the center of this region describes the mean fluid velocity. The spread of this region describes the random deviations from the mean. In a typical random diffusion process, this width is given by $(\Delta x)^2\sim\delta t$, for short times. For longer times, the functional form of K is determined in a complicated way, which is however of no concern to us in this paper.

Now we can show that Eqs. (10a) and (10b) imply that $F(\mathbf{x},t)$ must approach a constant, with the passage of time. To do this, let us consider a series of times, $t_1, t_2, t_3 \cdots t_n, t_{n+1} \cdots$. We apply (10a) and (10b) from one element of the series of times to the next. Thus

$$M(t_n) < M(t_{n-1}), \tag{11a}$$

$$m(t_n) > m(t_{n-1}). \tag{11b}$$

It is clear that $M(t_n)$ and $m(t_n)$ must each approach constant limits. For $M(t_n)$ is always decreasing and yet remains greater than some fixed number, $m(t_s)$, where t_s is any element of the series such that $t_n > t_s$. Similarly $m(t_n)$ is always increasing and yet less than $M(t_s)$. Now there are just two possibilities: (a) The two constant limits are different; (b) they are the same. We easily see that alternative (a) is self-contradictory. To do this, we denote the two limits by M and m, respectively. Then $M-m=\lim[M(t_n)-m(t_n)]$. But by (11a) and (11b), we have

$$M-m<\lim_{n\to\infty}[M(t_{n-1})-m(t_{n-1})] =\lim_{n\to\infty}[M(t_n)-m(t_n)]=M-m.$$

Because this is a contradiction, alternative (b) must hold. Then $F(\xi,t)$ must approach a constant limit, and $P(\mathbf{x},t)$ must approach $a\rho_0(\mathbf{x},t)$, where a is a constant. If, as happens in quantum theory, the integral of $\rho_0(\mathbf{x},t)$ is normalized to unity, then since by definition the integral of P is also normalized to unity, we must have $a=1$, and

$$P(\mathbf{x},t)\to\rho_0(\mathbf{x},t). \tag{12}$$

5. APPLICATION TO DIRAC EQUATION AND EXTENSION TO MANY-PARTICLE PROBLEM

We may apply the preceding results to the causal interpretation of the Dirac equation,[16] where, as in the Schrödinger equation, we have a stream velocity, $\mathbf{v}_0=\psi^*\alpha\psi/\psi^*\psi$, and a conserved density, $\rho_0=\psi^*\psi$. If we assume a fluid of the same kind as that treated in Sec. 4, and replace $\nabla S/m$ by $\psi^*\alpha\psi/\psi^*\psi$ and $|\psi|^2$ by $\psi^*\psi$, then according to the results of Sec. 4, the probability density will ultimately approach $\psi^*\psi$.

Our results can also be extended very readily to the case of many particles. We first discuss this extension in a purely formal way. We have a wave function, $\psi(\mathbf{x}_1,\mathbf{x}_2\cdots\mathbf{x}_N,t)$, defined in a $3N$-dimensional configuration space. Writing $\psi=R\exp(iS/\hbar)$, we have a set of $3N$ velocity fields, $\mathbf{v}_n=\nabla_n S(\mathbf{x}_1,\mathbf{x}_2,\cdots\mathbf{x}_N,t)$, where ∇_n refers to differentiation with respect to the coordinates of the nth particle. We have a conservation equation in the configuration space.[17] We may now assume that each particle follows the line of flow given by $\mathbf{v}_n(\mathbf{x}_1,\mathbf{x}_2,\cdots\mathbf{x}_N,t)$. Thus, our model is formally just a $3N$-dimensional extension of the model given previously. Hence, if we assume random fluctuations of the $3N$-dimensional velocity field, we shall obtain the result that the probability density in configuration space, $P(\mathbf{x}_1,\mathbf{x}_2,\cdots\mathbf{x}_N,t)$, approaches $|\psi(\mathbf{x}_1,\mathbf{x}_2,\cdots\mathbf{x}_N,t)|^2$.

To obtain a possible physical picture of the meaning of this model, we may use the causal interpretation of the N-particle problem recently proposed by de Broglie.[18] De Broglie has shown that the usual formulation in terms of a wave function in the $3N$-dimensional configuration space can be replaced by an equivalent formulation, according to which each particle is accompanied by its own 3-dimensional wave field, which depends on the precise locations of the other $(N-1)$ particles. Since each wave field satisfies its own Schrödinger's equation, the preceding demonstration still applies.

The above model would imply that each particle moves in its own fluid, and that the fluids interpenetrate each other. For the case of equivalent particles, however, de Broglie has suggested that all particles can be regarded as moving in a common three-dimensional fluid, the velocity of which, at any point $\mathbf{x}$, is dependent on the locations of all the particles, $\mathbf{x}_n$. Thus, we would merely need as many fluids as there are types of particles.

6. ON THE RELATION BETWEEN THE THEORY OF MEASUREMENTS AND FLUCTUATIONS IN THE ψ FIELD

We have demonstrated that with time, the limiting distribution, $P=|\psi|^2$, will be established for any functional form of $K(\xi,\xi',t,t')$, at least within a region which is such that $K(\xi,\xi',t,t')$, does not vanish for any pair of points ξ' and ξ in the region in question. But without a further specification of the $K(\xi,\xi',t,t')$, the rate of approach to the limiting distribution cannot be estimated.

The very fact that no conclusion drawn from the assumption that $P=|\psi|^2$ has as yet been contradicted experimentally, suggests, however, that at least to a fairly high degree of approximation, P is equal to $|\psi|^2$ in all quantum-mechanical systems which have thus far been investigated. Hence, we are led in our model to assume that the existing fluctuations are at least rapid enough to insure the approximate maintenance of the relation, $P=|\psi|^2$ in the very wide variety of systems which has thus far been studied.

In connection with the theory of measurements, however, there arises an important case in which the rate of approach to the equilibrium distribution must be quite slow, if the theory as a whole is to be consistent. This is the case of two wave packets separated by a classical order of distance, throughout which the mean density $|\psi|^2$ is completely negligible.

To show why this case is important, let us recall

[16] D. Bohm, Progr. Theoret. Phys. (Japan) **9**, 273 (1953).
[17] See reference 2, Paper I, Eq. (16).
[18] See reference 11; also Compt. rend. **235**, 1345, 1372 (1953).

briefly some results of the theory of measurements given in a previous paper.[2]

It was shown that in a measurement process, the interaction between measuring apparatus and observed system breaks the wave function into a series of classically separated packets, corresponding to the various possible results of the measurement. The particle, however, enters one of the packets and thereafter remains in it. It is important that the particle remain in this packet; because if it does, the other packets will never play any physical role, so that they can thereafter be neglected and the complete wave function replaced by a simplified one corresponding to the actual result of the measurement. Thus, we understand how a measurement can come to have a definite result, despite the spread of the wave function over a range of possibilities.

Now, if the introduction of a random fluctuation of the ψ field led to an appreciable diffusion of the particle from one of these classically separated packets to another, the above definiteness of the result of a measurement would be destroyed. It is essential therefore for the over-all consistency of the theory that the probability that the particle diffuse across a large region where $\rho_0(\mathbf{x},t)$ is very small shall be negligible.[19]

It is easy to see, however, that almost any reasonable assumptions concerning the fluctuations will lead to this result. For the mean current of particles is $\langle \rho \mathbf{v} \rangle_{\mathrm{Av}}$. Now ρ is everywhere of the order of magnitude of $\rho_0(\mathbf{x},t)$, which is by hypothesis very small in the region between the wave packets. Thus a large probability of a fluctuation that would carry a particle across this space would mean an enormous fluctuation velocity in this region. The mere assumption that fluctuation velocities do not differ by large orders of magnitude in different parts of the fluid is therefore sufficient to insure that the probability of diffusion across this space be very small.

7. CONCLUSION

The essential result of this paper has been to show that the probability density $P=|\psi|^2$ follows from reasonable assumptions concerning random fluctuations of the ψ field. Now, it has already been demonstrated[2] that once the probability distribution $P=|\psi|^2$ has, for any reason whatever, been set up in a statistical ensemble of quantum-mechanical systems, then the results predicted for all measurement processes will be precisely the same in the causal interpretation as in the usual interpretation. The difference between the two points of view, however, is this: in the usual interpretation, the irregular statistical fluctuations in the observed results[20] obtained in general when we make very precise measurements on *individual* atomic systems are assumed, so to speak, to be fundamental elements of reality, since it is supposed that they cannot be analyzed in more detail, and that they cannot be traced to anything else.[21] In the model that we have proposed here, however, the statistical fluctuation in the results of such measurements are shown to be ascribable consistently to an assumed deeper level of irregular motion in the ψ field.

In this paper we have proposed as a possible picture of this deeper level the more specific model of a fluid, undergoing a random fluctuation of its velocity and density about certain mean values determined from Schrödinger's equation, and having in it an inhomogeneity that follows the local stream velocity of the fluid. Of course, this proposal has not yet reached a definitive stage, since we have given only a very general description of the assumed fluctuations and of the properties of the inhomogeneity. Nevertheless, such a model, incompletely defined in character as it is, already suggests a number of interesting questions.

For example, the fluid may have vortex motion. In another paper[22] it will be shown that such vortex motion provides a very natural model for the nonrelativistic wave equation of a particle with spin (the Pauli equation). Work now in progress indicates that a generalization of such a treatment to relativity may yield a model of the Dirac equation.

Another interesting problem to be studied is the possible effects of the assumption of nonlinear equations for the ψ field, which could, as we have seen in Sec. 2, explain the existence of the irregular fluctuations that lead to $P=|\psi|^2$. Such nonlinear equations can lead to many qualitatively new results. For example, it is known that they have a spectrum of stable solutions having localized pulse-like concentrations of field,[23] which could describe inhomogeneities such as we

[19] Note that the slowness of this particular type of diffusion does not interfere with the validity of the relation $P=|\psi|^2$, for the wave function as a whole (i.e., over a whole set of wave packets). For the relation $P=|\psi|^2$ will already have been established by random fluctuations before the measurement took place; and as we have seen, once established, the relationship persists and is not thereafter altered by the fluctuations no matter what happens. But what we have been discussing is *another* probability; namely, the probability that if a particle has entered a given packet, it will within a given time diffuse to another packet. It is this probability that is negligible.

[20] Let us recall that as discussed in reference 5, Sec. 3, there exist real observable large-scale phenomena obtained in a measurement process, which depend on the properties of *individual* atoms (e.g., clicks of a Geiger counter, tracks in a Wilson chamber, etc.)

[21] For example, they cannot in general be ascribed to the uncontrollable actions of the measuring apparatus, as demonstrated by Einstein, Rosen, and Podolsky, Phys. Rev. **47**, 774 (1933) and also D. Bohm, *Quantum Theory* (Prentice Hall Publications, New York, 1951), p. 614. As Bohr has made clear [Phys. Rev. **48**, 696 (1935)] the measuring apparatus plus observed object must be regarded as a single indivisible system which yields a statistical aggregate of irregularly fluctuating observable phenomena. It would be incorrect, however, to suppose that these fluctuations originate in anything at all. They must simply be accepted as fundamental and not further analyzable elements of reality, which do not come from anything else but just exist in themselves. For a complete discussion of this problem, see, *Albert Einstein, Philosopher-Scientist*, Paul Arthur Schilpp, Editor (Library of Living Philosophers, Evanston, 1949).

[22] Bohm, Tiomno, and Schiller (to be published).

[23] Finkelstein, LeLevier, and Ruderman, Phys. Rev. **83**, 326 (1951).

have been assuming in this paper. Such pulse-like concentrations of field would also tend, for many types of field equations, to follow the local stream velocity.[11] The transitions between different possible forms of the inhomogeneous pulse-like part of the solution, combined with transitions between various modes of vibration in the rest of the fluid, could perhaps describe changes from one type of particle to another. Thus, we see that at least in its qualitative aspects, the model seems to have possibilities for explaining some of the kinds of phenomena that are actually found experimentally at the level of very small distances.

The authors would like to express their gratitude to the Conselho Nacional de Pesquisas of Brazil and the Section des Relations Culturelles of France, which provided grants that made this research possible.

This page is deliberately left blank.

Reprinted from *Foundations of Physics*, Vol. 13, No. 2, pp. 253-285, Copyright (1983) with permission from Kluwer Academic/Plenum Publishers.

Dirac's Aether in Relativistic Quantum Mechanics

Nicola Cufaro Petroni[1] **and Jean Pierre Vigier**[2]

Received August 27, 1982

The introduction by Dirac of a new aether model based on a stochastic covariant distribution of subquantum motions (corresponding to a "vacuum state" alive with fluctuations and randomness) is discussed with respect to the present experimental and theoretical discussion of nonlocality in EPR situations. It is shown (1) that one can deduce the de Broglie waves as real collective Markov processes on the top of Dirac's aether; (2) that the quantum potential associated with this aether's modification, by the presence of EPR photon pairs, yields a relativistic causal action at a distance which interprets the superluminal correlations recently established by Aspect et al.; (3) that the existence of the Einstein–de Broglie photon model (deduced from Dirac's aether) implies experimental predictions which conflict with the Copenhagen interpretation in certain specific testable interference experiments.

1. INTRODUCTION

Among all great physicists who founded quantum theory, Professor P. A. M. Dirac stands apart with Einstein and de Broglie. Indeed, once he had given (in a famous book[(1)]) the best known axiomatic presentation of the Copenhagen interpretation of this theory, he never stopped exploring new "strange" ideas, even when they were likely to destabilize an interpretation he had himself put in orbit with his crucial discoveries in electron-positron theory. In a paper written in his honor it is thus only fitting that one should discuss two of Dirac's famous "strange" ideas, i.e.,

- his departure from a pointlike model of particles to justify the propagation of possible superluminal interactions;
- his contribution to the revival (in a new form, of course) of the old aether concept.

Since we want to concentrate essentially on the second idea, we shall only briefly recall the first idea as a possible basis for an interpretation of the experimental confirmation of nonlocal correlations in EPR experiments in photon pair emitted in the singlet state.[(2)]

Clearly the idea that extended particles are nonlocal in nature, i.e., that they can propagate in their interior superluminal interactions and/or information goes back to Dirac. He was the first to notice that if one treats the classical extended electron as a point charge imbedded in its own radiating electromagnetic field, the equations obtained are of the same form as those already in current use, but that in their physical interpretation the finite size of the electron reappears in a new sense: the interior of the electron being a region of space through which signals can be transmitted faster than light. Physically this can be understood as follows. If we send out a pulse from a point A and a receiving apparatus for electromagnetic waves is set up at a point B, and if we suppose that there is an extended electron on the straight line joining A to B, then the disturbed electron will be radiating appeciably at a time a/c before the pulse has reached its center, so that this emitted radiation will be detectable at B at a time $2a/c$ earlier than when the pulse, which travels from A to B with the velocity of light, arrives (here, of course, a is the electron radius). In this way a signal could be sent from A to B faster than light through the interior of an electron.

This possibility of superluminal transmission of signals, of course, is a problem of this model of extended electron in the same sense as the nonlocal correlations in an EPR experiment. As we will discuss later (see Section 3), in order to preserve the Einsteinian causality we must use the concept of relativistic action at a distance, as developed in the predictive mechanics.[(3)] Indeed we will be able to explain causally the nonlocal correlations by means of a nonlocal quantum potential which satisfies the compatibility conditions of the relativistic action at a distance.

This idea has engineered a long set of researches starting for example with Yukawa's bilocal particle model[(4)] and Bohm and Vigiier's liquid droplet model.[(5)] The essential point is that, independently of the internal motions which yield a classical model of spin,[(6)] it has generally been demonstrated by Souriau *et al.*[(7)] that any extended particle model yields an internal rotation of the particle's center of matter density around its center of mass with the exact frequency of de Broglie's relation $\nu_0 = m_0c^2/h$. Of course, such extended particle models have received (until now) no direct experimental support. They open nevertheless interesting paths of research since:

- they offer the possibility to interpret the particle's newly discovered quantum numbers (T, Y, C, B, L,...) in terms of internal oscillations[8];
- they can contain (as suggested before) nonlocal hidden variables which can be utilized to support the nonlocal character of the quantum potential and lead to a causal action-at-a-distance interpretation of nonlocal correlations of EPR paradox.

Let us now come to the second idea, i.e., the reintroduction by Dirac of new possible aether models. As we shall see, this might well turn out to be one of Dirac's main contributions to the new era opened (in the author's opinion) by Aspect's confirmation of the real existence of superluminal correlations in the physical world.[2] In Dirac's own words[9]:

"In the last century, the idea of an universal and all pervading aether was popular as a foundation on which to build the theory of electromagnetic phenomena. The situation was profoundly influenced in 1905 by Einstein's discovery of the principle of relativity, leading to the requirement of a four-dimensional formulation of all natural laws. It was found that the existence of an aether could not be fitted in with relativity, and since relativity was well established, the aether was abandoned.

Physical knowledge has advanced very much since 1905, notably by the arrival of quantum mechanics, and the situation has again changed. If one reexamines the question in the light of present-day knowledge, one finds that the aether is no longer ruled out by relativity, and good reasons can now be advanced for postulating an aether.

Let us consider in its simpler form the old argument for showing that the existence of an aether is incompatible with relativity. Take a region of space-time which is a 'perfect vacuum,' that is, there is no matter in it and also no fields. According to the principle of relativity, this region must be isotropic in the Lorentz sense—all directions within the light cone must be equivalent to one another. According to the aether hypothesis, at each point in the region there must be an aether, moving with some velocity, presumably less than the velocity of light. This velocity provides a preferred direction within the light-cone in space-time, which direction should show itself up in suitable experiments. Thus we get a contradiction with the relativistic requirement that all directions within the light cone are equivalent.

This argument is unassailable from the 1905 point of view, but at the present time it needs modification, because we have to apply quantum mechanics to the aether. The velocity of the aether, like other physical variables, is subject to uncertainty relations. For a particular physical state, the velocity of the aether at a certain point of space-time will not usually be a well defined quantity, but will be distributed over various possible values according to a probability law obtained by taking the square of the modulus of a wave function. We may set up a wave function which makes all values for the velocity of the aether equally probable. Such a wave function may well represent the perfect vacuum state in accordance with the principle of relativity ....

Let us assume the four components v_μ of the velocity of the aether at any point of space-time commute with one another. Then we can set up a representation with the wave functions involving the v's. The four v's can be pictured as defining a point on a three-dimensional hyperboloid in a four-dimensional space, with the equation:

$$v_0^2 - v_1^2 - v_2^2 - v_3^2 = 1, \qquad v_0 > 0 \tag{1}$$

A wave function which represents a state for which all aether velocities are equally probable must be independent of the v's, so it is a constant over the hyperboloid (1). If we form the square of the modulus of this wave function and integrate over the three-dimensional surface (1) in a Lorentz-invariant manner, which means attaching equal weights to elements of the surface which can be transformed into one another by a Lorentz transformation, the result will be infinite. Thus this wave function cannot be normalized."

In other words, Dirac has bypassed all former relativistic objections to a static aether's existence by introducing a chaotic random moving subquantal aether behavior: a step subsequently revived and developed by Bohm and Vigier,[5,10] de Broglie,[11] Sudarshan *et al.*,[12] Cufaro Petroni and Vigier.[13]

To stress and clarify this essential point, we shall briefly recall a few evident results in a simplified case. One can see that Dirac's aether can be easily connected with the original "negative energy sea," which still remains the essential basis for the second quantization formalism as well as for all subsequent field theories. Indeed this negative energy sea can be considered as the first reintroduction of a material vacuum in relativistic quantum mechanics. As one knows,[14] Dirac's original vacuum is characterized (for spin-1/2 particles) by the fact that all positive energy states are not filled whereas all negative energy states are filled. In order to turn this vacuum into Dirac's aether it must be made covariant, i.e., not detectable with a Michelson and Morley experiment. As stated by Dirac,[9] we can satisfy such a condition if we consider that the four-momenta of the particles of Dirac's vacuum are uniformly distributed on the lower mass hyperboloid (see Fig. 1). Indeed, with a Lorentz transformation the equation of the hyper-

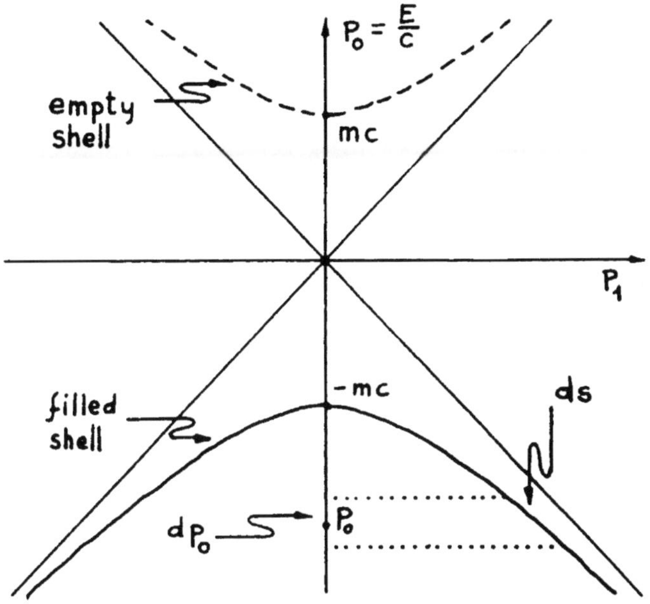

Fig. 1. Two-dimensional representation of Dirac's aether in momentum space: four-momenta are uniformly distributed on the lower filled mass shell.

boloid remains the same and, if the state distribution was uniform along this spacelike surface, i.e., if

$$dN = K\sqrt{|ds^2|} \tag{2}$$

(where dN is the number of states in a section ds of the hyperboloid and K is a constant), the new observer will see the same uniform distribution of states on his hyperboloid.

Of course, the distribution in energy is not constant in this case. We can compute this distribution starting from the obvious statement that in a section dp_0 of the p_0 axis (around a point p_0) we have a number of states $\rho(p_0)\,dp_0$ which equals the number of states in the corresponding ds on the hyperboloid (we fix here $p_1 \geqslant 0$, $p_0 \leqslant -mc < 0$), so that

$$K\,ds = dN = \rho(p_0)\,dp_0 \tag{3}$$

and hence

$$\rho(p_0) = K\frac{\sqrt{|ds^2|}}{dp_0} \tag{4}$$

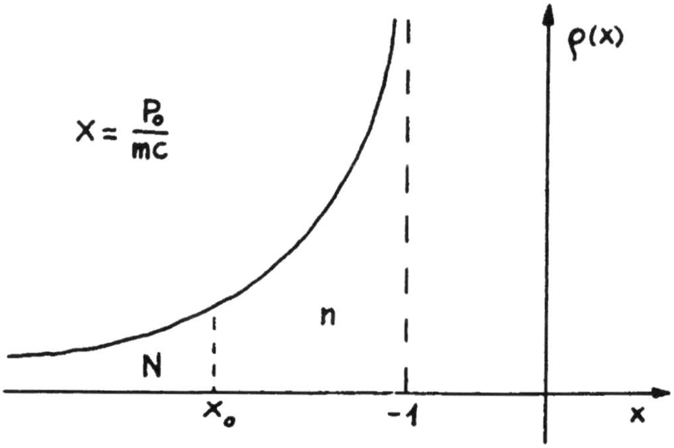

Fig. 2. Plot of the density ρ of states along the energy axis in Dirac's aether.

so that from the explicit expression of ds we have

$$\rho(p_0) = K\sqrt{\left(\frac{dp_1}{dp_0}\right)^2 - 1} = \frac{Kmc}{\sqrt{p_0^2 - m^2c^2}} \tag{5}$$

We have plotted the curve $\rho(x)$ with $x = p_0/mc$ in Fig. 2, and we can remark that our density diverges for $x \to -1$ and tends to zero for $x \to -\infty$. Nevertheless, we can prove that, if we take a fixed $x_0 \in]-\infty, -1[$, the number n of states between x_0 and -1 is always finite; on the contrary, the number N of the remaining states between $-\infty$ and x_0 always diverges. Indeed, we have

$$\begin{aligned} n &= \int_{x_0}^{-1} \rho(x)\, dx = \operatorname{arc\,cosh}(x_0) \\ N &= \int_{-\infty}^{x_0} \rho(x)\, dx = +\infty \end{aligned} \tag{6}$$

This proves that in Dirac's aether distribution the weight of the almost light-like four-momenta must be predominant.

The main poblem now raised by this exposition is: How does Dirac's aether interact with a positive energy particle put in it? Beyond the precise mechanism of this interaction, what about the conservation laws? We can make here some remarks: It is clear that the theory of Dirac's equation requires only that all negative energy levels must be filled with just one particle for each level. Then, if we consider the four-momentum of this particle (for example of energy $\bar{E}$), we see that we have an infinity of possibilities for the p^μ direction (at least two for the two-dimensional case,

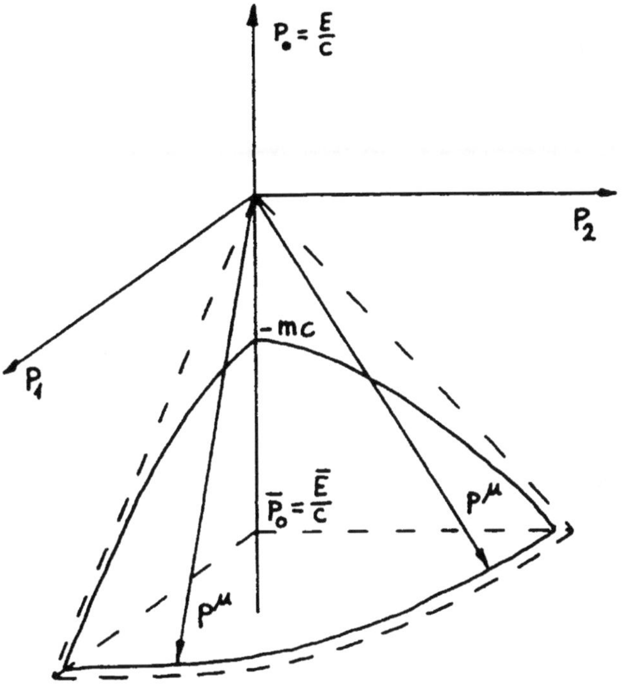

Fig. 3. Three-dimensional representation of Dirac's aether in momentum space: for each $\bar{p}_0$ value there is an infinity of equiprobable possible directions of the corresponding four-momentum p^μ.

but infinite in the other cases; see Fig. 3). So, if the level $\bar{E}$ is filled by only one particle, its p^μ is not completely determined and is uniformly distributed on the corresponding section of the hyperboloid. In this case, even if all the energy levels are filled, a vacuum particle can interact with a positive energy particle by exchanging momentum but no energy at all (with the obvious exception of the case of pair creation or annihilation). Hence we can say that in our subquantal medium a positive energy particle can travel without loss of energy (without "friction") but it can change the direction of its momentum **p** by interacting with the aether particles or produce pair creations and annihilations. As we will see in the subsequent section, these are exactly the possible interactions we need in order to describe the quantum statistics of the Klein–Gordon equation as a stochastic process.

Of course this transition from "hole theory" to Dirac's aether is too simple. In order to interpret the relativistic wave equations of quantum mechanics, Sudarshan *et al.*[12] and Vigier[14] had to complexify Dirac's aether model, i.e., introduce aether models built as superfluid states of particle–antiparticle pairs. In such model, the de Broglie waves are considered as real collective motions in which localized soliton-like energy-carrying particles[15] are surrounded by real physical "pilot waves" which interpret interference phenomena as well as nonlocal quantum correlations in configuration space. One is thus now confronted with the possible confrontation of the offsprings of Dirac's aether with present experimental possibilities.

The preceding discussion determines the plan of our paper. In the second section we shall illustrate in a simplified model how random stochastic jumps at the velocity of light yield (in a particle–antiparticle mixture) the basic relativistic second order equation of wave mechanics, i.e., the Klein–Gordon equation, which, as we will see, contains action at a distance tied to the quantum potential of Bohm[10] and de Broglie.[11] In the third section we shall discuss in terms of such a causal action at a distance the nonlocal quantum interactions which result from the experiments of Aspect.[2] In the fourth section we shall confront the conflicting predictions of the Copenhagen interpretation and of the stochastic interpretation of quantum mechanics in a particular situation in which Bohr's wave-packet-collapse concept conflicts with Maxwell's (i.e., Einstein's and de Broglie's) theory of light.

2. STOCHASTIC DERIVATION OF THE RELATIVISTIC QUANTUM EQUATIONS

According to our plan, we now utilize the concept of Dirac's chaotic aether (which assumes that the particles imbedded in it undergoes random jumps at the velocity of light) as a physical basis for the construction of the so-called stochastic interpretation of quantum mechanics. From this standpoint, the probabilistic character of quantum mechanics is not an irreducible limit of human knowledge but (following Einstein and de Broglie) appears as a natural consequence of the random character of the irregular deviations from the deterministic movement of a classical particle induced by the action of Dirac's chaotic aether. The explicit derivation of the relativistic quantum equations from such a stochastic process is, of course, very important in this type of model, because it materializes the link between the quantum and subquantum features of the microscopic world, so that all the correct predictions of the quantum mechanics can be reproduced in principle in this stochastic interpretation. This, evidently, realizes a hidden-variable theory, but (as we will later see) not a local one, as required by Bell's theorem.[16]

From the beginning of this line of thought[5,10] many demonstrations were published in the nonrelativistic[17] as well as in the relativistic[18] domain, both for spinless and spinning particles.[19] In such stochastic models the nature of the subjacent chaotic medium was not always clearly defined. In the authors' opinion, we are now left with only two lines of research in which this problem is clearly discussed, i.e., (1) the stochastic electrodynamics,[20] which consider charged particles imbedded in covariant electromagnetic vacuum; (2) the stochastic model based on Dirac's aether. In the later case one can deduce[13] from the features of this chaotic relativistic aether the fact that our particle must jump at the velocity of light, and (as we will also see later[19]) this is a fundamental characteristic in deducing the relativistic quantum equations. To show this explicitely we will give here an example of the derivation of the relativistic quantum equation for spinless particles (the Klein–Gordon equation) based on the hypothesis that the stochastic jumps are made at the velocity of light.[21]

To simplify our demonstration, we will limit ourselves to the case in which we assume that the random walks occur on a square lattice in a two-dimensional space-time (see Fig. 4) with coordinates x^0, x^1. We will describe a limit process where in each step we will suppose that our particle, starting from an arbitrary point $P_0(x^0, x^1)$, can only make jumps of fixed length, always at the velocity of light. Of course, this prescription completely determines the lattice of all possible particle positions. On such a lattice the particle can follow an infinity of possible trajectories. In our calculation we will consider first a lattice with fixed dimensions: Indeed for each jump we pose

$$\Delta x^0 = t\tau, \qquad \Delta x^1 = s\tau \qquad (t, s = \pm 1) \tag{7}$$

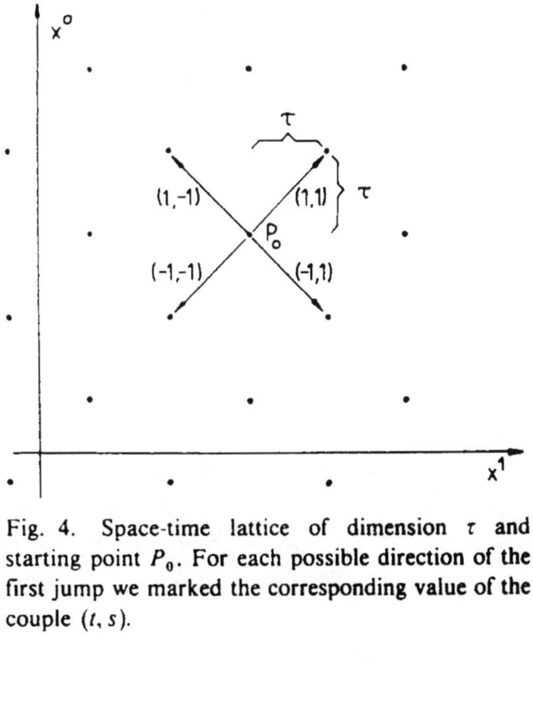

Fig. 4. Space-time lattice of dimension τ and starting point P_0. For each possible direction of the first jump we marked the corresponding value of the couple (t, s).

so that for the velocity we always have (for $\hbar = c = 1$)

$$v = \frac{\Delta x^1}{\Delta x^0} = \frac{s}{t} = \pm 1 \tag{8}$$

Here τ is the parameter which fixes the lattice dimensions: Of course, in order to recover the quantum equations, we will consider later the limit $\tau \to 0$. Moreover, it is clear from (7) and Fig. 4 that on this lattice we also consider the possibility of trajectories running backward in time: We will interpret them as trajectories of antiparticles running forward in time, following the usual Feynman interpretation.[22]

In order to describe random walks on this lattice, let we consider the following Markov process on the set of the four possible velocities of each jump: We define two sets of stochastic variables $\{\varepsilon_j\}$, $\{\eta_j\}$, with $j \in N$, in such a way that the only possible values of each ε_j and η_j are ± 1, following this prescription:

$$\varepsilon_j = \begin{cases} 1 \\ -1 \end{cases} \text{if in the } (j+1)\text{th jump the sign of velocity} \begin{cases} \text{doesn't change} \\ \text{changes} \end{cases}$$

$$\eta_j = \begin{cases} 1 \\ -1 \end{cases} \text{if in the } (j+1)\text{th jump the direction}$$

$$\text{of the time} \begin{cases} \text{doesn't change} \\ \text{changes} \end{cases}$$

with respect to the preceding jth jump. It means that the realization of the signes of ε_j, η_j determines one of the four possible directions of the $(j+1)$th jump on the ground of the direction of the jth jump, as we can see in Fig. 5.

Of course, a sequence $\{\varepsilon_j, \eta_j\}$, with $j \in N$, of values of these stochastic variables completely determines one of the infinite possible trajectories, except for the first jump, because there is no "preceding" jump for it. Thus, starting from $P_0(x^0, x^1)$, in the first jump we can get one of the four possible points $P_1(x^0 + t\tau, x^1 + s\tau)$, and after N jumps, as we can easily see by direct calculation, one of the points $P_N(x^0 + tT_N, x^1 + sD_N)$, where

$$\begin{aligned} T_N &= \tau(1 + \eta_1 + \eta_1\eta_2 + \cdots + \eta_1\eta_2 \cdots \eta_{N-1}) \\ D_N &= \tau(1 + \varepsilon_1\eta_1 + \varepsilon_1\varepsilon_2\eta_1\eta_2 + \cdots + \varepsilon_1\varepsilon_2 \cdots \varepsilon_{N-1}\eta_1\eta_2 \cdots \eta_{N-1}) \end{aligned} \tag{9}$$

We come now to the problem of the assignment of a statistical weight to each trajectory. In order to do that, we introduce for each $(j+1)$th jump a probability for each realization of the signs of the corresponding jth couple ε_j, η_j. In Table 1 we have listed these probabilities for a general ε_j, η_j couple.

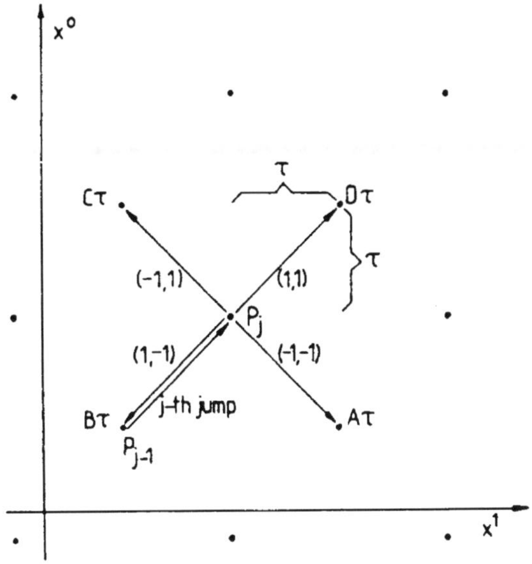

Fig. 5. An example of the four possible successions of two jumps. For each possible $(j+1)$th jump we marked the value of the couple (ε_j, η_j) and the corresponding probability.

Moreover, we suppose that A, B, C, D are constant over all the space-time.

Among these four constants we can also posit the usual relation

$$(A+B+C+D)\tau = 1 \tag{10}$$

In order to derive the Klein–Gordon equation, we consider a function $f(x^0, x^1)$ defined over all the space-time and, generally speaking, with complex values, and then we define the following set of functions:

$$F_N^{t,s}(x^0, x^1) = \langle f(P_N) \rangle = \langle f(x^0 + tT_N, x^1 + sD_N) \rangle \tag{11}$$

Table I. Probabilities for the Four Possible Successions of Two Jumps

ε_j	η_j	Probability
-1	-1	$A\tau$
1	-1	$B\tau$
-1	1	$C\tau$
1	1	$D\tau$

Here $\langle\cdot\rangle$ indicates an average for all the possible points P_N attained, following trajectories constituted by N jumps, starting from $P_0(x^0, x^1)$ with a first jump made in the direction fixed by (t, s).

In fact, it is clear that the terminal point P_N is not uniquely determined by the initial point P_0 and the number of jumps N, because the possibility to choose different trajectories of N jumps. Of course, on the average, the statistical weight of each P_N is calculated from the probabilities associated with the trajectories which lead to P_N, as stated in the previous section. We remark finally that, because the arbitrariness of the starting point P_0, the function $F_N^{t,s}$ is defined over all space-time.

We can start to make this average from the first jump so that, remembering (9) and (11):

$$\begin{aligned}F_N^{t,s}(x^0, x^1) &= \langle f[x^0 + t\tau + t\tau\eta_1(1 + \eta_2 + \cdots + \eta_2 \cdots \eta_{N-1}),\\ &\qquad x^1 + s\tau + s\tau\varepsilon_1\eta_1(1 + \varepsilon_2\eta_2 + \cdots + \varepsilon_2 \cdots \varepsilon_{N-1}\eta_2 \cdots \eta_{N-1})]\rangle\\ &= \langle f(x^0 + t\tau + t\eta_1 T_{N-1}, x^1 + s\tau + s\varepsilon_1\eta_1 D_{N-1})\rangle\\ &= D\tau F_{N-1}^{t,s}(x^0 + t\tau, x^1 + s\tau) + A\tau F_{N-1}^{-t,s}(x^0 + t\tau, x^1 + s\tau)\\ &\quad + B\tau F_{N-1}^{-t,-s}(x^0 + t\tau, x^1 + s\tau) + C\tau E_{N-1}^{t,-s}(x^0 + t\tau, x^1 + s\tau) \qquad (12)\end{aligned}$$

and using (10), that is $D\tau = 1 - (A + B + C)\tau$, we get

$$\begin{aligned}F_N^{t,s}(x^0, x^1) &= F_{N-1}^{t,s}(x^0 + t\tau, x^1 + s\tau)\\ &\quad + A\tau[F_{N-1}^{-t,s}(x^0 + t\tau, x^1 + s\tau) - F_{N-1}^{t,s}(x^0 + t\tau, x^1 + s\tau)]\\ &\quad + B\tau[F_{N-1}^{-t,-s}(x^0 + t\tau, x^1 + s\tau) - F_{N-1}^{t,s}(x^0 + t\tau, x^1 + s\tau)]\\ &\quad + C\tau[F_{N-1}^{t,-s}(x^0 + t\tau, x^1 + s\tau) - F_{N-1}^{t,s}(x^0 + t\tau, x^1 + s\tau)] \qquad (13)\end{aligned}$$

We pass now to the limit $N \to \infty$ (and τ fixed): If we indicate with $F^{t,s}$ the functions for $N \to \infty$ we have, from (13),

$$\begin{aligned}F^{t,s}(x^0, x^1) &= F^{t,s}(x^0 + t\tau, x^1 + s\tau)\\ &\quad + A\tau[F^{-t,s}(x^0 + t\tau, x^1 + s\tau) - F^{t,s}(x^0 + t\tau, x^1 + s\tau)]\\ &\quad + B\tau[F^{-t,-s}(x^0 + t\tau, x^1 + s\tau) - F^{t,s}(x^0 + t\tau, x^1 + s\tau)]\\ &\quad + C\tau[F^{t,-s}(x^0 + t\tau, x^1 + s\tau) - F^{t,s}(x^0 + t\tau, x^1 + s\tau)] \qquad (14)\end{aligned}$$

and then

$$
\begin{aligned}
&-[F^{t,s}(x^0+t\tau,x^1)-F^{t,s}(x^0,x^1)]/t\tau \\
&\quad=(s/t)[F^{t,s}(x^0,x^1+s\tau)-F^{t,s}(x^0,x^1)]/s\tau \\
&\qquad+[F^{t,s}(^0+t\tau,x^1-s\tau)-F^{t,s}(x^0,x^1+s\tau)]/t\tau \\
&\qquad-[F^{t,s}(x^0+t\tau,x^1)-F^{t,s}(x^0,x^1)]/t\tau \\
&\qquad+(A/t)[F^{-t,s}(x^0+t\tau,x^1+s\tau)-F^{t,s}(x^0+t\tau,x^1+s\tau)] \\
&\qquad+(B/t)[F^{-t,-s}(x^0+t\tau,x^1+s\tau)-F^{t,s}(x^0+t\tau,x^1+s\tau)] \\
&\qquad+(C/t)[F^{t,-s}(x^0+t\tau,x^1+s\tau)-F^{t,s}(x^0+t\tau,x_1+s\tau)] \qquad (15)
\end{aligned}
$$

In the limit $\tau \to 0$, when our lattice tends to recover all of space-time, we get the following set of four partial differential equations (one for each possible value of the couple t, s of the first jump):

$$
\begin{aligned}
-\frac{\partial F^{t,s}}{\partial x^0}=\frac{s}{t}\frac{\partial F^{t,s}}{\partial x^1}+\frac{A}{t}(F^{-t,s}-F^{t,s})+\frac{B}{t}(F^{-t,-s}-F^{t,s}) \\
+\frac{C}{t}(F^{t,-s}-F^{t,s}) \qquad (16)
\end{aligned}
$$

where we neglected the arguments (x^0, x^1) of the functions.

If we define now the following four linear combinations of the four functions $F^{t,s}$:

$$
\begin{cases}
\phi=F^{1,1}+F^{-1,-1}+F^{1,-1}+F^{-1,1} \\
\psi=F^{1,1}+F^{-1,-1}-F^{1,-1}-F^{-1,1} \\
\chi=-F^{1,1}+F^{-1,-1}-F^{1,-1}+F^{-1,1} \\
\omega=-F^{1,1}+F^{-1,-1}+F^{1,-1}-F^{-1,1}
\end{cases} \qquad (17)
$$

we can build a new equivalent set of equations by combining Eqs. (16):

$$
\begin{cases}
\dfrac{\partial\phi}{\partial x^0}+\dfrac{\partial\psi}{\partial x^1}=-2(A+B)\chi \\[2ex]
\dfrac{\partial\psi}{\partial x^0}+\dfrac{\partial\phi}{\partial x^1}=-2(C+B)\omega \\[2ex]
\dfrac{\partial\chi}{\partial x^0}+\dfrac{\partial\omega}{\partial x^1}=0 \\[2ex]
\dfrac{\partial\omega}{\partial x^0}+\dfrac{\partial\chi}{\partial x^1}=-2(A+C)\psi
\end{cases} \qquad (18)
$$

By derivation and successive linear combination of equations (18), we have:

$$\begin{cases} \Box\phi = -2(A+B)\dfrac{\partial\chi}{\partial x^0} + 2(C+B)\dfrac{\partial\omega}{\partial x^1} = -2(A+2B+C)\dfrac{\partial\chi}{\partial x^0} \\ \Box\psi = -2(C+B)\dfrac{\partial\omega}{\partial x^0} + 2(A+B)\dfrac{\partial\chi}{\partial x^1} = -2(A+2B+C)\dfrac{\partial\omega}{\partial x^0} \\ \quad - 4(A+B)(A+C)\psi \\ \Box\chi = 2(A+C)\dfrac{\partial\psi}{\partial x^1} = -2(A+C)\dfrac{\partial\phi}{\partial x^0} - 4(A+C)(A+B)\chi \\ \Box\omega = -2(A+C)\dfrac{\partial\psi}{\partial x^0} = 2(A+C)\dfrac{\partial\phi}{\partial x^1} + 4(A+C)(C+B)\omega \end{cases} \tag{19}$$

(where $\Box$ is a two-dimensional d'Alembert operator); and if we pose

$$-B = \frac{A+C}{2}, \qquad 2(A^2 - C^2) = m^2 \tag{20}$$

we finally get

$$\begin{cases} (\Box + m^2)\psi = 0 \\ \Box\phi = 0 \\ (\Box + m^2)\chi = -2(A+C)\dfrac{\partial\phi}{\partial x^0} \\ (\Box + m^2)\omega = 2(A+C)\dfrac{\partial\phi}{\partial x^1} \end{cases} \tag{21}$$

We now make the following remarks:

(a) We can interpret the first equation of (21) as a Klein–Gordon equation. The function ψ, which satisfies this Klein–Gordon equation, is the average of a function f over all the possible final points reached following all the possible trajectories of infinite jumps. In this average, as we can see from (17), we consider also the first jump by supposing that the four possibilities for the signes of t, s are equiprobables.

(b) The functions χ, ϕ, ω which satisfies the remaining equations in (21) have no direct physical interpretation and seem to us to constitute only a formal tool in the deduction of the equation for the complete average ψ. However, we see that in (21) the equation for ψ is not coupled at all with the other equations for χ, ϕ, ω, so that the solution of the Klein–

Gordon equation is absolutely independent of the solutions of the rest of the system.

(c) The previous derivation of (21) from (18) shows that each solution $(\phi, \chi, \psi, \omega)$ of (18) is a solution of (21), but it is possible to show that not all the solutions of (21) are solutions of (18). Indeed, for example, we can verify by direct calculation that

$$\begin{aligned} &\psi = \exp[ip \cdot x] \quad (\text{with } p^2 = m^2) \\ &\chi = \phi = \omega = 0 \end{aligned} \tag{22}$$

is a solution of (21), but it is not a solution of (18). Therefore, it is important to analyze the following question: We proved the statement "the function ψ defined as a stochastic average in (17) and satisfying the system (18) always is a solution of a Klein–Gordon equation." What about the inverse statement "all the solutions of a Klein–Gordon equation are interpretable as stochastic averages satisfying a system like (18)?" We will show here that this inverse statement holds in the following sense: If ψ is an arbitrary solution of the Klein–Gordon equation always, we can determine the functions χ, ϕ, ω in such a way that $(\phi, \chi, \psi, \omega)$ is a solution of (18). In fact, if ψ is an arbitrary solution of the Klein–Gordon equation in (21), we can choose ϕ as an arbitrary solution of $\Box\phi = 0$, and then we determine χ and ω as follows:

$$\begin{aligned} \chi &= \frac{1}{C-A}\left[\frac{\partial\phi}{\partial x^0} + \frac{\partial\psi}{\partial x^1}\right] \\ \omega &= \frac{1}{A-C}\left[\frac{\partial\psi}{\partial x^0} + \frac{\partial\phi}{\partial x^1}\right] \end{aligned} \tag{23}$$

It is only a question of calculation to show now that our $(\phi, \chi, \psi, \omega)$ is a solution of (18) (with $-B = (A + C)/2$) and of (21).

(d) We are confronted here with an old problem characteristic of relativistic quantum mechanics,[23] namely the existence of negative probabilities. Indeed, we can immediately see from (20) that A, B, C, D cannot be simultaneously positive if we want to get the system (21). If, for example, we choose $A, C < 0$, we have

$$B = -\frac{A+C}{2} > 0 \quad \text{and} \quad D = \frac{1}{\tau} - (A + B + C) = \frac{1}{\tau} - \frac{A+C}{2} > 0$$

so that the probabilities of the inversion of the sign of the velocity (A, C) have an opposite sign with respect to the probabilities (B, D) of the noninversion of the sign of the velocity (see Fig. 5). This choice of the signs is also coherent with the definition (17) of ψ as an average, where $F^{1,1}$ and $F^{-1,-1}$ are considered with the same probability but with opposite sign with respect to $F^{1,-1}$ and $F^{-1,1}$. Of course, we have no final answer to the question "what is a negative probability?": We can only quote a proposition for his interpretation[24] in which the negative sign of the probability distribution is interpreted as reflecting the opposite "charge" values of antiparticles in a particle-antiparticle distribution. We further remark that this is a problem which arises every time we are dealing with particles and antiparticles, and hence that it would be very strange not to meet it here where the possibility of trajectories running backward in time on our lattice are interpreted with the presence of pair creation and annihilation.[22] On the other hand, it is clear that if we had not assumed the possibility of the trajectories running backward in time (i.e., the antiparticle behavior) all our statistics would be different since it is possible to show[25] that one obtains in this case a classical diffusion equation that one can not reduce to the quantum Klein–Gordon equation.

3. DETERMINISTIC NONLOCAL INTERPRETATION OF THE ASPECT-RAPISARDA EXPERIMENTS ON THE EPR PARADOX

In this section we are going to utilize the hydrodynamical-stochastic interpretation of the quantum mechanics, physically based on the real existence of a chaotic Dirac's aether, as a starting point for a deterministic interpretation of the recent results of the Aspect–Rapisarda experiments on correlated photon pairs. As is well-known, the paradoxical features of the quantum mechanical description of correlated systems first discussed by Einstein, Podolsky, and Rosen[26] are now experimentally tested, in the form established by Bohm[27] for discrete variables. Recent discussions have convinced physicists that an essential property of the EPR paradox lies in the nonlocal character of quantum correlations, which seem to be in striking contradiction with Einstein's relativistic causal description of nature—and imply causal anomalies.[28] Indeed, in the EPR paper the hypothesis of the noninteraction between two correlated system at a great distance is essential in order to achieve the demonstration of the incompleteness of the quantum mechanics,[29] and Bell's theorem[16] states that there is a measurable difference between the predictions of quantum mechanics and any local hidden-variable theory for correlated particles.

To illustrate this, we briefly recall a typical experimental set-up to check Bell's inequalities. Let us consider a pair of photons (1 and 2) issuing from a cascade source S in a single state of polarization, so that they move in opposite directions parallel to the same x axis. These photons are successively detected through two linear polarizers (L and N) with polarization directions A and B perpendicular to the x axis (see Fig. 6). We know[1] that a photon impinging upon a linear polarizer either passes or is stopped, thus answering yes or no (1 or 0) to the question: "Is your linear polarization found parallel or perpendicular to the direction $A(B)$ of the polarizer $L(N)$?" We can thus compute the probability of the four possible answers to the composite question: "Does the photon 1 pass the polarizer L and the photon 2 the polarizer N?" For this calculation we need only the initial and final states $|i\rangle$ and $|f\rangle$ of our composite system, so that, denoting by (1, 1), (1, 0), (0, 1), and (0, 0) the probabilities of the four possible answers, we compute the probabilities as $|\langle i|f\rangle|^2$.

Of course, if our initial state is

$$|i\rangle = (1/\sqrt{2})(|y_1\rangle|y_2\rangle + |z_1\rangle|z_2\rangle) \tag{24}$$

in terms of state vectors polarized along two orthogonal axes y and z in an $x = \text{const.}$ plane, the final state is, for example, for the case (1, 1),

$$|f\rangle = (\cos A\,|y_1\rangle + \sin A\,|z_1\rangle)(\cos B\,|y_2\rangle + \sin B\,|z_2\rangle) \tag{25}$$

so that we have

$$(1, 1) = |\langle i|f\rangle|^2 = (1/2)(\cos A\,\cos B + \sin A\,\sin B)^2 = (1/2)\cos^2\alpha \tag{26}$$

if $\alpha = A - B$. In an analogous way we get immediately

$$(0, 0) = (1/2)\cos^2\alpha \quad (1, 0) = (0, 1) = (1/2)\sin^2\alpha \tag{27}$$

The crux of the new situation now lies in Bell's proof[16] that this quantum mechanical predictions on the two photon coincidences cannot result from

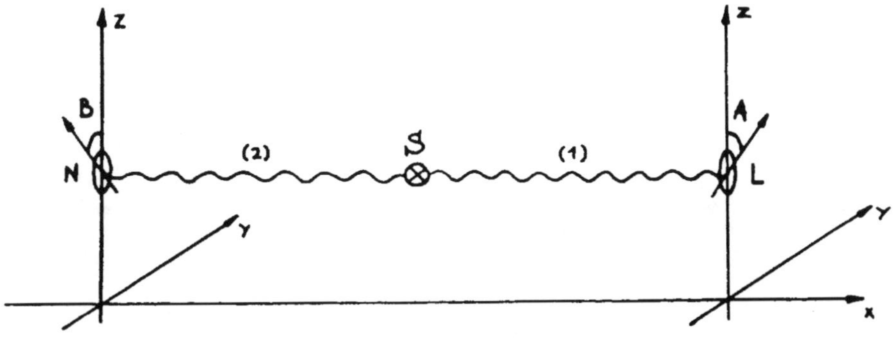

Fig. 6. Schematic view of an EPR–Bohm experiment.

the correlation functions obtained in a local hidden-variable theory. The same result was attained, with another example, in a recent paper by Feynman.[23]

As one knows, despite the almost general confirmation of such quantum predictions in EPR experiments[30] (i.e., of the violation of Bell's inequality) a supplementary device with four photon coincidences was needed to definitely prove the nonlocal character of this quantum correlation.[31] This set-up essentially rests on the use of calcite crystals acting as random switches on the photon paths which orient them, with a 1/2-probability, in the ordinary (O) or extraordinary (E) rays. The photon thus pick at random four possible paths and are subsequantly detected through two pairs of linear polarizers L, L', N, N' (see Fig. 7). The recent result of this experiment,[2] obtained by Aspect's group, confirms the quantum mechanical prediction with great precision for separation of 12 m between the polarizers $L(L')$ and $N(N')$. If the forthcoming Rapisarda experiment also confirms this result, we will be faced with the problem of the interpretation of nonlocal correlations in the microscopic domain.

Two remarks can be made at this stage of the discussion:

(1) There is no possibility left by these experiments, to construct a local hidden-variable theory for quantum mechanics,[16] but it is still possible in principle to build a nonlocal one coherent with a characteristic feature of the hydrodynamical-stochastic interpretation—since the quantum potential for correlated systems is always nonlocal.[32] The problem, as we will now see, is how to construct a coherent causal covariant nonlocal theory.

(2) There is no possibility (as claimed by the authors in another paper[33]) for a observer in L to use this EPR experiment to send macroscopic superluminal signals to the observer in N, because they are always dealing with coincidence experiments. However, we can deduce[33] from

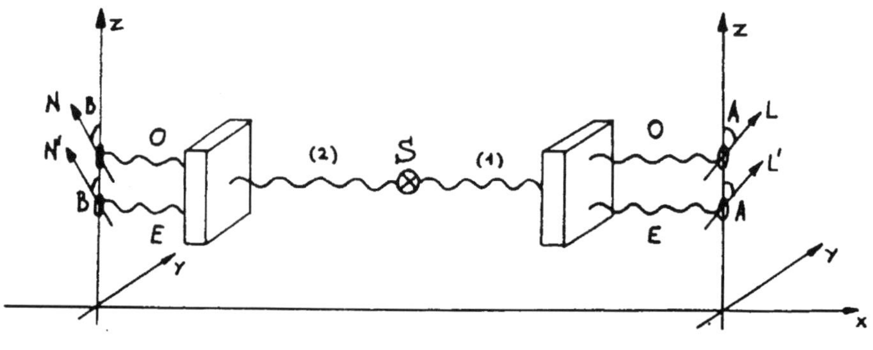

Fig. 7. Schematic view of the Aspect–Rapisarda experiment.

an "*a posteriori*" analysis of the experimental results the existence of past superluminal "exchange of information" between the two photons, in the sense that we can explain by a closed causal chain the existence of correlations and any other variation of that induced by operations on the polarizers only by a sort of nonlocal link between spatially separated events.

In this section we will extend the analysis of the nonlocal character of the quantum potential to the case of spinning particles, in order to show that, for correlated systems, also the spins (and polarizations) are nonlocally connected.

We start with a non-zero mass photon model ($m_\gamma \neq 0$). This is justified: (1) by the well-known fact[34] that the zero-mass limit of a nonzero mass spin-1 Proca particle cannot be physically distinguished from a Maxwell wave, since the so-called transverse waves just correspond to $J_3 = \pm 1$ (i.e., to opposite circular polarizations), while the longitudinal solutions $J_3 = 0$ (pratically decoupled from transverse waves when $m_\gamma \to 0$) describes the Coulomb field when $m_\gamma \to 0$; (2) by the theoretical result that (with $m_\gamma \neq 0$) one has found a classical counterpart (i.e., the Weyssenhoff particle) to the photon field,[35] so that one can determine a classical counterpart of spin for isolated "classical" photons which is distributed[36] in the hydrodynamical representation of the Proca wave equation.

Both in the usual quantum mechanical theory[37] and in the stochastic interpretation of quantum mechanics[38] a system of two correlated photons ($m_\gamma \neq 0$) can be represented by a second rank tensor $A_{\mu\nu}$. As one knows, this compound state of two spin-1 particles can be split [as a consequence of the group representation relation $D(1) \otimes D(1) = D(2) \oplus D(1) \oplus D(0)$] into a symmetric part $A_{\underline{\mu\nu}}$, a skew part $A_{\mu\nu}$, and a trace $A_{\mu\mu}$, representing respectively the $J = 2$, $J = 1$, and $J = 0$ compound states. Since the aforementioned experiments utilize 0-1-0 singlet states cascades, we shall limit ourselves to the $A_{\mu\mu}$, $D(0)$, $J = 0$ singlet case.

Denoting by 1 and 2 the two photons (with coordinates x_1^μ and x_2^μ), we represent our compound state by a scalar field

$$\Phi(x_1, x_2) = A_{1\mu}(x_1) A_2^\mu(x_2) = \exp[R(x_1, x_2) + iS(x_1, x_2)]$$

where $\hbar = c = 1$

As one knows,[3,39] such a scalar field satisfies the system of relations

$$\begin{cases} (\Box_1 + \Box_2 - 2m_\gamma^2)\Phi = 0 \\ (\Box_1 - \Box_2)\Phi = 0 \end{cases} \tag{28}$$

or, equivalently,

$$\begin{cases}(\Box_1 - m_\gamma^2)\Phi = 0 \\ (\Box_2 - m_\gamma^2)\Phi = 0\end{cases} \tag{29}$$

along with the transverse gauge conditions $\partial_{1\mu} A_1^\mu = \partial_{2\mu} A_2^\mu = 0$, the second relation (28) representing the causality constraint in the so-called predictive mechanics with action at a distance.[3] In this case, the Lagrangian of our pair will be

$$\mathcal{L} = m_\gamma^2 \Phi^* \Phi + \partial_{1\mu} \Phi^* \partial_1^\mu \Phi + \partial_{2\mu} \Phi^* \partial_2^\mu \Phi \tag{30}$$

A classical relativistic hydrodynamical analysis[36,40] then allows one to build the energy-momentum tensor for each single photon (from now on, because of the $1 \leftrightarrow 2$ symmetry, we will calculate only the quantities relative to the photon 1), i.e.,

$$\begin{aligned} t_{1\mu\nu} &= \frac{\partial \mathcal{L}}{\partial(\partial_1^\nu A_1^\lambda)} \partial_{1\mu} A_1^\lambda + c.c. - \mathcal{L}\,\delta_{\mu\nu} \\ &= \partial_{1\mu} \Phi^* \partial_{1\nu} \Phi + \partial_{1\mu} \Phi \partial_{1\nu} \Phi^* - \mathcal{L} \delta_{\mu\nu} \end{aligned} \tag{31}$$

From Belinfante's tensor,[40]

$$\begin{aligned} f_{1\mu\nu\lambda} &= \frac{\partial \mathcal{L}}{\partial(\partial_1^\lambda A_1^\rho)} \mathcal{I}_{\mu\nu}^{\rho\sigma} A_{1\sigma} + c.c. \\ &= \tfrac{1}{2}(\partial_{1\lambda} \Phi^*)(A_{2\mu} A_{1\nu} - A_{1\mu} A_{2\nu}) + c.c. \end{aligned} \tag{32}$$

where $\mathcal{I}_{\mu\nu}^{\rho\sigma} = \frac{1}{2}(\delta_{\rho\mu}\,\delta_{\sigma\nu} - \delta_{\rho\nu}\,\delta_{\sigma\mu})$, the spin density tensor becomes (if u_i^μ are the unitary four-velocity of the photons)

$$\tfrac{1}{2} S_{1\mu\nu} = -u_1^\lambda f_{1\mu\nu\lambda} = (A_{1\mu} A_{2\nu} - A_{1\nu} A_{2\mu})\, u_1^\lambda\, \partial_{1\lambda} \Phi^* + c.c. \tag{33}$$

and the spin vector can be written

$$S_{1\mu} = \frac{i}{2} \varepsilon_{\nu\alpha\beta\mu} u_1^\nu S_1^{\alpha\beta} \tag{34}$$

Moreover, denoting now by a dot the derivative along a current line, we can show that, because of the $t_{1\mu\nu}$ symmetry, we have[40]

$$\dot{S}_{1\mu\nu} = \partial_{1\lambda}(u_1^\lambda S_{1\mu\nu}) = t_{1\mu\nu} - t_{1\nu\mu} = 0 \tag{35}$$

From this results that:[41]

(1) $S_{1\mu}$ has a constant length in the sense that $\dot{S}_1^2 = 0$; indeed we see that[40] $S_1^2 = S_{1\mu}S_1^\mu = \frac{1}{2}S_{1\alpha\beta}S_1^{\alpha\beta}$ because of the properties of $\varepsilon_{\alpha\mu\nu\beta}$, the antisymmetry of $S_{1\mu\nu}$, $u_1^\mu u_{1\mu} = -1$, and $u_{1\mu}A_1^\mu = u_{1\mu}A_2^\mu = 0$. Hence $\dot{S}_1^2 = 0$, because we showed that $\dot{S}_{1\mu\nu} = 0$.

(2) The derivative of S_1 is $\dot{S}_1 = (i/2)\varepsilon_{\nu\alpha\beta\mu}S_1^{\alpha\beta}(u_1^\lambda \partial_{1\lambda} u_1^\nu)$, so that it depends in a nonlocal way from the $A_2(x_2)$ contained in $S_1^{\mu\nu}$.

(3) The elements of the photon pairs interact permanently not by exchanging tachyons but through action at a distance, which reflects the disturbance of Dirac's covariant stochastic aether.[13,14] In present experiments the photons are "holding hands" over 12 meters an any disturbance of one is carried superluminally to the other by a phase-like disturbance of the stochastic quantum potential—which includes a quantum torque.

Despite the presence of an action at a distance, it is possible to show[41] that this system is relativistically deterministic, in the sense that we shall now show:

- The system of two $J = 1$ particles can be solved in the forward (or backward) time direction in the sense of the Cauchy problem.
- The paths of the two particles are time-like.
- The formalism is invariant under the Poincaré group $P = T \otimes \mathscr{L}^\uparrow$.

Indeed, writing $P_i^\mu = \partial S/\partial q_{i\mu}$ $(i = 1, 2)$, we can split internal from external variables by writing $P^\mu = p_1^\mu + p_2^\mu$; $y^\mu = (1/2)(p_1^\mu - p_2^\mu)$; $Q^\mu = (1/2)(q_1^\mu + q_2^\mu)$; $z^\mu = q_1^\mu - q_2^\mu$; q_i^μ, and p_i^μ representing pairs of canonical variables. Splitting (29) into real and imaginary parts, we obtain, for the real parts,

$$\begin{cases} (1/2)\,\partial_{1\mu}S\partial_1^\mu S + U_1 = (1/2)\,m_\gamma^2 \\ (1/2)\,\partial_{2\mu}S\partial_2^\mu S + U_2 = (1/2)\,m_\gamma^2 \end{cases} \tag{36}$$

where we have $U_i = -(1/2)(\Box_i R + \partial_i^\mu R \partial_{i\mu} R)$. This separation can be performed in the rest frame of the center of mass of the two photons, where we consider the case of an eigenstate of P_μ, i.e., $\Phi = \varphi(z^\mu)\exp[ik_\mu(x_1^\mu + x_2^\mu)/2]$, where k^μ is a constant timelike vector. In that case we have $(\partial_1^\mu + \partial_2^\mu)R = 0$ and $(\partial_1^\mu + \partial_2^\mu)S = k^\mu$, so that, subtracting Eq. (36), we get $k^\mu(\partial R/\partial z^\mu) = 0$, and hence R only depends on $z_\perp^\mu = z^\mu - (z_\nu k^\nu)k^\mu/k^2$. In order to satisfy the condition $\{y \cdot P, U\} = 0$ for the existence of causal timelike world lines,[3,39] we must now make the substitution $z_\perp^\mu \to \tilde{z}^\mu = z^\mu + (z_\nu P^\nu)P^\mu/P^2$, so that $(\partial_1^\mu + \partial_2^\mu)R = 0$ and $U_1 = U_2 = U(\tilde{z}_\mu)$. In that case the relations (36) represent a pair of causally bound photons connected by a causal action at a distance. Moreover:

(1) The causality condition $P \cdot y = 0$ implies that the Poisson bracket of the two photon Hamiltonians $\{H_1, H_2\}$ is zero, i.e., that their corresponding proper times τ_1 and τ_2 are independent.

(2) $q_i^\mu = x_i^\mu$ in the rest frame of the center of mass Σ_0 ($k_j = 0$).

(3) Subtracting Eq. (36) with $U_1 = U_2$, we get[13] $P \cdot y = 0$, so that $\dot{P}_\mu = 0$, where the dot denotes the operation $(1/2)(\partial/\partial\tau_1 + \partial/\partial\tau_2)$. This yields $P \cdot \dot{y} = 0$, which shows that no energy can be exchanged between the photons in Σ_0, so that no causal anomaly results from this particular type of action at a distance.

(4) We have $\dot{p}_i \cdot P = 0$, so that the paths of both photons remain timelike.

(5) The formalism shows that[42] our causal covariant action at a distance is instantaneous only in Σ_0, and its velocity η can thus be calculated in any other frame Σ by the $\Sigma_0 \rightarrow \Sigma$ corresponding Lorentz transformation. In the particular case of the Aspect–Rapisarda experiment, this immediately yields $\eta = 7{,}57c$ in the laboratory frame.

This analysis implies that the hydrodynamical-stochastic interpretation of the quantum mechanics based on the physical existence of a chaotic Dirac's aether can provide all the essential elements needed to build a nonlocal hidden-variable theory, since the nonlocal quantum potentials and quantum torques satisfy the compatibility conditions[3] required by the predictive mechanics in order to have a relativistically deterministic theory. This also implies that both EPR paradox and the experimentally confirmed violation of Bell's inequality can be completely interpreted in a model that does not imply mysterious retrodictions[43] or *a priori* limitations of our comprehension[44]—since quantum mechanics appears as a statistical manifestation of a subquantum classical, relativistic and deterministic world in which there is also place for actions at a distance whose physical basis is the nonlocal quantum potential or, in other words, the physical existence of de Broglie's waves on Dirac's aether.

The importance of causal action at a distance is now evident. Despite the fact that we are dealing with a nonlocal theory, we claim that there is no possibility left for causal anomalies.[39] Indeed, a perfectly deterministic nonlocal theory is not at all a theory in which we can send superluminal signals in contradiction to relativity and causality,[28] since the existence of such "signals" requires the existence of a "free will," i.e., of somebody who "decides" at a given time to send something to somebody else. If, as claimed in our model, absolutely everything (bodies, men, "free will," etc.) are completely determined, all events are fixed somewhere in space-time, so that we cannot properly speak of "signals." In this scheme, the world is thus describable by a causal ensemble of particle in mutual interaction, the causality implied in it being absolute. The measuring processes themselves and the observers satisfy the same causal laws and are real physical processes with antecedents in time. The measuring process (observer plus apparatus plus observed object) can now be considered as a set of particles which belong to an overall causal process, so that the intervention of a measurement contains no extranatural "free will" or "observer consciousness." Quantum measuring devices now act as spectral analysers[10] which split into subpackets the real de Broglie's waves associated with particles: The particle entering in one of them according to its random causal motion.[45] In brief, there is no "free will" signal production and thus no possible causal paradoxes: Nothing exists beyond the motion and interactions of material particles in a random stochastic aether.

4. HOW DOES A PHOTON INTERFERE WITH ITSELF?

We conclude this paper with a brief discussion of the present theoretical and experimental status of Professor Dirac's initial views on the nature of quantum mechanics illustrated in the first pages of his famous book on quantum mechanics.[1] As every physicist knows, this book contain the deepest and clearest exposition ever made of the basic concepts underlying the Copenhagen interpretation of quantum mechanics. It is thus very important that the physical *gedanken* experiment discussed by him (based on the theory of light in single photon eases) are now about to become testable directly (a natural consequence of technical progress in the field of detection of single photons) and that explicit, realisable (in the author's opinion) experiments are now proposed and discussed in the literature in order to test the validity of the said concepts.[46]

Dirac starts his discussion of the principles of quantum mechanics by a discussion of the principle of superposition of states which he analyzed in the case of isolated photons both for polarization and interference. Since the present experiments are really built to test the validity and signifiance of his analysis for interference, we shall quote him at some length:[1]

"We shall discuss the description which quantum mechanics provides of the interference of photons. Let us take a definite experiment demonstrating interference. Suppose we have a beam of light which is passed through some kind of interferometer, so that it gets split up into two components and the two components are subsequently made to interfere. We may, as in the preceding section, take an incident beam consisting of only a single photon and inquire what will happen to it as it goes through the apparatus. This will present to us the difficulty of the conflict between the wave and corpuscular theories of light in an acute form.

Corresponding to the description that we have in the case of the polarization, we must now describe the photon as going partly into each of the two components into which the incident beam is split. The photon is then, as we may say, in a translational state given by the superposition of the two translational states associated with the two components. We are thus led to a generalization of the term 'translational state' applied to a photon. For a photon to be in definite translational state it need to be associated with one single beam of light, but may be associated with two or more beams of light which are the components into which one original beam has been split. Translational states are thus superposable in a similar way to wave functions.

Let us consider now what happens when we determine the energy in one of the components. The result of such a determination must be either the whole photon or nothing at all. Thus the photon must change suddenly from being partly in one beam and partly in the other to be entirely in one of the beams. This sudden change is due to the disturbance in the translational state of the photon which the observation necessarily makes. It is impossible to predict in which of the two beams the photon will be found. Only the probability of either result can be calculated from the previous distribution of the photon over the two beams.

One could carry out the energy measurement without destroying the component beam by, for example, reflecting the beam from a movable mirror and observing the recoil. Our description of the photon allows us to infer that, *after* such an energy measurement, it would not be possible to bring about any interference effects between the two components. So long as the photon is partly in one beam and partly in the other, interference can occur when the two beam are superposed, but this possibility disappears whe the photon is forced entirely into one of the beams by an observation. The other beam then no longer enters into the description of the photon, so that it counts as being entirely in one beam in the ordinary way for any experiment that may subsequantly be performed on it.

On these lines quantum mechanics is able to effect a reconciliation of the wave and corpuscular properties of light."

This justifies Dirac's famous sentence: "The new theory, which connect the wave function with probabilities for one photon, gets over the difficulty by making each photon go partly into each of the two components. Each photon then interferes only with itself. Interference between two different photons never occurs."

To summarize, this analysis evidently rests (1) on the idea that individual photons interfere only with themselves; (2) on the assumption that one cannot tell through which branch of the interference device the photon goes (i.e., through which slit it passes in the Young hole experiment), since any such detection in one branch would collapse the probability wave of the other branch thus annihilating the interference pattern—even when built by photons coming one by one in independent wave packets; (3) on the description of photons as either waves *or* particles—never the two simultaneously.

As one knows, the only alternative interpretation for the photon case rests on Einstein's[47] and de Broglie's[48] suggestion that individual photons are waves *and* particles, i.e., that there are real Maxwell waves (practically devoid of energy and momentum) which carry (pilot) localized nondispersive concentrations of energy-momentum which correspond to individual photons. In an interference device, for example, the real wave goes through both branches (both slits in the double slit experiment) the photon going throught one slit only. An individual photon is thus influenced by the wave of both slits in the interference observation region—so that it is distributed according to Maxwell's wave superposition principle. This yields in this case the quantum mechanical prediction—since Maxwell's wave are then equivalent to the probabilistic ψ field of quantum mechanics.

Clearly the only experimental way to distinguish between these two interpretation would be:

(a) to discover a means for detecting through which branch (slit) the photon goes without destroying the subsequent interference region;

(b) to utilize such a means to construct a specific precise experimental set-up in which the two preceding interpretations yield conflicting testable predictions.

Let us first discuss point (a). Curiously, the discovery of a possible mean to follow a photon path without destroying its interference properties rests on a typical consequence of wave mechanics itself, i.e., the possibility of duplicating photons by using a 3-photon resonance mechanism initially suggested by Bassini, Cagnac, *et al.*[49] and developed by Gozzini[50]—since the use of such a photon duplicator on one of the interference branches would tell us (by absorbing one of them) by which path it has gone, while the remaining one could still be used for interference detection. The principle of Gozzini's duplicator is simple. Before it, all known laser amplifiers were difficult to use due to parasitic light, specially when one wants to act with highly directional light. Moreover the "copies" of an indident photon are generally emitted in a sample of excited cells, i.e., are not in phase with the exciting photon, even when inserted in a coherent wave packet. The Pisa duplicator rests on the idea that one can stimulate with three photons the transition from the level $3^2S_{1/2}$ to the level $3^2P_{1/2}$ of the sodium (separated by an energy E_{10}) by irradiating sodium vapor with two lasers of frequency ν_1 and incident photons ν_2 such that $2\nu_1 - \nu_2 = E_{10}/h$, according to the

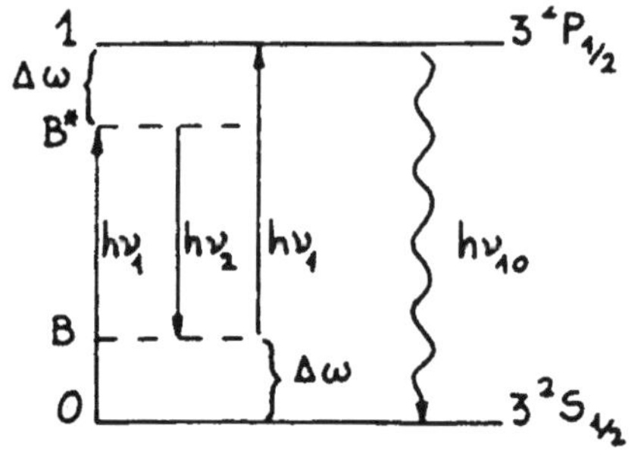

Fig. 8. Scheme of level transitions in Gozzini's duplicator.

scheme of Fig. 8, where the incident absorbed photons are represented by ↑ and the emitted photons by ↓. If this process satisfies the relations $\nu_1 = \nu_0 - \Delta\nu$ and $\nu_2 = \nu_0 - 2\Delta\nu$, ν_0 being the resonance frequency, it is then possible to induce a transition through two intermediate virtual states B and B^*: The absorption of two photons of frequency ν_1 combined with an incoming stimulating photon ν_2 induces the production of two photons $h\nu_2$ of equal phases (since theoretically built in the laser-like process $B^* \to B$) and one luminescence photon $h\nu_0$. This duplication process presents the great interest of eliminating any Doppler contribution, since it does not depend on the sodium atom's momentum if the three photons satisfy the geometry of Figs. 9 and 10, where the relation $\hbar\mathbf{k}_1 + \hbar\mathbf{k}_1 - \hbar\mathbf{k}_2 = 0$ implies total momentum conservation—so that all excited atoms enter resonance independently of their velocity. Since one can operate with the set-up of Fig. 11, one can localize the process at the point P, eliminate the fluorescence $h\nu_0$ with a Fabry–Perot device and, by pulsing the incoming $h\nu_2$ packets, individualize the time of copy creation in the duplicator.

Of course, following an argument of Selleri[51] (who has played a pioneer role in this type of proposals[46]), the use of such a duplicator as path detector implies a simple preliminary test to check that the duplication process is really associated with a passage of a photon $h\nu_2$ through the duplicator. It goes as follows: let us consider (see Fig. 12) the arrival of

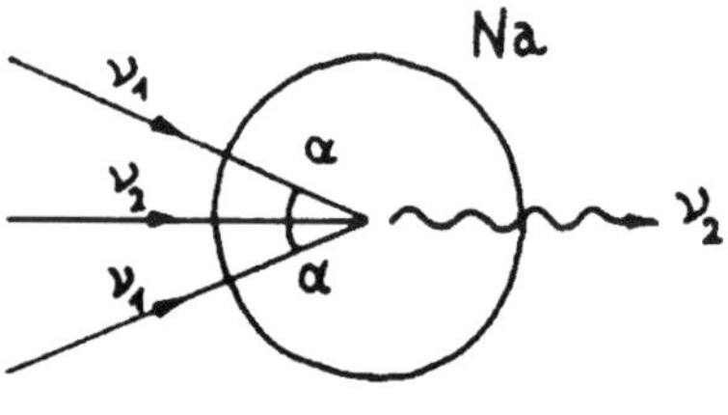

Fig. 9. Geometry of the duplicator's interactions.

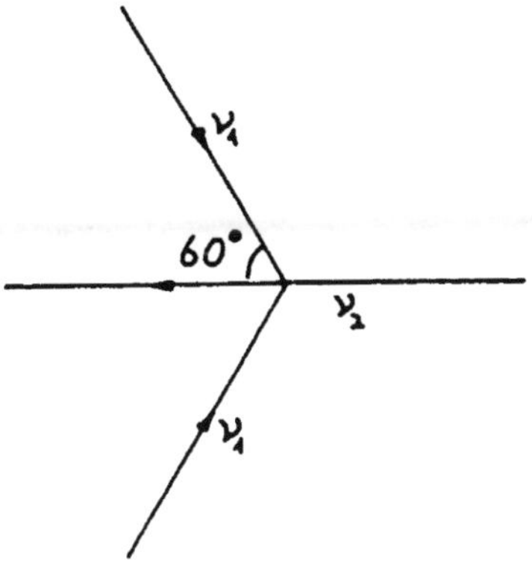

Fig. 10. Memomentum conservation in Gozzini's duplicator.

photons $h\nu_2$ one by one on a semitransparent mirror M of transmission coefficient 1/2. As one knows,[52] two photo multipliers $PM1$ and $PM2$ then necessarily detect anticoincidences, since the photon enters the reflected *or* the transmitted beam. If one then introduces Gozzini's duplicator on one of the beams (say the transmitted), two possibilities arise, i.e.:

(a) Coincidences appear, which would imply that the duplicator D is excited only by an empty wave.

(b) Anticoincidences persist, which show that D is only excited when hit by a photon $h\nu_2$.

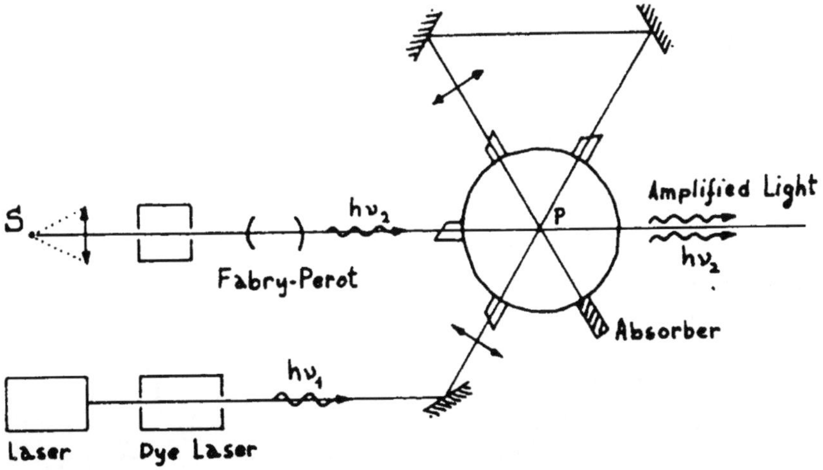

Fig. 11. Schematic representation of Gozzini's photon duplicator D. Two photons $h\nu_1$ emitted in the laser criss-cross at P with a sodium molecule and a photon $h\nu_2$ emitted at S.

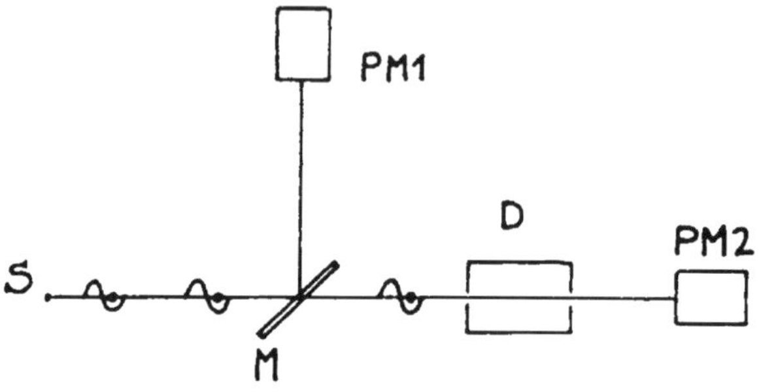

Fig. 12. Representation of Selleri's set-up to test the relation of Gozzini's duplicator D with de Broglie's waves. The semitransparent mirror M (with transmission coefficient 1/2) acting on photons coming one by one lies on the transmitted path. The existence (nonexistence) of anticorrelations between the photomultipliers $PM1$ and $PM2$ ensures the triggering (nontriggering) of D by the passage of a real photon on the transmitted path.

If case (a) is true (an unlikely possibility in the author's opinion), this would imply a direct argument for the real existence of the Einstein–de Broglie waves. If case (b) is verified (in conformity with the usual laser theory) then the appearence of two $h\nu_2$ photons is correlated with the impact of one $h\nu_2$ on the duplicator, and no photon exists in the reflected wave.

With the duplicator in hand we can now discuss possible set-ups which satisfy B. Various proposals have been made to that effect. The latest, by Garuccio, Rapisarda, and Vigier,[46] rests on the assumption (which can also be checked by experiment[46]) that the two outgoing photons $h\nu_2$ have the same frequency and *the same phase* as the incoming photon $h\nu_2$. This assumption, theoretically justified by the similarity of the $B^* \to P$ decay with the usual laser mechanism, is of course not established experimentally and it is quite possible that it would turn out that the two outgoing photons present a random phase fluctuation with respect to the incoming wave packet, so that the use of the duplicator apparently prevents the use of the outgoing $h\nu_2$ photons in interference devices. It is thus important (still in our opinion) that Andrade e Silva, Selleri, and Vigier[53] have been able to construct a proposal which modifies the Garuccio, Rapisarda, and Vigier proposal in such a way that one can compare the antagonistic predictions of the Copenhagen and the causal stochastic interpretations of quantum mechanics, i.e., show, independently of the phases of the duplicator's photons, that in the limit of one photon only the Bohr–Dirac model of purely probabilistic waves yields a different prediction from the Einstein–de Broglie real Maxwell wave model—so that their merit can be assessed by experiment.

We shall rediscuss this later proposal here not only to satisfy B, but also to show how the Einstein–de Broglie model, which rests on Dirac's

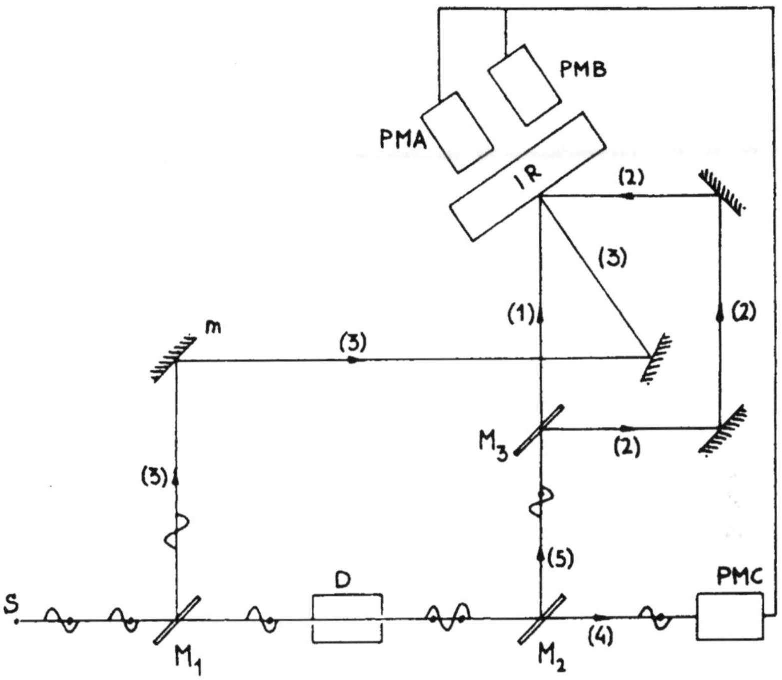

Fig. 13. Representation of the Andrade e Silva, Selleri, and Vigier set-up. M_1, M_2, and M_3 are semitransparent mirrors with transmission coefficient 1/2. *PMA* and *PMB* are photo multipliers connected with half-wave receivers coinciding with the maximum and the minimum of the interference fringes of the interference pattern of paths (1) and (2). They are put in coincidence with *PMC* so that one is sure that two photons have effectively emerged from the duplicator *D*.

aether, yields a new interpretation of Dirac's famous statement that each photon interferes only with itself. As in the Garuccio, Rapisarda, and Vigier proposal,[46] one starts (see Fig. 13) from a set of successive packets (which are issued from an incoherent source) which impinge on a semitransparent mirror M_1. If the experiment has confirmed assumption (b) (i.e., if in this experiment no coincidences have been observed), the appearance of correlated photons in *PMC* and in *PMA* and *PMB* implies that the duplicator *D* has been excited by a transmitted photon from M_1—and that no energetic photon is propagating on the M_1 reflected path (3). Following Dirac, this implies that no probability wave exists along the reflected path (3). No use can further be made of path (3). On the contrary, following Einstein and de Broglie (and also Maxwell's concept of unquantized real existing light waves), a real energy empty pilot wave propagates along (3) which can be reflected by two mirrors into an interference region *IR*, on which also converge the two beams (1) and (2) generated by the further splitting (by a semitransparent mirror M_3) of one of the two photon beams generated by a semitransparent mirror M_2 which splits (in 1/3 of the cases[46] selected by the *PMA*, *PMB*, and *PMC* coincidence) the two photons issuing from D into the respective paths (4) and (5). Following Dirac, this experiment clearly predicts interferences (independent of the phases produced in D) which can be detected (following Pfleegor and Mandel's device[54]) on a pile of half wave detectors connected with the photo multipliers *PMA* and *PMB*. Following Einstein, de Broglie, and Maxwell, the device predicts something else, since we can write for the overall intensity I observed in *IR*

$$I = I_1 + I_2 + I_3 + 2\sqrt{I_1 I_2} \cos \delta_{12} \tag{37}$$

where I_i is the intensity of the ith beam and δ_{ij} is the relative phase shift of the ith and jth beam. In the preceding relation we have suppressed terms in $\cos \delta_{13}$ and $\cos \delta_{23}$, since their phase shifts assume different values in different events. As stressed by Andrade e Silva, Selleri, and Vigier,[53] this implies that the term I_3 is always present (unless one suppresses m) and has an observable effect on the fringe visibility parameter

$$v = 2\sqrt{I_1 I_2}/(I_1 + I_2 + I_3) \tag{38}$$

which can be measured by dividing the coefficient of $\cos \delta_{12}$ by the nonoscillating term in the interference region. Of course, one could further check the existence of an empty pilot wave on (3) by a stroboscoping device on path (3). In other words, we are now

(1) in a position to check the existence of the Einstein–de Broglie–Maxwell wave;

(2) in a situation where the concept of wave packet collapse (really induced by D in our case) yields for the Copenhagen School predictions which contradict the one-photon limit of Maxwell's theory of light.

The answers to the experiment will be interesting to observe. If it confirms Maxwell, then one can only conclude that the presence of real pilot waves which accompany real photons justify Dirac's statement that photons interfere only with themselves. This statement, however, does not preclude the possibility, used by de Broglie and Andrade e Silva,[55] that for coherent beams waves which belong to different photons can interfere, as shown in their interpretation of the Pfleegor and Mandel experiment.

5. CONCLUSIONS

We conclude this paper with some remarks. As a consequence of the results of Aspect's experience (i.e., as a consequence of the violation of Bell's inequality) we are now confronted with the following question: How can we interpret the correlations between spacelike separated events? We think that two different attitudes are possible:

1. Assume that the violation of Bell's inequality proves the existence of correlations between spatially separated events and *hence* the existence of interactions (or signals) exchanged between such events. In this case the problem is to know if the correct interpretation of this new fact lies (a) in a signal exchange made via Feynman's zig-zag (with all the consequences of the possibility to really travel backward in time), as claimed by Wigner[56] and Costa de Beauregard,[43] or (b) in a completely deterministic theory based on the relativistic action at a distance of a quantum potential interpreted in the frame of Dirac's aether, as explained in Section 3.
2. Assume a no-problem attitude in the sense that, from a standpoint based only on the directly observed facts, the violation of Bell's inequality cannot directly prove the existence of a signal exchange between two spatially separated events: the necessity to choose a causal chain on a nonlocal correlation is no reason to assume the existence of nonlocal interactions. This no-problem attitude, which reflects Bohr's attitude toward the EPR paradox,[44] draws its justification from the fact, already remarked, that we cannot use an EPR mechanism to send any superluminal macroscopic signal.

It is an open question which of these two attitudes is the correct one. This makes clear that the Aspect–Rapisarda experiment, despite the importance of finally completely testing the existence of the quantum nonlocal correlations, is not a crucial epistemological experiment in the sense that it does not completely impose a choice between the various standpoints. For this reason, we think that this experiment (comparable in its importance to Michelson's experiment) only opens a new era of theoretical and experimental research. The future choice really depends on the results of the proposed experiments on the direct testing of the existence of the Broglie's waves on Dirac's aether, since only these new forthcoming results can shed new light on the old question of the real nature of the ψ field in wave mechanics.

ACKNOWLEDGMENTS

The authors are grateful to Profs. Gozzini and Selleri for many suggestions and information which helped the preparation of this work. One of us (N.C.P.) want to thank the Italian M.P.I. for a grant which made thi paper possible.

REFERENCES

1. P. A. M. Dirac, *The Principles of Quantum Mechanics* (Oxford, London, 1958).
2. A. Aspect, P. Grangier, and G. Roger, *Phys. Rev. Lett.* **47**, 460 (1981); *Phys. Rev. Lett* **49**, 91 (1982).
3. Ph. Droz-Vincent, *Ann. Inst. H. Poincaré*, **27**, 407 (1977); *Phys. Rev.* **D19**, 702 (1979) *Ann. Inst. H. Poincaré*, **33**, 377 (1980).
4. H. Yukawa, *Proceding International Conference of Elementary Particles* (Kyoto, 1965).
5. D. Bohm and J. P. Vigier, *Phys. Rev.* **96**, 208 (1954); *Phys. Rev.* **109**, 1882 (1958).
6. C. Fenech, M. Moles, and J. P. Vigier, *Lett. N. Cim.* **24**, 56 (1979).
7. F. Halbwachs, J. M. Souriau, and J. P. Vigier, *J. Phys. Rad.* **22**, 26 (1981).
8. Ph. Guèret, M. Moles, P. Merat, and J. P. Vigier, *Lett. Math. Phys.* **3**, 47 (1979); N Cufaro Petroni, Z. Maric, Dj. Zivanovic, and J. P. Vigier, *J. Phys. A* **14**, 501 (1981); N Cufaro Petroni, Z. Maric, Dj. Zivanovic, and J. P. Vigier, *Lett. N. Cim.* **29**, 565 (1980)
9. P. A. M. Dirac, *Nature* **168**, 906 (1951).
10. D. Bohm, *Phys. Rev.* **85**, 166, 180 (1952); *Phys. Rev.* **89**, 458 (1953).
11. L. de Broglie, *La thermodynamique de la particule isolée* (Gauthier-Villars, Paris, 1964).
12. K. P. Sinha, C. Sivaram, and E. C. G. Sudarshan, *Found. Phys.* **6**, 65 (1976); *Found Phys.* **6**, 717 (1976); *Found. Phys.* **8**, 823 (1978).
13. J. P. Vigier, *Astron. Nachr.* **303**, 55 (1982); N. Cufaro Petroni and J. P. Vigier, *Causa Action at a Distance and a new Possible Deduction of Quantum Mechanics from Genera Relativity: The Many-Body Problem*," *Preprint* (Institut H. Poincaré, Paris, 1982).
14. J. P. Vigier, *Lett. N. Cim.* **29**, 467 (1980).
15. Ph. Guèret and J. P. Vigier, *Nonlinear Klein–Gordon Equation Carrying a Non-dispersive Soliton-like Singularity*," Preprint (Institut H. Poincaré, Paris, 1982); Ph. Guèret and J P. Vigier, *Soliton Model of Einsteinian Nadelstrahlung in Real Physical Maxwel Waves*," Preprint (Institut H. Poincaré, Paris, 1982); Ph. Guèret and J. P. Vigier, *De Broglie's Wave-Particle Duality in the Stochastic Interpretation of Quantum Mechanics. A Testable Physical Assumption*, Preprint (Institut H. Poincaré, Paris, 1982).
16. J. S. Bell, *Physics* **1**, 195 (1965); *Rev. Mod. Phys.* **38**, 447 (1966); J. F. Clauser, M. A. Horne, A. Shimony, and R. A. Holt, *Phys. Rev. Lett.* **23**, 880 (1969).
17. E. Nelson, *Phys. Rev.* **150**, 1079 (1966); L. de la Peña Auerbach, *J. Math. Phys.* **10**, 1620 (1969); L. de la Peña Auerbach and A. M. Cetto, *Found. Phys.* **5**, 355 (1975).
18. W. Lehr and J. Park, *J. Math. Phys.* **18**, 1235 (1977); J. P. Vigier, *Lett. N. Cim.* **24**, 258. 265 (1979); F. Guerra and P. Ruggiero, *Lett. N. Cim.* **23**, 529 (1979); N. Cufaro Petroni and J. P. Vigier, *Int. J. Th. Phys.* **18**, 807 (1979); Kh. Namsrai, *Found. Phys.* **10**, 353, 731 (1980).
19. D. Bohm, R. Schiller, and J. Tiomno, *Suppl. N. Cim.* **1**, 48, 67 (1955); L. de la Peña Auerbach, *J. Math. Phys.* **12**, 453 (1971); N. Cufaro Petroni and J. P. Vigier, *Phys. Lett.* **A73**, 289 (1979); *Phys. Lett.* **A81**, 12 (1981); "Stochastic Interpretation of Relativistic Quantum Equations," in *Old and New Questions in Physics, Cosmology, Philosophy, and Theoretical Biology: Essays in Honor of Wolfgang Yourgrau*, Alwyn van der Merwe, ed. (Plenum, New York, 1983).
20. T. H. Boyer, *Phys. Rev.* **D11**, 790, 809 (1975); L. de la Peña Auerbach and A. M. Cetto,

Found. Phys. **8**, 191 (1978); *Int. J. Quant. Chem.* **12**, Suppl. 1, 23 (1978); P. Claverie and S. Diner, *Int. J. Quant. Chem.* **12**, suppl. 1, 41 (1978); L. de la Peña Auerbach and A. Jàuregui, *Found. Phys.* **12**, 441 (1982).

21. N. Cufaro Petroni and J. P. Vigier, *Random Motions at the Velocity of Light and Relativistic Quantum Mechanics*," Preprint (Insituto di Fisica, Bari, 1982).
22. R. P. Feynman, *Phys. Rev.* **76**, 749, 769 (1949).
23. J. D. Bjorken and S. D. Drell, *Relativistic Quantum Mechanics* (McGraw-Hill, New York, 1964); E. P. Wigner, *Phys. Rev.* **40**, 749 (1932); J. E. Moyal, *Proc. Camb. Phil. Soc.* **45**, 99 (1949); P. A. M. Dirac, *Proc. Roy. Soc. A* **180**, 1 (1942); W. Pauli, *Rev. Mod. Phys.* **15**, 175 (1943); R. P. Feynman, *Rev. Mod. Phys.* **20**, 367 (1948); *Int. J. Theor. Phys.* **21**, 467 (1982).
24. J. P. Vigier and Ya. P. Terletskij, *Sov. Phys. J.E.T.P.* **13**, 356 (1961).
25. A. Avez, *Interprétation Probabiliste d'équations aux derivées Partielles Hyperboliques Normales*," Preprint (Journèes Relativistes, Paris, 1976).
26. A. Einstein, B. Podolsky, and N. Rosen, *Phys. Rev.* **47**, 777 (1935).
27. D. Bohm, *Quantum Theory* (Prentice-Hall, Englewood Cliffs, N.J., 1951); D. Bohm and Y. Aharonov, *Phys. Rev.* **108**, 1070 (1957).
28. C. Møller, *The Theory of Relativity* (Oxford, London, 1972).
29. V. Augelli, A. Garuccio, and F. Selleri, *Ann. Fond. L. de Broglie* **1**, 154 (1976).
30. S. J. Freedman and F. Clauser, *Phys. Rev. Lett.* **28**, 938 (1972); R. A. Holt and F. M. Pipkin, Harvard, Preprint (1974), unpublished; G. Faraci, S. Gutkowsky, S. Notarrigo, and A. R. Pennisi, *Lett. N. Cim.* **9**, 667 (1974); L. Kasday, J. Ullman, and C. S. Wu, *N. Cim.* **B25**, 663 (1975); J. F. Clauser, *Phys. Rev. Lett.* **36**, 1223 (1976); E. S. Fry and R. C. Thompson, *Phys. Rev. Lett.* **37**, 465 (1976); A. R. Wilson, J. Lowe, and D. K. Butt, *J. Phys.* **G2**, 613 (1976); M. Bruno, M. D'Agostino, and C. Maroni, *N. Cim.* **B40**, 42 (1977).
31. A. Aspect, *Phys. Lett.* **A67**, 117 (1975); *Progr. Sci. Cult.* **1**, 439 (1976); *Phys. Rev.* **D14**, 1944 (1978); F. Falciglia, G. Iaci, and V. A. Rapisarda, *Lett. N. Cim.* **26**, 327 (1979); L. Pappalardo and V. A. Rapisarda, *Lett. N. Cim.*. **29**, 221 (1980); A. Garuccio and V. A. Rapisarda, *N. Cim.* **A65**, 269 (1981).
32. D. Bohm and B. Hiley, *Found. Phys.* **5**, 93 (1975); C. Philippidis and D. Bohm, *The Aharonov–Bohm Effect and the Quantum Potential*, preprint (Birbeck College, London, 1982).
33. N. Cufaro Petroni and J. P. Vigier, *Lett. N. Cim.* **25**, 151 (1979).
34. L. de Broglie, *La Mechanique Ondulatoire du Photon* (Gauthier-Villars, Paris, 1940); L. Bass and E. Schrödinger, *Proc. Roy. Soc. Lond.* A232, 1 (1955); S. Deser, *Ann. Inst. H. Poincaré* **16**, 79 (1972); M. Moles and J. P. Vigier, *C. R. Acad. Sci. Paris* **B276**, 697 (1973).
35. F. Halbwachs, F. Piperno, and J. P. Vigier, *Lett. N. Cim.* **33**, 311 (1982).
36. A. Garuccio and J. P. Vigier, *Lett. N. Cim.* **30**, 57 (1981).
37. L. D. Landau and E. M. Lifshitz, *Teoria Quantistica Relativistica* (Editori Riuniti, Roma, 1978).
38. N. Cufaro Petroni and J. P. Vigier, *Lett. N. Cim.* **26**, 149 (1979); *Phys. Lett.* **A88**, 272 (1982); Kh. Namsrai, *J. Phys.* **A14**, 1307 (1981); *Sov. J. Part. Nucl.* **12**, 449 (1981).
39. N. Cufaro Petroni, Ph. Droz-Vincent, and J. P. Vigier, *Lett. N. Cim.* **31**, 415 (1981).
40. F. Halbwachs, *Théorie Relativiste des Fluides à Spin* (Gauthier-Villars, Paris, 1960).
41. N. Cufaro Petroni and J. P. Vigier, *Causal Action at a Distance Interpretation of the Aspect–Rapisarda Experiment*, Preprint (Institut H. Poincaré, Paris, 1982).
42. A. Garuccio, V. A. Rapisarda, and J. P. Vigier, *Lett. N. Cim.* **32**, 451 (1981).
43. O. Costa de Beauregard, *N. Cim.* **B42**, 41 (1977); *Ann. Fond. L. de Broglie* **2**, 231

(1977); *Phys. Lett.* **A67**, 171 (1978); *N. Cim.* **B51**, 267 (1979); *N. Cim. Lett.* **17**, 551 (1980).

44. N. Bohr, *Phys. Rev.* **48**, 696 (1935).
45. M. Cini, M. De Maria, G. Mattioli, and F. Nicolo, *Found. Phys.* **9**, 479 (1979).
46. F. Selleri, *Lett. N. Cim.* **1**, 908 (1969); F. Selleri and J. P. Vigier, in *Old and New Questions in Physics, Cosmology, Philosophy, and Theoretical Biology: Essays in Honor of Wolfgang Yourgrau, Alwyn van der Merwe*, ed. (Plenum, New York, 1983); A. Garuccio and J. P. Vigier, *Found. Phys.* **10**, 797 (1980); A. Garuccio, K. Popper, and J. P. Vigier, *Phys. Lett.* **A86**, 397 (1981); A. Garuccio, V. A. Rapisarda, and J. P. Vigier, *Phys. Lett.* **A90**, 17 (1982); J. and M. Andrade e Silva, *C. R. Acad. Sci. Paris*, **290**, 501 (1980).
47. A. Einstein, *Ann. der Phys.* **17**, 132 (1905); *Ann. der Phys.* **18**, 639 (1905); *Zeit. Phys.* **18**, 121 (1917).
48. L. de Broglie, *Ann. Phys.* **3**, 22 (1925).
49. B. Cagnac, G. Grynberg, and F. Biraben, *Jour. de Phys.* **34**, 845 (1973); G. Grynberg, F. Biraben, M. Bassini, and B. Cagnac, *Phys. Rev. Lett.* **37**, 283 (1976).
50. A. Gozzini, in *Proceedings of the Symposium on Wave-Particle Dualism* (Reidel, Dordrecht, 1983).
51. F. Selleri, *Ann. Fond. L. de Broglie*, **7**, 45 (1982).
52. L. Mandel and K. Dajenais, *Phys. Rev. Lett.* **18**, 2217 (1978).
53. J. Andrade e Silva, F. Selleri, and J. P. Vigier, *Some Possible Experiments on Quantum Waves*, preprint (Institut H. Poincaré, Paris, 1982).
54. R. L. Pfleegor and L. Mandel, *Phys. Rev.* **159**, 1084 (1967).
55. L. de Broglie and J. Andrade e Silva, *Phys. Rev.* **172**, 1284 (1968).
56. E. P. Wigner, *Symmetries and Reflections* (MIT press, Mass., 1971).

Reprinted from *Lettere al Nuovo Cimento*, Vol. 24, No. 8, pp. 258-264, Copyright (1979)
with permission from the Società Italiana di Fisica

Superluminal Propagation of the Quantum Potential in the Causal Interpretation of Quantum Mechanics.

J. P. VIGIER

Equipe de Recherche Associée au C.N.R.S. no. 533
Institut Henri Poincaré - 11 rue Pierre et Marie Curie, 75231 Paris Cedex 05

(ricevuto il 9 Novembre 1978)

The theoretical controversy and subsequent experiments which started with Bell's discovery (¹) that local hidden variable theories (LHV) imply testable consequences which differ from the standard predictions of quantum mechanics (QM) have evidently reached a critical stage. Its experimental issue might prove as important for the future evolution of physics as the negative result of the Michelson-Morley experiment. Indeed the theoretical discussion started by BELL (²) has shown that a positive experimental test of the truth of quantum-mechanical measurement predictions in the case of correlated particles (in latter versions of the Einstein-Podolsky-Rosen type of experiments) has far reaching consequences which go beyond Bohr's initial statements on the conservation of macroscopic causality. They imply a destruction of the Einsteinian concept of material causality in the evolution of Nature, since they would establish the physical reality of nonlocal interactions between spacelike separated instruments o fmeasurement, *i.e.* two polarizers measuring the relative-spin orientations of a pair of correlated particles emitted in the singlet state.

Indeed preliminary results which favour Bohr's interpretation of QM will be definitely established by the experiment of ASPECT (³) if its results confirm Bohr's prediction.

The present situation can be summarized as follows:

a) if the present trend of experiments persists (as believed by many people including the author of this letter) to support (except in special conditions which we shall discuss later) the experimental predictions of quantum mechanics even in the case of spacelike separated measurements;

b) if one accepts the Copenhagen interpretation of QM as a complete theory describing the behaviour of pointlike particles;

(¹) J. S. BELL: *Physics*, **1**, 195 (1964).

(²) See for example the discussion between BELL, D'ESPAGNAT, SHIMONY and COSTA DE BEAUREGARD in *Epistemological Letters* from 1970 to 1978.

(³) A. ASPECT: *Phys. Lett.*, **54** A, 117 (1975); *Progr. in Sci. Culture*, **1**, 439 (1976); *Phys. Rev. D*, **14**, 1944 (1976).

d) if one recalls the fact that trigger mechanism can transform (by cascade types of processes) individual microscopic phenomena into macroscopic events in space-time, one must:

A) discard the LHV version of the causal interpretation in the initial form given by DE BROGLIE (4), BOHM (5) and various authors including myself (6) as well as Bohm's non local interpretation (7,8).

B) drop causality in the physical world since (to utilize Stapp's statement (9)) « if the statistical predictions of quantum theory are true in general and if the macroscopic world is not radically different from what is observed, then what happens macroscopically in one space-time region must in some cases depend on variables that are controlled by experimenters in far away spacelike separated regions ». This statement valid for fermions and bosons (photons) (10) results from the new established fact that quantum mechanical predictions necessarily result from the use of state vector swhich are the superposition of the products of the eigenvectors corresponding to the two opposite polarizations (*i.e.* to mixtures of the second kind) which imply this spacelike connection.

The devastating character of a result in favour of QM is its contradiction with (*c*)) since it evidently implies a noncausal nonmaterial perception of information between spacelike separated domains.

At this stage only two interpretations seem to be left. The first is to maintain (*c*)) and according to Costa de Beauregard accept the possibility of Feynman zigzags along the light cone with advanced potentials. This allows information to travel backward into time ... a concept on which he has based a belief into psychokinesis and paranormal phenomena (11). The second is to accept real material propagation of particles (of the tachyon type) between spacelike separated regions, a concept which evidently contradicts the conservation of energy and Carnot's principle.

The aim of the present letter is to show that a third interpretation is possible based on a slightly modified version of the model of the causal interpretation of QM in terms of a fluid with regular fluctuations developed by Bohm and the author (12) (B.V.). This new model:

A) preserves Einstein's concept of causality (*c*)) in the sense that individual particles now considered as extended timelike hypertubes (13,14), move along timelike paths and can only transmit superluminal information localized within their internal structure;

---

(4) L. DE BROGLIE: *La physique quantique restera-t-elle indéterministe* (Paris, 1953).
(5) D. BOHM: *Phys. Rev.*, **85**, 166, 180 (1952).
(6) J. P. VIGIER: *Structure des microobjets dans l'interprétation causale de la théorie des quanta* (Paris, 1956).
(7) D. BOHM and B. J. HILEY: *Found. Phys.*, **5**, 93 (1974).
(8) F. SELLERI and G. TARROZI: *Extension of the domain of validity of Bell's inequality*, University of Bari, preprint (1978).
(9) H. P. STAPP: *Nuovo Cimento*, **40** B, 191 (1977).
(10) J. F. CLAUSER, M. A. HORNE, A. SHIMONY and R. A. HOLT: *Phys. Rev. Lett.*, **26**, 880 (1969).
(11) O. COSTA DE BEAUREGARD: *Epistemological Lett.*, **18** (January 1978).
(12) D. BOHM and J. P. VIGIER: *Phys. Rev.*, **96**, 208 (1954).
(13) D. BOHM and J. P. VIGIER: *Phys. Rev.*, **109**, 882 (1958).
(14) F. HALBWACHS, J. M. SOURIAU and J. P. VIGIER: *Journ. Phys. Radium*, **22**, 26 (1961).

B) carries nonlocal superluminal information which is not transmitted by individual particles (limited to timelike motions) but result from the superluminal propagation of a real physical collective excitation (*i.e.* a density wave) which propagates like a phase phenomenon (analogous to the successive lighting of electric bulbs on a christmas tree) on the top of a continuous thermostat of such extended elements ... which corresponds to the material vacuum which supports the real physical ψ waves associated to individual particles in the B.V. particular version of the causal interpretation of quantum mechanics.

In this paper we shall limit our demonstration to the simple case of a spin zero isolated particle for which we shall establish that information starting on the ψ wave's boundary (such as the opening or closing of a slit in the double slit Young hole interference experiment) reacts with superluminal velocity (via the quantum potential) on the particle motions which move with infraluminal group velocities along the lines of flow of the said ψ waves. As will be shown in a subsequent letter this saves causality (and allows a detailed realistic causal interpretation) in the EPR experiments.

The interpretation of macroscopic local interactions (at least up to distances of the order of the coherent lengths of wave packets) which seem to appear in the Einstein-Podolsky-Rosen type of experiments thus implies a new theoretical step. We must modify the causal model of quantum statistics introduced by Bohm and Vigier (12) by adding three new assumptions:

I) the fluid elements (and the particles) which follow the lines of flow of the fluid with irregular fluctuations are built from extended «rigid» elements in the sense discussed later;

II) the stochastic fluctuations occur at the velocity of light;

III) the fluid is a mixture of extended particles (13,14) (and antiparticles); the latter being mathematically equivalent to particles moving backward in time (15,16).

With the help of these assumptions it has been shown independently (using different methods) by Lehr and Park (17) and by Vigier (18) that the corresponding statistics is correctly described by the Klein-Gordon equation. This is interesting since it shows that one can deduce one of the basic equations of wave mechanics from the stochastic theory of ultra-relativistic Brownian motion in space-time.

Assumption *A*) is now easily established. Indeed the idea that extended particles are nonlocal in nature *i.e.* that they can propagate in their interior superluminal interaction and/or information is not new in the literature. Dirac (19) was the first to notice that if one treats the classical extended electron as a point charge imbedded in its own radiating e.m. field the equations obtained are of the same form as those already in current use, but that in their physical interpretation the finite size of the electron reappears in a new sense: the interior of the electron being a region of space through which signals can be transmitted faster than light. Physically this can be understood as follows. If we send out a pulse from a point *A* and a receiving apparatus for electromagnetic waves is set up at a point *B* and if we suppose there is an extended electron

---

(15) M. Flato, G. Rideau and J. P. Vigier: *Nucl. Phys.*, **61**, 250 (1965).
(16) Ya. P. Terletski and J. P. Vigier: *Žurn. Ėksp. Teor. Fiz.*, **13**, 356 (1961).
(17) W. Lehr and J. Park: *Journ. Math. Phys.*, **18**, 1235 (1977).
(18) J. P. Vigier: *Model of quantum statistics in terms of a fluid with irregular stochastic fluctuations propagating at the velocity of light: a derivation of Nelson's equations*, preprint Inst. H. Poincaré (July 1978), to be published.
(19) P. A. M. Dirac: *Proc. Roy. Soc.*, **167** A, 148 (1938).

on the straight line joining A to B then the disturbed electron will be radiating appreciably at a time a^{-1} before the pulse has reached its centre so that this emitted radiation will be detectable at B at a time a^{-1} earlier than when the pulse which travels from A to B with the velocity of light arrives. In this way a signal can be sent from A to B faster than light so that it is possible for a signal to be transmitted faster than light through the interior of an electron, the whole theory being of cause perfectly Lorentz invariant and causal in Einstein's sense since no particle travels faster than light.

The same result can be obtained from the relativistic generalization of the motion of extended rigid body. As one knows there is no such thing as a perfectly classical extended rigid body in relativity since the distance between two arbitrary subelements depends on the spacelike cross-section of the timelike hypertube defined by the observer. However if we consider in any such arbitrary spacelike region a given chain of contiguous subelements and assume that the corresponding strings cannot cross at any time another chain built with different elements we arrive at the relativistic concept of an elastic solid *i.e.* to the closest possible analogue of a rigid body in classical theory. This has been utilized by GUTKOWSKY *et al.* ([20]) to construct the classical counterpart of Dirac's electron. If we then add to this concept the idea that we are dealing with a spherically shell of matter with a current $J_\mu = \sigma v_\mu$ such that with $\boldsymbol{v}\cdot\boldsymbol{v} = -c^2$:

1) σ vanishes everywhere at any fixed time except at the surface of sphere of centre z and radius r where $\sigma \neq 0$ is constant;

2) r is constant;

3) v_μ depends on time but at any time is constant on the surface of the sphere;

4) conditions a), b), c) hold with respect to every proper inertial frame of the charged sphere and with respect to a particular external inertial frame, we fall exactly on the well-known rigidity conditions introduced by Born ([21]) in relativity theory. In this sense such a rigid body has only three degrees of freedom: the remaining three degrees of freedom of the spherical shell being determined as shown by Pounder ([22]), by requiring that Born's rigidity condition are satisfied on the surface.

Born's rigid shell evidently implies transmission of superluminal interaction and/or information since such a shell travelling at (or very close to) the velocity of light implies knowledge and interactions which crosses the surface of the light cone. Indeed knowledge of the position A on the shell (or of its centre z implies as shown in fig. 1 knowledge of B on the opposite end of a diameter and a collision (or interaction) at A' which switches the hypertube (and the path of z) back into the forward light cone (into A'' and B'') implies a deflection of the path of B' beyond the light cone's surface, *i.e.* a nonlocal modification of the physical situation at B'. Of course such nonlocal interactions can be neglected in classical relativity theory since the actual size of the particles' cross-sections have been shown to be very small (perhaps of the order of Planck's length, *i.e.* $\sim 10^{-32}$ cm so that the introduction of extended structures can be approximated by the statement that their centre-of-mass and centre-of-matter density associated with extended particle models are restricted to the forward light cone.

To demonstrate B) we shall not discuss the detail of the demonstration of ref. ([17,18]) here but just recall the elements necessary to show that they imply that the associated quantum potential Q of stochastic origin propagates within our fluid with superluminal velocity; a fact overlooked in ref. ([17,18]).

(20) D. GUTKOWSKY, M. MOLES and J. P. VIGIER: *Nuovo Cimento*, **39** B, 193 (1977).
(21) M. BORN: *Ann. der Phys.*, **30**, 1 (1909).
(22) J. R. POUNDER: *Comm. of Dublin Inst. for Advanced Studies*, **11** A, 1 (1954).

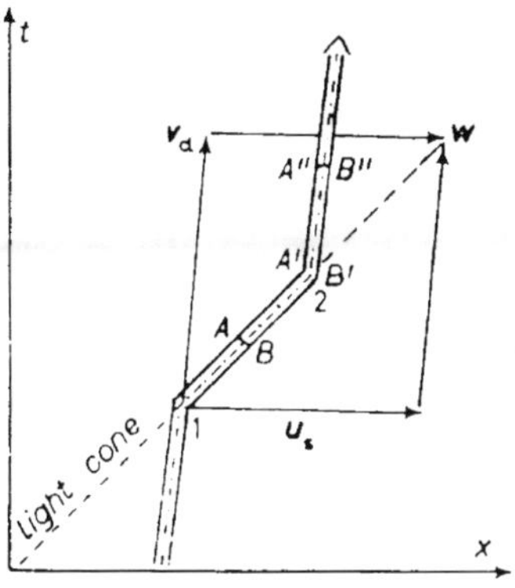

Fig. 1. – In this model a particle whose boundaries are denoted by $''$ and centre motion by —·—·— moves along the drift average four velocity v_d. It undergoes stochastic jumps at the velocity of light w from point 1 to point 2, after which it reintegrates the drift flow. A and B (as $A'B'$ and $A''B''$) represent the opposite extremities of a particle's diameter.

To show this, we just recall that the particles (and relativistic fluid elements) follow the fluid average lines of flow characterized by a four-vector drift velocity $\boldsymbol{v}_d$ parallel to ∇S with $\psi = \varrho^{\frac{1}{2}} \exp[iS/\hbar]$ and $\Box\psi = -(m^2c^2/\hbar^2)\psi$. This corresponds to the ψ wave's group velocity and defines an average conserved drift current $\boldsymbol{j} = \varrho \boldsymbol{v}_d$ which has been shown ([12]) to represent a limiting equilibrium distribution for particles carried by the fluid. The stochastic jumps on the light cone which pass particles (or fluid elements) from one line of flow to another are represented by zero-length four vectors $\boldsymbol{w}$ (with $\boldsymbol{w}\cdot\boldsymbol{w} = 0$) which can be decomposed into the sum of two vectors, *i.e.* $\boldsymbol{w} = \boldsymbol{v}_d + \boldsymbol{u}_s$ with $\boldsymbol{v}_d\boldsymbol{v}_d = -c^2$ and $\boldsymbol{u}_s\boldsymbol{u}_s > 0$. If the spacelike vector $\delta\boldsymbol{x}$ then represents ([18]) spacelike stochastic jumps which carry our extended fluid elements from one drift current timelike hypertube into another, one of the main points of the demonstration of ref. ([17,18]) is that if we denote by $\boldsymbol{u}_s$ the corresponding spacelike stochastic velocity one finds

$$\boldsymbol{u}_s = \frac{\langle(\delta x)^2\rangle}{2\,\Delta\tau}\frac{1}{\varrho}\nabla\varrho = D\frac{\nabla\varrho}{\varrho}\,; \tag{1}$$

where the symbol $\langle\,\rangle$ denotes averages taken over elementary domains, $\Delta\tau$ the proper time differential along the drift lines of flow parallel to $\boldsymbol{v}_d$ and D the relativistic generalization of the diffusion coefficient *i.e.* $D = \langle(\delta x_\mu)(\delta x_\mu)\rangle/2\,\Delta\tau$. One deduces there from that the particles (fluid elements) are submitted to a drift force of stochastic origin, *i.e.* the quantum potential $Q = \log M$ of the causal interpretation ([23]). This appears in the drift fluids' Hamilton-Jacobi equation of motion,

$$\nabla S\cdot\nabla S = -M^2c^2 = -[m^2 + (\hbar^2/c^2)(\Box\varrho^{\frac{1}{2}}/\varrho^{\frac{1}{2}})]$$

so that we obtain ([23])

$$\mathrm{d}(M\boldsymbol{v}_d)/\mathrm{d}\tau = c\nabla M\,,$$

([23]) L. de Broglie: *Une tentative d'interprétation causale et non linéaire de la mécanique ondulatoire* (Paris, 1956); J. P. Vigier: *Compt. Roy. Acad. Sci.*, **166**, 598 (1968).

which yields finally per unit of « mass » M the stochastic force F, *i.e.*

$$F = \nabla Q\,, \tag{2}$$

where $Q = -c^2 \log M$ is a function of ϱ only.

We know also that $\mathrm{d}\boldsymbol{v}_d/\mathrm{d}\tau$ is spacelike since $\boldsymbol{v}_d \cdot \boldsymbol{v}_d = -c^2$.

The essential conclusion of relation (2) (though not explicitly formulated in ref. ([17,18,23])) is that the gradient $\nabla Q = \nabla M/M$ is spacelike (since parallel to the spacelike drift acceleration) so that $\boldsymbol{v}_d \cdot \nabla Q \geqslant 0$. This implies that the quantum potential (described as a function of the scalar density ϱ in our model) propagates with superluminal velocities within our drift current. This is easily understood since one has seen that

$$\mathrm{d}\varrho/\mathrm{d}\tau = \nabla(\varrho \nabla S) = 0\,, \tag{3}$$

so that ϱ remains constant along the drifting tubes of flow.

The last step of our demonstration is to show how this superluminal propagation of Q is a natural consequence of the nonlocal character of our fluid elements.

As an example, we discuss the propagation of a defect in the ϱ distribution in the spacelike direction. For clarity, we shall limit ourselves to the simplified case of a two-dimensional description since it can evidently be generalized. We start with a set of hypertubes (labelled 1, 2, 3 ... in fig. 2) which follow the drift lines of flow before they

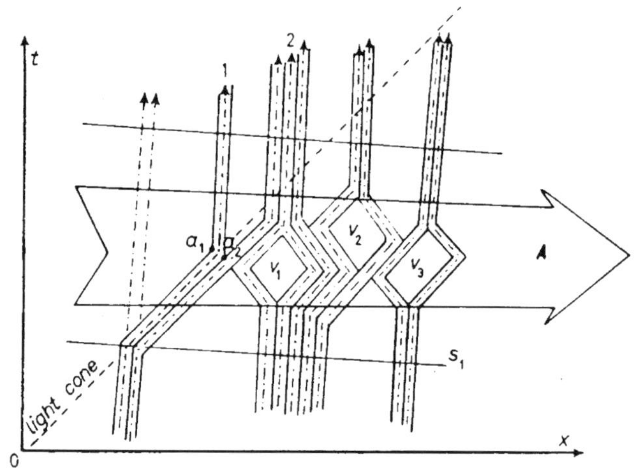

Fig. 2. – Two dimensional schematic model of a spacelike decompression density wave. The stochastic movements (parallel to the light cone represented by the lines – – –) followed by a return to the average timelike drift motions —·—·— open successive holes v_1, v_2, v_3 ... which propagate along the spacelike arrow A. All particles have timelike motions.

reach the constant phase surface S_1. These particles undergo quantum jumps at S at the velocity of light from one drift line to another as a consequence of collisions. If one then follows particle 1 and if for some external reason (modification of the boundary condition, etc.) it shifts its motion in the timelike direction in a_1 then a_2 (which is spacelike separated from a_1) interacts with its neighbourhood and leaves open a hypervolume v_1 equivalent to a ϱ defect. This can be filled by a space displacement of particle 2 which in turns can be repeated and one sees that the successive set of hypervolumes v_1, v_2, v_3, etc. move along a spacelike arrow A and correspond to a decompression wave equivalent to a spacelike gradient of ϱ in this direction. The same reasoning applies of course to a ϱ excess (*i.e.* to a pressure wave) so that both mechanisms are analogous to the mechanism of propagation of sound waves (or pressure waves) in a fluid except that the propagation is now superluminal. It explains how distant boundary modifications can influence individual particle behaviour and statistics. The interconnectedness between boundary conditions and particle, carried by a superluminal modification of the ψ wave, is comparable to the long-range correlation of helium atoms in a superfluid state. In a stable situation (with stable boundaries) the stochastic jumps are random but if they are modified then a co-ordinated movement propagates superluminaly over the chaos, modifying the orientation and density of the regular drift motion.

We shall conclude this letter with two restrictive remarks:

The first is that even if this model explains how an individual particle « knows » about the modification of neighbouring boundary conditions (such as the opening of a distant slit) it must be enlarged to spinning particles and to the case of correlated particles (represented in configuration space) in order to interpret EPR experiments. This implies of course (as noted by Bohm and Hiley ([7])) superluminal propagation of a many-body quantum potential: a question which will be discussed in a subsequent work. In principle however it is not astonishing that even limited internal departure from LHV should lead as shown by FLATO *et al.* ([24]) to a breakdown of Bell's inequalities.

The second remark is that the price to pay for this superluminal propagation of Q in the ψ waves in the existence of a random subquantum level of matter (the BV vacuum ([12]) de Broglie's hidden thermostat ([25])) invariant in Dirac's sense ([26]) which modifies the ψ waves in a way comparable to superconductivity ([7]). This opens new lines of research (including the real physical range of our collective superluminal interactions) which could be tested in future E.P.R. experiments.

([24]) M. FLATO, C. PIRON, J. GREA, D. STERNHEIMER and J. P. VIGIER: *Helv. Phys. Acta*, **48**, 219 (1975).
([25]) L. DE BROGLIE: *La thermodynamique de la particule isolée* (Paris, 1964).
([26]) P. A. M. DIRAC: *Nature*, **168**, 906 (1951).

This page is deliberately left blank.

Reprinted from *Lettere al Nuovo Cimento*, Vol. 24, No. 8, pp. 265-272, Copyright (1979) with permission from the Società Italiana di Fisica

Model of Quantum Statistics in Terms of a Fluid with Irregular Stochastic Fluctuations Propagating at the Velocity of Light: a Derivation of Nelson's Equations.

J. P. VIGIER

Equipe de Recherche Associée, au C.N.R.S. no. 533
Institut Henri Poicaré, 11 rue Pierre et Marie Curie - 75231 Paris Cedex 05

(ricevuto il 9 Novembre 1978)

Within the frame of the general discussion on the principles and physical content of quantum mechanics (QM) one the most interesting branches since 1952 deals with the possible stochastic nature of its associated statistics. An increasing set of results ([1-3]) have now established striking formal similarities with classical models of stochastic theory such as Markov processes ([4,5]).

Two basic obstacles remain however, which have prevented until now the completion of the main statistical interpretation of QM in terms of real physical stochastic motions.

The first obstacle is the existence of a wrong sign (from the classical point of view) in the stochastic version of Newton's second law: a sign which is clearly necessary to derive Schrödinger-type wave equations. For example in the notations of de la Peña and Cetto ([3]) Newton's law takes the form

$$m(D_c v + D_s u) = F^+ \tag{1}$$

for Brownian motion: in contrast with the form given by NELSON ([2]) *i.e.*

$$m(D_c v - D_s u) = F^+ , \tag{2}$$

from which he has deduced (combined with the continuity equation) a remarkable derivation of Schrödinger's equation.

The second obstacle is the relativistic generalization of these stochastic models. Indeed HAKIM ([6]) has shown that it is not enough to write a relativistic generalization

---

([1]) D. BOHM and J. P. VIGIER: *Phys. Rev.*, **96**, 208 (1954).
([2]) E. NELSON: *Phys. Rev.*, **150**, 1079 (1966).
([3]) L. DE LA PEÑA and A. M. CETTO: *Found. Phys.*, **5**, 355 (1975).
([4]) I. FENYES: *Zeits. Phys.*, **132**, 81 (1952); W. WEIZEL: *Zeits. Phys.*, **134**, 264 (1953); N. WIENER and A. SIEGEL: *Phys. Rev.*, **91**, 1551 (1953).
([5]) F. GUERRA and P. RUGGIERO: *Phys. Rev. Lett.*, **31**, 1022 (1973).

of (2) since if $\Delta t \to 0$ the only value for the diffusion constant ν_0 (in $(dx)^2 \simeq 2\nu_0 dt$) compatible with relativistic invariance is $\nu_0 = 0$. As a consequence LEHR and PARK (7) have been led to add to eq. (2) two supplementary axioms *i.e.* *a*) the discretization of time in the stochastic model; *b*) the attribution of the speed of light *c* to the stochastic particle between interactions with the thermostat. Under these conditions they do indeed recover the Klein-Gordon equation provided antiparticles are considered as particles moving backward in time.

The aim of the present letter is to derive Nelson's equation and quantum statistics from a relativistic generalization of the hydrodynamical model of QM developed by MADELUNG (8), TAKABAYASI (9) and extended to spinning particles by various authors (10).

This classical relativistic model generalizes the nonrelativistic stochastic hydrodynamical model of QM of Bohm and Vigier on terms of a fluid with irregular fluctuations (1). It contains three new physical features.

I) the fluid elements (and the particles) which follow the lines of flow of the fluid with irregular fluctuations are built from extended elements in the sense discussed by Bohm (11) and Souriau (12).

II) The stochastic fluctuations occur at the velocity of light.

III) The fluid is a mixture of extended particles (and antiparticles): the latter being mathematically equivalent to particles moving backward in time (13,14).

The existence of such fluctuations (which induce in the particle a Markov type of Brownian motion) has been shown (1) to lead any initial distribution of the particles in the fluid into a limiting equilibrium distribution const $\cdot\, \varrho(x_\mu(\tau))$ proportional to the fluid's average conserved drift density $\varrho(x_\mu(\tau))$. This means that the fluctuations of our Madelung fluid induce on our particles stochastic jumps at the velocity of light (from one line of flow to another) and that such jumps can be decomposed into the regular drift motion $\boldsymbol{v}_d$ plus an apparent spacelike random part $\boldsymbol{u}_s$ with $\boldsymbol{v}_d = dx_\mu(\tau)/d\tau$, τ representing the proper time along the drift lines: so that $\boldsymbol{v}_d \cdot \boldsymbol{v}_d = -c^2$.

Indeed any velocity $\boldsymbol{w}$ represented by a point P (with $w_\mu w_\mu = 0$) of the light cone can be decomposed into the sum of two four-velocities $\boldsymbol{v}_d$ and $\boldsymbol{u}_s$ *i.e.* $\boldsymbol{w} = \boldsymbol{v}_d + \boldsymbol{u}_s$ with $\boldsymbol{u}_s \cdot \boldsymbol{u}_s \geqslant 0$. Since the three indeprendent components of $\boldsymbol{w}$ determine the four components of $\boldsymbol{u}_s$. As a consequence if one considers a particle of the preceding type it undergoes two independent types of motions: *a*) regular motions along the fluid's drift lines of flow with the fluids own velocity $\boldsymbol{v}_d$ *b*) stochastic jumps in any direction with the velocity of light with a four velocity $\boldsymbol{w}$ satisfying $\boldsymbol{w} \cdot \boldsymbol{w} = 0$.

To establish (*a*)) let us first recall that a particle or a regular fluid element (which can be compared with the stochastic particle and the thermostat's elements in the usual Brownian motion) are now represented in four dimensional space-time by time like hypertubes instead of timelike lines. These hypertubes can be naturally assumed to have a minimum spacelike radius $\bar{r}/2$ which yields the minimum distance $\bar{r}$ which separates two continuous particles in any spacelike section passing through their centre of mass. Independently of the stochastic jumps our drifting fluid is thus comparable

(7) W. LEHR and J. PARK: *Journ. Math. Phys.*, **18**, 1235 (1977).
(8) E. MADELUNG: *Zeits. Phys.*, **40**, 332 (1926).
(9) T. TAKABAYASI: *Prog. Theor. Phys. (Japan)*, **8**, 143 (1952); **9**, 187 (1953).
(10) Summarized in F. HALBWACHS: *Théorie des fluides à spin* (Paris, 1960).
(11) D. BOHM and J. P. VIGIER: *Phys. Rev.*, **109**, 882 (1958).
(12) F. HALBWACHS, J. M. SOURIAU and J. P. VIGIER: *Journ. Phys. Radium*, **22**, 26 (1961).
(13) M. FLATO, G. RIDEAU and J. P. VIGIER: *Nucl. Phys.*, **61**, 250 (1965).
(14) YA. P. TERLETSKI and J. P. VIGIER: *Žurn. Èksp. Teor. Fiz.*, **13**, 356 (1961).

with a timelike set of extended fibers and the minimum time needed to pass from one of these hypertubes to the next is thus $r/c = \Delta t$ since the jumps occur at the velocity of light. This implies that the proper-time variable which corresponds to adjacent events in our stochastic model have nonzero minimum temporal separation $\overline{\Delta\tau}$.

The second step is just to generalize to our relativistic model the average velocities utilized by de la Peña and Cetto (3) to discuss the nonrelativistic theory of classical and quantum-mechanical systems. Let us start (fig. 1) from a four dimensional volume limited on the side by the fluid's regular lines of flow and, at both extremities, by two spacelike constant phase surfaces (15) S_1 and S_3. If the domain is small enough such surfaces are separated by an interval $2\Delta\tau$: an interval $\pm\Delta\tau$ separating S_1 and S_3 from a median section S_2. Of course $|\Delta\tau| \gg \overline{\Delta\tau}$.

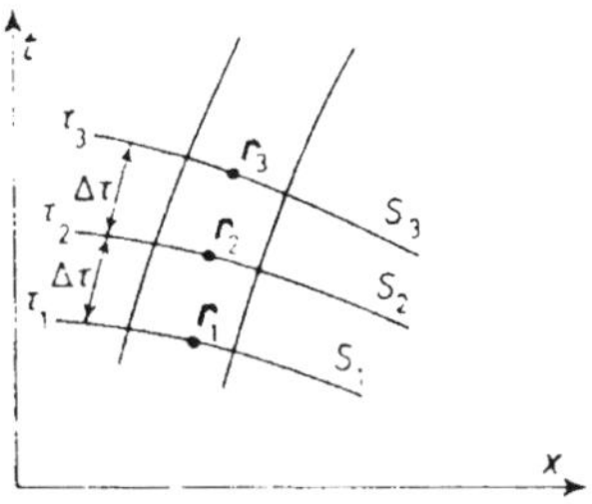

Fig. 1.

As a consequence of the assumed stochastic equilibrium we can treat on the same footing the fluid behaviour and an ensemble of similarly prepared particles characterized by the density $\varrho(\boldsymbol{x}, \tau)$ in configuration space where $\boldsymbol{x}$ represents a point in four dimensional space-time.

We shall now establish that the preceding model leads to the correct quantum-mechanical statistics (governed in our simplified case by the Klein-Gordon equation) in the simple case of a charged scalar particle. The simplification is justified since the introduction of spin complicates, but does not modify significantly, the various steps of our demonstration.

We can describe the average local motions of the elements of the ensemble by the selection of all particles that at proper time $\tau = \tau_2$ are contained in a small four-dimensional volume element around the point $\boldsymbol{r} = \boldsymbol{r}_2$ with co-ordinates $(r_2)_\mu$. This is necessary in our model, since if one starts from a particle in its local drift rest frame (*i.e.* the frame in which the neighbouring fluid element is practically at rest) its stochastic jumps along the light cone can bring it into any neighbouring line of flow: both in the forward and backward proper time direction. As a consequence our general stochastic model implies the use of a four-dimensional stochastic space-time volume element to recover all possible stochastic jumps of each drifting particle. We have thus made the new theoretical step of introducing along with the average space positions the new concept of an average time in a four dimensional volume element.

In order to describe the global motion of this element we select the particles that at proper time τ_2 are contained on a small section (space-volume element) of S_2 limited by the hypertubes boundary. According to our model it is possible to distinguish two different kinds of motion of this volume element during a short interval $\Delta\tau$. Besides its

(15) J. P. Vigier: *Compt. Rend.*, **266**, 598 (1968).

motion as a whole in the hypertube (which preserves the fluid's scalar density ϱ) the element will suffer variations of ϱ due to the stochastic jumps which move matter from one line of flow to another and will bring fluid across the hypertubes' boundary. Generalizing de la Peña and Cetto (3)'s ideas we can obtain a simplified description in terms of two quasilocal statistical velocities. If we take any one of the particles of our volume element and call $\boldsymbol{r}_1$ and $\boldsymbol{r}_2$ its average mean position at $\tau_1 = \tau_2 - \Delta\tau$ and $\tau_1 = \tau_2 + \Delta\tau$ we can calculate the average of $\boldsymbol{r}_3 - \boldsymbol{r}_2$ over the subensemble defined by the particles which belong to our small volume element. We call these average values the mean and denote them with $\langle\ \rangle$. We thus write

$$\text{(3)} \qquad \boldsymbol{r}_3 - \boldsymbol{r}_2 = \langle \boldsymbol{r}_3 - \boldsymbol{r}_2 \rangle + \delta_+ \boldsymbol{r} \quad \text{and} \quad \boldsymbol{r}_2 - \boldsymbol{r}_1 = \langle \boldsymbol{r}_2 - \boldsymbol{r}_1 \rangle + \delta_- \boldsymbol{r} .$$

Since one must assume (in our model) the homogeneity, isotropy and time independence of our stochastic mechanism the change variable $\delta_\pm r_i$ must satisfy $\langle(\delta_+ r_i)\rangle = \langle(\delta_- r_i)\rangle$ so what we can omit the indexes from such expressions and write in general $\langle \delta \boldsymbol{r}_i \rangle = 0$.

We can now derive from (3) two different velocities *i.e.*

$$\boldsymbol{b}_+(2) = ((\boldsymbol{r}_3 - \boldsymbol{r}_2)/\Delta\tau) \quad \text{and} \quad \boldsymbol{b}_-(2) = ((\boldsymbol{r}_2 - \boldsymbol{r}_1)/\Delta\tau) ,$$

whose mean values

$$\boldsymbol{v}_+(2) \equiv \langle \boldsymbol{b}_+(2) \rangle = \langle ((\boldsymbol{r}_3 - \boldsymbol{r}_2)/\Delta\tau) \rangle \quad \text{and} \quad \boldsymbol{v}_-(2) = \langle \boldsymbol{b}_-(2) \rangle = \langle ((\boldsymbol{r}_2 - \boldsymbol{r}_1)/\Delta\tau) \rangle$$

are the relativistic generalization of the mean forward and backward velocities. From these one can derive the regular fluid's velocity $\boldsymbol{v}_d$ and a stochastic belocity $\boldsymbol{u}_s$ through the relations

$$\text{(4)} \qquad \boldsymbol{v}_d(2) \equiv \langle ((\boldsymbol{r}_3 - \boldsymbol{r}_1)/2\Delta\tau) \rangle = \tfrac{1}{2}(\boldsymbol{v}_+ + \boldsymbol{v}_-)$$

$$\text{(5)} \qquad \boldsymbol{u}_s(2) \equiv \langle [(\boldsymbol{r}_3 - \boldsymbol{r}_2) - (\boldsymbol{r}_2 - \boldsymbol{r}_1)]/2\Delta\tau \rangle = \tfrac{1}{2}(\boldsymbol{v}_+ - \boldsymbol{v}_-)$$

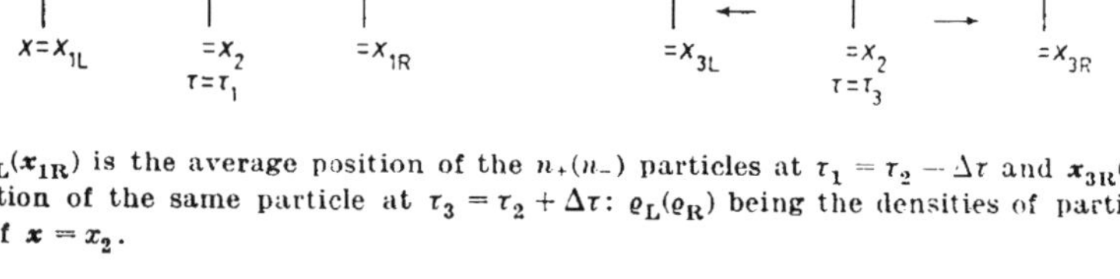

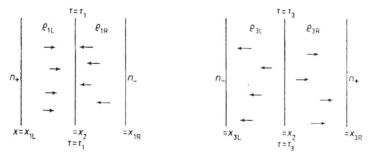

Fig. 2. – $x_{1L}(x_{1R})$ is the average position of the $n_+(n_-)$ particles at $\tau_1 = \tau_2 - \Delta\tau$ and $x_{3R}(x_{3L})$ is the average position of the same particle at $\tau_3 = \tau_2 + \Delta\tau$: $\varrho_L(\varrho_R)$ being the densities of particles to the left (right) of $x = x_2$.

Now the stochastic velocity $\boldsymbol{u}_s$ can be determined in any spacelike direction by calculating the flow between τ_1 and τ_3 of all elements which cross a drift timelike plane passing through $\boldsymbol{r}_2$ and orthogonal to a spacelike direction x. Indeed let us consider (see fig. 2) an ensemble of fluid elements (particles) which are at τ_2 in the neighbourhood of x. If $\varrho_{1L}(\varrho_{1R})$ then represents the scalar densities in the neighbourood of $x_{1L}(x_{1R})$ at $\tau = \tau_1$ we see that these densities are related to n_+ and n_- through

$$n_+ = (x_{3R} - x_2)\,\varrho_{3R} = (x_2 - x_{1L})\,\varrho_{1L} \qquad \text{and} \qquad n_- = (x_{1R} - x_2)\,\varrho_{1R} = (x_2 - x_{3L})\,\varrho_{3L}\,.$$

This yields

$$x_1 + x_3 - 2x_2 =$$
$$= (1/(n_+ + n_-))[-\varrho_{1L}(x_2 - x_{1L})^2 + \varrho_{1R}(x_{1R} - x_2)^2 + \varrho_{3R}(x_{3R} - x_2)^2 - \varrho_{3L}(x_2 - x_{3L})^2]$$

which can be averaged over the ensemble. Since each of the parentheses then become $\langle(\delta x)^2\rangle$ we can write to the first approximation (with $n_+ + n_- = 2\varrho(\tau_2)\Delta x$):

$$\boldsymbol{u}_s = \frac{\langle x_1 + x_3 - 2x_2\rangle}{2\,\Delta\tau} = \frac{\langle(\delta x)^2\rangle}{2\,\Delta\tau}\frac{1}{\varrho}\nabla\varrho = D\frac{\nabla\varrho}{\varrho}\,, \tag{6}$$

if we define as usual the diffusion coefficient as $D = \langle(\delta r_i)^2\rangle/2\,\Delta\tau$ and neglect higher-order terms in $\Delta\tau$. D is always > 0 since our quantum jumps are spacelike.

This is exactly the relativistic generalization of Einstein's definition [16] of the stochastic velocity in Brownian motion. We have further $\boldsymbol{v}_\pm = \boldsymbol{v}_d \pm \boldsymbol{u}_s$ which connect out forward (particle) and backward (antiparticle) velocities with the fluids regular drift velocity $\boldsymbol{v}_d$ and its stochastic velocity $\boldsymbol{u}_s$.

The second step is to associate the two velocities needed to describe our motion to four accelerations required to describe the forward and backward changes of these velocities. To do this we require the existence of our minimum proper time interval $\Delta\tau$ which allows us to define the four accelerations

$$\begin{cases} \boldsymbol{b}_+(3) - \boldsymbol{b}_+(2) = \boldsymbol{a}_+^+ + \delta_+\boldsymbol{b}_+\,, \\ \boldsymbol{b}_-(3) - \boldsymbol{b}_-(2) = \boldsymbol{a}_-^+ + \delta_+\boldsymbol{b}_-\,, \\ \boldsymbol{b}_+(2) - \boldsymbol{b}_+(1) = \boldsymbol{a}_+^- + \delta_-\boldsymbol{b}_+\,, \\ \boldsymbol{b}_-(2) - \boldsymbol{b}_-(1) = \boldsymbol{a}_-^- + \delta_-\boldsymbol{b}_-\,, \end{cases} \tag{7}$$

which evidently lead to systematic drift and stochastic derivative operators. Indeed if we define as D_d and D_s the following operations on a general function $f(\boldsymbol{r})$ of the stochastic variable $\boldsymbol{r}$, *i.e.*

$$D_d f(\boldsymbol{r}_2) = \langle[f(\boldsymbol{r}_3) - f(\boldsymbol{r}_1)]/2\,\Delta\tau\rangle \qquad \text{and} \qquad D_s f(\boldsymbol{r}_2) = \langle[f(\boldsymbol{r}_1) + f(\boldsymbol{r}) - 2f(\boldsymbol{r}_2)]/2\,\Delta\tau\rangle\,,$$

which are evidently related with the forward (D^+) and backward (D^-) derivative operators through the relation: $D^\pm = D_d \pm D_s$ we see they thus correspond to scalar (proper time type) derivatives in timelike and spacelike directions ... and yield the drift and stochastic velocities through $\boldsymbol{v}_d = D_d\boldsymbol{r}$ and $\boldsymbol{u}_s = D_s\boldsymbol{r}$: where the dummy index 2 has been omitted. This generalizes $v_\mu = \mathrm{d}x_\mu/\mathrm{d}\tau$ and lead to the preceding mean accelerations through the expressions $\boldsymbol{a}_\pm^+ = D^+\boldsymbol{v}_\pm$ and $\boldsymbol{a}_\pm^- = D^-\boldsymbol{v}_\pm$.

---

[16] A. Einstein: *Investigations on the Theory of Brownian Movement* (New York, N. Y., 1956).

Moreover a development in Taylor series yields

$$(8)\qquad \begin{cases} D_{\rm d} f = \dfrac{\partial f}{\partial \tau} + (\boldsymbol{v}_{\rm d} \cdot \nabla) f + \dots, \\ D_{\rm s} f = (\boldsymbol{u}_{\rm s} \cdot \nabla) f + D(\nabla \cdot \nabla) f + \dots, \end{cases}$$

where the diffusion coefficient D is given as before by the relation

$$\langle \delta r_i \cdot \delta r_j / 2\,\Delta\tau \rangle = D\delta_{ij}\,,$$

in the drift rest frame: diffusion in time representing, as before, particle-antiparticle transition: δr_i and δr_j denoting any pair of Cartesian components of $\delta_{\pm}\boldsymbol{r}$ ... which are assumed to be statistically independent if $i \neq j$.

The third (essential) step is to derive the covariant generalization of Nelson's equation, in our model. To do that we recall that any detailed description must start from the general equation

$$m\ddot{\boldsymbol{r}} = \boldsymbol{f}_{\rm d} + \boldsymbol{f}_{\rm s}\,,$$

where $\boldsymbol{f}_{\rm d}$ represents the drift spacelike forces and $\boldsymbol{f}_{\rm s}$ the purely random effects the $\dot{}$ denoting proper-time derivatives. The corresponding statistical theory must, according to our model, start from the ensemble of particles which at any proper time τ_2 lie in the neighbourhood of $\boldsymbol{r}_2$. The mean of the preceding relation thus becomes

$$(9)\qquad m\langle\ddot{\boldsymbol{r}}\rangle = \boldsymbol{F}_{\rm d} + \boldsymbol{F}_{\rm s} = \boldsymbol{F}, \qquad \text{where } \boldsymbol{F}_{\rm d} = \langle \boldsymbol{f}_{\rm d}\rangle \text{ with } \boldsymbol{F}_{\rm s} = \langle \boldsymbol{f}_{\rm s}\rangle = 0.$$

Since the mean value of $\ddot{\boldsymbol{r}}$ is taken over the same ensemble utilized to define our average velocities and accelerations in the preceding steps, it must be expressed as a linear combination of $\boldsymbol{a}^{\pm}_{\pm}$. To determine these combinations, we remark that $\langle\ddot{\boldsymbol{r}}\rangle$ and $\langle f_{\rm d}\rangle$ can be split into two parts *i.e.* a part $\langle\ddot{\boldsymbol{r}}\rangle^{+}$ (or $\langle f_{\rm d}\rangle^{+}$) which is invariant under proper time reversal *i.e.* $\tau_3 - \tau_2 \rightarrow \tau_1 - \tau_2$ and a part $\langle\ddot{\boldsymbol{r}}\rangle^{-}$ (or $\langle f_{\rm d}\rangle^{-}$) that changes sign under this discrete symmetry which changes $\boldsymbol{v}_{\rm d}$ but conserves $\boldsymbol{u}_{\rm s}$. Combining equation (9) with its counterpart obtained through a proper-time reversal operation we obtain the new set of equations

$$(10)\qquad m\langle\ddot{\boldsymbol{r}}\rangle^{\pm} = \boldsymbol{F}^{\pm}_{\rm d}\,.$$

We now make the final step in our demonstration of Nelson's equation (2) by examining the implications of eq. (10). The first implication is the importance of the proper-time relation $m\langle\ddot{\boldsymbol{r}}\rangle^{+} = \boldsymbol{F}^{+}_{\rm d}$ which evidently represents the stochastic generalization of Newton's law for our model. Indeed the usual four-dimensional acceleration $\ddot{\boldsymbol{x}}$ of a classical point $\boldsymbol{x}$ satisfies $\dot{\boldsymbol{x}}\ddot{\boldsymbol{x}} = 0$ (since $\dot{\boldsymbol{x}} \cdot \dot{\boldsymbol{x}} = -c^2$) and is invariant under proper-time reversal. The same holds for our stochastic case since: *a*) the drift acceleration $\dot{\boldsymbol{v}}_{\rm d}$ is orthogonal to $\boldsymbol{v}_{\rm d}$; *b*) the stochastic spacelike velocity $\boldsymbol{u}_{\rm s}$ is locally orthogonal to $\boldsymbol{v}_{\rm d}$ so that the corresponding stochastic accelerations (which vanish on the average since $\langle \boldsymbol{F}_{\rm s}\rangle = 0$) are thus always orthogonal to $\boldsymbol{v}_{\rm d}$.

The second implication is that $\langle\ddot{\boldsymbol{r}}\rangle^{+}$ must be expressed by just the linear combination of relations (7) which are proper-time–inversion invariant *i.e.* $(\boldsymbol{a}^{+}_{-} + \boldsymbol{a}^{-}_{+})$ or $(\boldsymbol{a}^{+}_{+} + \boldsymbol{a}^{-}_{-})$ or a linear combination therefrom.

The third implication is that a mean acceleration (which corresponds mathematically to second-order proper-time derivatives) should be defined physically only by the motions of fluid elements surrounding $\boldsymbol{r}_2$ *i.e.* enclosed within the four-dimensional volume element limited by S_1 and S_3 utilized to define mean quantities. We deduce therefrom and from the explicit form of the $\boldsymbol{a}$'s given in eq. (13), that the only quantity of this type invariant under $\tau \to -\tau$ is $(\boldsymbol{a}_-^+ + \boldsymbol{a}_+^-)$. Indeed the definition of $(\boldsymbol{a}_+^+ + \boldsymbol{a}_-^-)$ implies knowledge of the behaviour of fluid elements which lie outside our volume since it contains four-velocities of elements which are crossing S_3 and S_1 in the backward and forward directions *i.e.* are leaving this volume. Moreover one sees that the combination $\langle\ddot{\boldsymbol{r}}\rangle = (\boldsymbol{a}_-^+ + \boldsymbol{a}_+^-)$ evidently represents the relativistic definition of the sum of the mean accelerations of antiparticles $(\boldsymbol{a}_-^+)$ and particles $(\boldsymbol{a}_+^-)$ passing through $\boldsymbol{r}_2$ at $\tau = \tau_2$:

As a consequence we must write relation (10) in the form

$$\tfrac{1}{2} m(\boldsymbol{a}_-^+ + \boldsymbol{a}_+^-) = \boldsymbol{F}^+ , \tag{11}$$

which is exactly the relativistic generalization of the form given by de la Peña and Cetto ([3]) to Nelson's equation. Clearly eq. (11) contains particle-antiparticle symmetry.

The same argument applies to the — part of (10). Indeed the only combinations of $a_\pm^\pm$ that change sign under proper-time reversal are $(\boldsymbol{a}_+^+ - \boldsymbol{a}_-^-)$ and $(\boldsymbol{a}_-^+ - \boldsymbol{a}_+^-)$ and the second only is exclusively defined by the motion of fluid elements between S_1 and S_3. We thus have $\frac{1}{2} m(\boldsymbol{a}_-^+ - \boldsymbol{a}_+^-) = F^-$ which satisfies the continuity equation and is compatible with the introduction of the Lorentz force for charged fluid elements. Moreover these relations can be rewritten with the help of the definitions of D_d and D_s into the form

$$m(D_\mathrm{d} \boldsymbol{v}_\mathrm{d} - D_\mathrm{s} \boldsymbol{u}_\mathrm{s}) = \boldsymbol{F}^+ \tag{12a}$$

and

$$m(D_\mathrm{d} \boldsymbol{u}_\mathrm{s} + D_\mathrm{s} \boldsymbol{v}_\mathrm{d}) = \boldsymbol{F}^- . \tag{12b}$$

In eq. (12*b*) both sides tend (as they should) to zero in the nonstochastic limit.

The last step of our demonstration is, of course, the derivation of the integrated stochastic equations which result from (11) and (12). This can evidently be done in two ways. The first is to start from the drift rest frame at $\boldsymbol{r}_2$ and define as usual Smoluchowski's densities ϱ and P_R. The interested reader can then check immediately that since we have demonstrated *a*) and Nelson's equation (11) one can just follow Lehr's and Park's demonstration (7) to recover Klein-Gordon's equation.

The second way (which we will choose instead since it throws some interesting new light on the physics of the problem) is so complete the relativistic generalization of de la Peña's work ([3]).

In order to integrate (12*a*) and (12*b*) we define the quantities

$$D_\mathrm{q} = D_\mathrm{d} + \varepsilon D_\mathrm{s}, \quad \boldsymbol{v}_\mathrm{q} = \boldsymbol{v}_\mathrm{d} + \varepsilon \boldsymbol{u}_\mathrm{s} \quad \text{and} \quad \boldsymbol{F}_\mathrm{q} = \boldsymbol{F}^+ + \varepsilon \boldsymbol{F}^- \tag{13}$$

with $\varepsilon = \pm i$.

Relations (12*a*) and (12*b*) can thus be combined into the complex eqs. (14) *i.e.* $m D_\mathrm{q} \boldsymbol{v}_\mathrm{q} = \boldsymbol{F}_\mathrm{q}$ which can be integrated if one assumes that $\boldsymbol{F}_\mathrm{q}$ is just the general Lorentz force applied to our fluid of spinless charged particles *i.e.*

$$(\boldsymbol{F}_\mathrm{q})_\mu = (e/c)(\partial_\mu A_\mu - \partial_\mu A_\mu)(v_\mathrm{q})_\mu$$

with $\nabla \cdot \boldsymbol{A} = \partial_\mu A_\mu = 0$.

Indeed if we then write the relation (15) *i.e.* $\boldsymbol{v}_q = \varepsilon D \cdot \nabla S_q - (e/mc)\boldsymbol{A}$, where $D = \hbar/2m$, ∇ and $\boldsymbol{A}$ denoting the four-vectors ∂_μ and A_μ and $S_q = \text{const}$ representing the surfaces orthogonal to the four velocity $\boldsymbol{v}_q$. If we then utilize the Taylor developments (8) and substitute (15) and (16) into (14) we obtain the general relation

$$\nabla(2\varepsilon m D\dot{S}_q + \tfrac{1}{2} m\boldsymbol{v}_q \cdot \boldsymbol{v}_q + \varepsilon m D \nabla \cdot \boldsymbol{v}_q) = 0 \,, \tag{16}$$

which admits as first integral eq. (17) *i.e.*

$$-2\varepsilon m\dot{S}_q = 2\varepsilon^2 m D^2[\nabla S_q \nabla S_q + \nabla \cdot \nabla S_q] - 2\varepsilon D(e/c)\boldsymbol{A}\cdot\nabla S_q - \varepsilon D(e/c)\nabla \boldsymbol{A} + (e^2/2mc^2)\boldsymbol{A}\cdot\boldsymbol{A} \,.$$

Introducing further the wave function $\varphi(\boldsymbol{r}, \tau) = \exp[\varepsilon mc^2\tau/2\hbar]\psi(\boldsymbol{r})$ *i.e.*

$$\varphi(\boldsymbol{r}, \tau) = \exp[\varepsilon mc^2\tau/2\hbar]\varrho^{\frac{1}{2}}(\boldsymbol{r})\exp[\varepsilon S_q(\boldsymbol{r})] \,.$$

we obtain from (7) the usual relativistic generalization of the Schrödinger equation, *i.e.*

$$2mD\varepsilon\dot{\varphi} = (1/m)[2mD\varepsilon\nabla - (e/c)\boldsymbol{A}]^2\varphi \,, \tag{18}$$

which reduces to the Klein-Gordon equation

$$(\partial_\mu - \varepsilon(e/c)A_\mu)^2\psi - (m^2c^2/\hbar^2)\psi = 0 \,. \tag{19}$$

Relation (19) yields ([15]) the relations

$$\mathrm{d}\varrho/\mathrm{d}\tau = \dot{\varrho} = 0 \qquad \text{and} \qquad \mathrm{d}(M\boldsymbol{v}_d)/\mathrm{d}\tau = -\nabla(Mc^2) \tag{20}$$

with

$$M^2 = \{m^2 - (\hbar^2/c^2)(\Box R/R)\} \,, \qquad \psi^*\psi = R^2 \qquad \text{and} \qquad \varrho = (M/m)R^2 \,.$$

* * *

The author wants to express his thanks to Profs. L. de Broglie, D. Bohm and M. Flato for long and helpful past discussions stressing the possible importance of Einstein's views on Brownian motion in the interpretation of QM. He is especially grateful to Prof. Luis de la Peña-Auerbach and A. M. Cetto not only for crucial suggestions but also for help in the preparation of this work. Without this help, it would not have been completed.

Reprinted from *Physical Review*, Vol. 109, No. 6, pp. 1882-1891, Copyright (1958) with permission from the American Physical Society.

Relativistic Hydrodynamics of Rotating Fluid Masses

DAVID BOHM, *Technion, Haifa, Israel*

AND

JEAN-PIERRE VIGIER, *Institut Henri Poincaré, Paris, France*

(Received April 23, 1957)

With the aid of the new notion of center of matter density, we give a relativistic treatment of the behavior of finite-size masses of rotating fluid. This treatment is based on an analysis of the relative motion of this center of matter density and the more familiar center of mass. In this way, we obtain a clear physical interpretation of the equations studied by Mathisson, Weysenhoff, and Möller. We also show that more general types of motions are possible, related to additional degrees of freedom of the relativistic fluid droplet. These degrees of freedom provide a framework for a theory of the quantum numbers of the elementary particles (isotopic spin, strangeness, etc.) which will be developed in a subsequent paper.

I. INTRODUCTION

IN a series of very interesting papers, Mathisson,[1] Möller,[2] Weysenhoff,[3] and Pryce[4] have developed a relativistic theory of the motions of rotating masses of matter. Their equations are deduced from the conservation of energy-momentum and angular-momentum tensors. From these conservation assumptions they demonstrate the possibility of qualitatively new types of motion resulting from the coupling of a mean velocity with the total angular momentum of the system. However, they do not make it clear to what this mean velocity refers. In fact, Möller suggested that these new motions are purely formal, or in other words, that the mean velocity defined in these theories refers only to the behavior of fictitious and purely mathematical "center of gravity" points.

Moreover, in all these papers, the deduction of the equations of motion is based in a very essential way on the assumption that the time-like components of a certain angular momentum vanish in the mean rest frame of the body, or in other words, that:

$$\mathfrak{M}_{\alpha\beta}u^{\beta}=0; \qquad (1)$$

where $\mathfrak{M}_{\alpha\beta}$ is the antisymmetric tensor for the total internal angular momentum, and u_α is the four-velocity with $u_\alpha u^\alpha=1$. This assumption however has not been justified by any specific physical arguments; so that it constitutes a further somewhat arbitrary mathematical restriction on the theory.

In the present paper, we shall give a relativistic treatment of the general problem of the behavior of a mass of conserved fluid that is in some kind of rotational motion. We shall begin by giving a clear physical interpretation of the meaning of the time components of the angular momentum. Then with the aid of the

[1] M. Mathisson, Acta Phys. Polon. **6**, 163 (1937).
[2] C. Möller, Ann. inst. Henri Poincaré **11**, 251 (1949).
[3] J. Weysenhoff, Acta Phys. Polon. **9**, 7 (1947).
[4] M. H. L. Pryce, Proc. Roy. Soc. (London) **A195**, 62 (1948).

new concepts of center of matter density, defined along with the already well-known concept of center of mass, we shall see that the ambiguity of the meaning of the motions described in the above quoted papers can be removed. Indeed, it will become clear that in our interpretation of the theory, the equations of Weysenhoff, Möller, etc., refer to the relative motion of the center of mass and center of matter density. Furthermore, we shall see that motions are possible which are more general than those treated by Weysenhoff and Möller, with the result that Eq. (1) need no longer be satisfied. We then obtain a set of equations for these more general cases, and we show that the motion cannot be fully determined without further physical assumptions replacing Eq. (1).

In another paper we develop an example of one of these more general theories, leading to a classical motion equation of the same form as the Dirac relativistic equation in quantum mechanics. When this classical equation is quantized in the usual way, one obtains a set of quantum numbers similar to those which have been proposed recently[5] for the elementary particles.

II. RELATIVISTIC FLUID MASSES

The difficulty of treating rotating masses in the theory of relativity is connected with the impossibility of defining a relativistic rigid body in a consistent way.[6] We may, however, overcome these difficulties of formulation by considering instead relativistic fluid masses, which are kept together by appropriate internal tensions that tend to hold these masses in some stable forms.[7] Relative to such stable forms, the body of fluid may be subjected to all kinds of internal movements, such as rotations, vibrations, creation and destruction of inner closed-vortex structures, etc., each corresponding to different possible physical motions.

At first sight, a general treatment of the problem of describing the behavior of such masses raises insuperable difficulties. If we attempted to treat of all the details of these possible complex motions, we would find ourselves blocked not only by mathematical difficulties, but also by the fact that we do not even know in general what the fluid equations are. Fortunately another point of view is possible if we are willing to restrict ourselves to an over-all average description. In this case, if we suppose that however complicated the motion may be, there is a conserved energy-momentum tensor density $T_{\mu\nu}$ (which of course contains the tensions that hold the body of fluid together), it is possible to define certain average properties of the motion, independently of the complex details that we ignore. These average properties can be treated mathematically and lead to a description of the general features of the motions of relativistic rotating fluid masses.

The assumption of a conserved energy-momentum tensor density takes the form

$$\partial^{\nu} T_{\mu\nu}=0. \tag{2}$$

We assume further that the energy-momentum tensor is symmetric (as has been the case for all fluids treated so far). This means that

$$T_{\mu\nu}=T_{\nu\mu}. \tag{3}$$

As a result, the angular-momentum tensor

$$L_{[\mu\nu]\lambda}=x_{\mu}T_{\nu\lambda}-x_{\nu}T_{\mu\lambda} \tag{4}$$

satisfies the conservation equation

$$\partial^{\lambda} L_{[\mu\nu]\lambda}=0 \tag{5}$$

(where the bracket, $[\mu\nu]$, indicates an antisymmetric pair of indices).

On the basis of Eq. (2) one can easily show[8] that if the fluid body is localized (so that $T_{\mu\nu}$ vanishes outside a space-like three-dimensional limited region) the total energy and momentum integrated over all space in any specified Lorentz frame are constants. In other words,

$$\mathcal{G}_{\mu}=\int T_{\mu 0}dV=\text{constant} \tag{6}$$

and

$$d\mathcal{G}_{\mu}/dt=0, \tag{7}$$

where dV represents the element of volume.

Moreover, it also follows from Eq. (2) that $\mathcal{G}_{\mu}$ transforms as a 4-vector under Lorentz transformations. On physical grounds we suppose that $\mathcal{G}_{\mu}$ is a time-like vector; otherwise there would have to be a Lorentz frame in which the fluid had momentum but no energy.

We shall assume further that we can define at each point of the fluid mass a 4-vector density j_{μ} satisfying the conservation equation

$$\partial^{\mu} j_{\mu}=0. \tag{8}$$

This 4-vector density can be written

$$j_{\mu}=Du_{\mu},$$

where u_{μ} ($u_{\mu}u^{\mu}=1$) represents the components of the local unitary 4-velocity and $D=(j^{\mu}j_{\mu})^{\frac{1}{2}}$ the invariant matter density.

[5] See for example, J. Schwinger, Phys. Rev. **104**, 1164 (1956) or B. D'Espagnat and J. Prentki, Nuclear Phys. **1**, 33 (1956). A promising attempt has been made to consider elementary particles as stable excited states of our model of fluid masses; see P. Hillion and J. P. Vigier, Compt. rend. **246**, 399, 564 (1958) and Hillion, Lochak, and Vigier, Compt. rend. **246**, 710, 896 (1958).

[6] Many authors have discussed this problem and have proposed various solutions, but these proposals are in any case very complicated and it is not yet clear whether they are completely free of contradictions. See, for example, J. L. Synge, in *Studies Presented to R. von Mises* (Academic Press, Inc., New York, 1954), p. 217.

[7] H. Poincaré [Acta Math. 7, 259 (1885)] has shown that rotating fluid masses with internal tensions tend to go into stable equilibrium forms, one of which is a rotating torus.

[8] Möller, reference 2. We use Möller's notation where Greek indices such as μ vary from zero to three, and Latin indices i vary from one to three only. However, we are using the metric $(1, -1, -1, -1)$.

The "matter density" j_0 is proportional to the quantity of matter in a given region, which could, for example, be the number of molecules, while the vector j_i represents the rate of flow of this matter across a unit area in the direction of the coordinate vector i. If the fluid consists of charged particles with a constant ratio of e/m, then the density of charge will be $\rho = j_0 e$. But whether the fluid is charged or not, there will be a set of quantities j_μ satisfying Eq. (8).

The quantities j_μ clearly will be important for the determination of the way in which the body of fluid will change its shape, size, position, and orientation in space. Indeed, one can in principle deduce all these properties on the basis of the fundamental hydrodynamic equations satisfied by the local flow velocity,[9] which is just

$$v_i(\mathbf{x},t) = j_i(\mathbf{x},t)/j_0(\mathbf{x},t).$$

Hence, if we wish to treat any of these properties of the motion of the fluid, we shall evidently have to study the behavior of the j_μ.

In the nonrelativistic limit, the mass density T_{00}/c^2 and the matter density j_0 are proportional. But in the relativistic domain these two quantities may be different. For if the fluid is in motion, the kinetic energy E contributes a term E/c^2 to the mass; and in any case the internal tensions which are contained in the tensor $T_{\mu\nu}$ may make a similar contribution.

The essential features of the distinction between mass density and matter density arising in the theory of relativity may be brought out most clearly in terms of the conceptions of center of mass and center of matter density. We shall discuss the center of matter density in the next section; and here we shall consider only the center of mass which is defined as

$$\mathcal{G}_0 X_i = \int T_{00} x_i dV, \tag{9}$$

where $\mathcal{G}_0 = \int T_{00} dV$ is the total energy of the body (the Latin subscript i refers to space-like indices).

The most interesting property of the center of mass is that it moves at a constant velocity proportional to the total momentum. To prove this, we write

$$\mathcal{G}_0 \frac{dX_i}{dt} = \int \frac{\partial T_{00}}{\partial t} x_i dV. \tag{10}$$

But by the conservation equation (2) we have

$$\mathcal{G}_0 \frac{dX_i}{dt} = -\int \frac{\partial T_{0j}}{\partial x_j} x_i dV = \int T_{0j} \frac{\partial x_i}{\partial x_j} dV = \mathcal{G}_i, \tag{11}$$

where we have integrated $\partial^j T_{0j}$ by parts and used the vanishing of T_{0j} outside the fluid body. This gives

$$dX_i/dt = \mathcal{G}_i/\mathcal{G}_0, \tag{12}$$

which is the usual relativistic relation between the velocity of a particle and its momentum.

If we integrate the time components of the angular-momentum density $L_{[\mu\nu]\lambda} = x_\mu T_{\nu\lambda} - x_\nu T_{\mu\lambda}$, over the total volume of the liquid droplet, we obtain the total angular momentum,

$$L_{\mu\nu} = \int\int\int L_{[\mu\nu]0} dV. \tag{13}$$

In the same way, we can show that $L_{\mu\nu}$ is a constant. For

$$\frac{d}{dt} L_{\mu\nu} = \int\int\int \frac{\partial}{\partial t} L_{[\mu\nu]0} dV = -\int\int\int \frac{\partial}{\partial x_i} L_{[\mu\nu]i} dV = 0, \tag{14}$$

since $L_{[\mu\nu]\lambda} = 0$ on the surface of the droplet. From the conservation of $L_{[\mu\nu]\lambda}$, it follows that $L_{\mu\nu}$ is a tensor.

The center of mass has a close relationship to the time-like components of the angular momentum. To see this, we obtain from Eqs. (13) and (4)

$$L_{i0} = \int L_{[i0]0} dV = \int (x_i T_{00} - x_0 T_{0i}) dV.$$

If we integrate at a constant value of the time coordinate x_0 we obtain [using Eqs. (12) and (6)]

$$L_{i0} = \mathcal{G}_0 X_i - \mathcal{G}_i X_0. \tag{15}$$

If we choose $X_0 = 0$ then the time component of the angular momentum is proportional to the center-of-mass coordinate. More generally, we have, by integrating Eq. (12),

$$\mathcal{G}_0 X_i = \mathcal{G}_i X_0 + \alpha_i,$$

where α_i is a constant. We then obtain

$$L_{i0} = \alpha_i.$$

Thus, we verify the constancy of the time component of the angular momentum which we have already derived from the conservation law directly.

The space components of the angular momentum are, of course, just the usual moments of the momenta:

$$L_{ij} = \int (x_i T_{j0} - x_0 T_{i0}) dV. \tag{16}$$

It can be seen from Eq. (15) that the center of mass varies from one Lorentz frame to another.[10] For example, let us choose the origin of our space-time coordinate system such that $X_0 = 0$ and $X_i = 0$ (so that the origin is at the center of mass and L_{0i} is zero). If X_0, X_i were a four vector, then evidently under a Lorentz transformation we would obtain for the new coordinates $X_0' = X_i' = 0$ and L_{0i}' would also have to be zero. To show that this cannot be true consider the

[9] A number of such sets of equations have already been proposed; see A. Lichnerowicz, *Théorie de la gravitation et de L'Electromagnétisme* (Masson et Cie, Paris, 1955), Chapt. III and IV.

[10] See p. 2 Papapetrou, Möller, etc.

infinitesimal Lorentz transformation (which consists only of a change of velocity with no rotation):

$$X_0' = X_0 - \epsilon_{0i} X_i,$$
$$X_i' = X_i - \epsilon_{i0} X_0,$$

with $\epsilon_{i0} = -\epsilon_{0i}$. We obtain immediately

$$L_{i0}' = L_{i0} + \epsilon_{0l} L_{il} = \epsilon_{0l} L_{il},$$

since $L_{i0}=0$. Now L_{il} is just the space-like part of the angular momentum. Hence, if the fluid body is spinning (so that $L_{il}\neq 0$), then L_{i0}' will not be zero.

We conclude from the above that the center-of-mass coordinates do not transform as a four-vector, and that in different Lorentz frames, the center of masses correspond to physically different points.

III. CENTER OF MATTER DENSITY

We shall now define the center of matter density. In analogy with what was done with the center of mass, one would be led to assume that this center (Y_μ) is given by

$$Y_i J_0 = \int j_0 x_i dV, \tag{17}$$

where $J_0 = \int j_0 dV$ is the total amount of matter in the droplet (which is a scalar constant because of the conservation equation). The time derivative dY_i/dt would then be given by

$$J_0 \frac{dY_i}{dt} = \int \frac{\partial j_0}{\partial t} x_i dV = -\int \left(\frac{\partial j_k}{\partial x_k}\right) x_i dV = \int j_i dV. \tag{18a}$$

Thus, the velocity w_k of the center-of-mass density is

$$w_k = \frac{dY_k}{dt} = \frac{\int j_k dV}{J_0} = \frac{J_k}{J_0}; \tag{18b}$$

where

$$J_k = \int j_k dV.$$

The difficulty with this definition is that the quantities J_0, J_k do not form a four-vector, because J_k depends on the volume element corresponding to the chosen Lorentz frame. (On the other hand, we recall that the velocity of the center of mass is proportional to a four-vector.) As a result, we cannot obtain for example an unambiguous definition of the frame in which the center of matter density is at rest.

We can remove this ambiguity by defining the total current J_k and the center of matter density Y_i according to Eq. (18), but in the special Lorentz frame Π_0 in which the center of mass is at rest[11] (so that $\mathcal{G}_i=0$). This frame does have a unique meaning because the total momentum is a four vector. Then if we wish to know J_i and Y_μ in another frame Π' we simply take their values in the frame Π_0 and Lorentz transform them according to the transformation laws of a four vector. In other words J_i and Y_μ are defined by (18) only by integration over the volume element associated with the frame Π_0.

To express this definition of (Y_μ) in more detail, we first denote by the superscript zero all quantities which refer to the frame Π_0 in which the total momentum $\mathcal{G}_i$ is zero. The velocity of the center of mass in the frame is then

$$w_k^0(t^0) = J_k^0(t^0)/J_0^0. \tag{19a}$$

From this we can define a four-velocity (where we choose units such that $c=1$):

$$v_\mu^0(t^0) = J_\mu^0(t^0)/D^0(t^0), \tag{19b}$$

with

$$D^0(t^0) = [-J_\mu^0(t^0) J^{0\mu}(t^0)]^{\frac{1}{2}},$$

(where $J_\mu = iJ_0$). The four-velocity v_μ^0 is then evidently unitary; that is $v_\mu^0 v^{0\mu} = 1$.

On the other hand, the four-velocity of the center of mass will be

$$u_\mu = \mathcal{G}_\mu / M_0, \tag{20}$$

with

$$M_0^2 = -\mathcal{G}_\mu \mathcal{G}^\mu.$$

To go to an arbitrary frame, for example the laboratory frame Σ, one simply makes a Lorentz transformation with the velocity u_μ. This transformation is defined by

$$x_\mu = \sum_\nu a_{\mu\nu} x_\nu^0, \tag{21a}$$

with

$$a_{ij} = \delta_{ij} + \frac{u_i u_j}{1+u_0},$$
$$a_{k0} = -a_{0k} = iu_k, \tag{21b}$$
$$a_{00} = u_0.$$

Then in the frame Σ the center of matter density has the coordinates

$$Y_k = Y_k^0(t^0) - \frac{u_k u_l Y_l^0(t^0)}{1+u_0} - u_k t^0, \tag{22a}$$

and

$$Y_0 = u_0 t^0 - u_k Y_k^0(t^0); \tag{22b}$$

where Y_0 represents the time coordinate of the center of matter density in the frame Σ. The above equations provide a parametric representation of the trajectory of the center of matter density in Σ, the parameter being t^0.

The velocity of matter density in Σ can be obtained by Lorentz transformation of (19b). This gives

$$v_i(Y_0) = \frac{1}{D_0}\left\{J_i^0 + \frac{\mathcal{G}_i(\mathcal{G}^\mu J_\mu^0 - M_0 J_0^0)}{\mathcal{G}_0(M_0+\mathcal{G}_0)}\right\}. \tag{23a}$$

$$v_0(Y_0) = -\frac{\mathcal{G}^\nu J_\nu^0}{M_0 D_0}. \tag{23b}$$

[11] We are here using a suggestion of T. Takabayasi.

Notice that the time parameter appearing in the above equations is still t^0. We could transform to the parameter $t=Y_0$ with the aid of Eq. (22b). However, it will be more convenient to use the proper time τ of the center of matter density as a parameter. Then τ is defined by the relation

$$\frac{d\tau}{dY_0}=\frac{1}{V_0}=\frac{M_0D_0(t^0)}{\mathcal{G}^\nu J_\nu^0(t^0)}.$$

IV. INTERNAL ANGULAR MOMENTUM

Thus far we have defined the angular momentum relative to points fixed in space. We wish now to define an internal angular momentum analogous to the nonrelativistic angular momentum relative to the center of mass. In the relativistic theory, the center of mass varies from one Lorentz frame to another. Moreover, the center of mass and the center of matter density are not in general the same. Thus there is an ambiguity with regard to the point relative to which the inner angular momentum of the fluid droplet ought to be defined. This ambiguity could be removed for example by choosing as the point relative to which the inner angular momentum is to be taken the center of mass (X_μ). This is in fact one of the possibilities that Möller[1] considered. However, as we shall see in Sec. 8 such a choice leads to results having little physical significance with regard to the motion of the droplet as a whole. We shall choose instead for this purpose the center of matter density (Y_μ); for this point reflects in a better way the average velocity of the droplet. If the angular momentum relative to this point is taken, then the resulting equations will, as we shall see, describe the fluctuating motion of the droplet as a whole, relative to that of the center of mass which moves at a constant velocity. Thus, a general description of the over-all motion of the fluid droplet is obtained.

In accordance with these considerations, we define the inner angular momentum of the fluid droplet as

$$\mathfrak{M}_{\mu\nu}=\int[(x_\mu-Y_\mu)T_{\nu0}-(x_\nu-Y_\nu)T_{\mu0}]dV. \quad (24)$$

We can express $\mathfrak{M}_{\mu\nu}$ in terms of $L_{\mu\nu}$ with the aid of Eq. (13). We have

$$L_{\mu\nu}=\int(x_\mu T_{\nu0}-x_\nu T_{\mu0})dV$$

$$=\int[(x_\mu-Y_\mu)T_{\nu0}-(x_\nu-Y_\nu)T_{\mu0}]dV$$

$$+\int(Y_\mu T_{\nu0}-Y_\nu T_{\mu0})dV,$$

since Y_ν is a constant in the integration, we then obtain

$$L_{\mu\nu}=\mathfrak{M}_{\mu\nu}+Y_\mu\mathcal{G}_\nu-Y_\nu\mathcal{G}_\mu. \quad (25)$$

As $L_{\mu\nu}$ is a tensor, it follows from (25) that $\mathfrak{M}_{\mu\nu}$ is also a tensor. By differentiating (25) with respect to the proper time τ and by noting that $d\mathcal{G}_\mu/d\tau=0$ we obtain

$$d\mathfrak{M}_{\mu\nu}/d\tau=\mathcal{G}_\mu v_\nu-\mathcal{G}_\nu v_\mu. \quad (26)$$

This is one of the basic equations postulated by Weysenhoff.

It is particularly instructive to consider Eq. (26) in the special frame Σ_0 in which $v_i=0$. This we shall call the rest frame of the particle, because it is the frame in which the center of matter density is at rest. In that frame we choose a set of axes such that $Y_i=0$ and $Y_0=0$ for the moment of interest. Then we have $dY_0/dt=1$. We then obtain for the time component of the angular momentum:

$$L_{i0}=\mathfrak{M}_{i0}+Y_i\mathcal{G}_0-Y_0\mathcal{G}_i$$
$$=\mathfrak{M}_{i0}-\mathcal{G}_i t,$$

since

$$L_{i0}=\mathcal{G}_0X_i-\mathcal{G}_i t.$$

This yields

$$\mathcal{G}_0X_i-\mathcal{G}_i t=\mathfrak{M}_{i0}-\mathcal{G}_i t,$$

and thus (27)

$$\mathfrak{M}_{i0}=X_i\mathcal{G}_0.$$

The above relation shows that in the frame Σ_0 the time component of inner angular momentum is proportional to the vector joining the center of mass and the center of matter density. This interpretation of $\mathfrak{M}_{i0}$ will be seen to play an important role in the further development of the theory.

V. BRIEF REVIEW OF WEYSENHOFF'S THEORY

We now proceed to give a brief review of the Weysenhoff theory, in order to lay the foundations for a discussion of its physical significance in terms of our fluid model.

The basic starting point is Eq. (26). Now Eq. (26) consists of six equations determining the antisymmetric tensor $d\mathfrak{M}_{\mu\nu}/d\tau$ in terms of $\mathcal{G}_\mu$ and v_μ. Moreover, there are four more equations coming from the conservation of the total momentum, *viz.*:

$$d\mathcal{G}_\mu/d\tau=0.$$

There are still, however, no equations to determine the time variation of the v_μ (of which only three are independent, since $v^\mu v_\mu=1$). In order to determine those equations, some further hypothesis is needed. Such a hypothesis is essentially a supplementary assumption connecting the center of mass and the center-of-matter density.

In the Weysenhoff theory, the supplementary assumption is

$$\mathfrak{M}_{\mu\nu}v^\nu=0. \quad (1)$$

By going to the rest frame of the particle Σ_0 (where $v^i=0$, $v_0=1$), we see that the above reduces to the three conditions,

$$\mathfrak{M}_{0i}=0. \quad (28)$$

Thus, Eq. (1) which at first sight seems to contain four conditions, is seen actually to contain only three. And by Eq. (27) it follows that, in this frame,

$$X_i=0.$$

Thus, Weysenhoff's assumption implies, in our theory, that in the rest frame Σ_0 the center of mass and the center of matter density coincide. It is clear from this that Weysenhoff's assumption serves to complete the definition of the equations of motion of the particle.

To obtain the equations of motion in detail, we first multiply (26) by v^ν. This gives (with $v^\nu v_\nu=1$):

$$\mathcal{G}_\mu=(\mathcal{G}_\nu v^\nu)v_\mu+v^\nu d\mathfrak{M}_{\mu\nu}/d\tau. \qquad (29)$$

Now because $\mathfrak{M}_{\mu\nu}v^\nu=0$ we have

$$v^\nu\frac{d\mathfrak{M}_{\mu\nu}}{d\tau}+\mathfrak{M}_{\mu\nu}\frac{dv^\nu}{d\tau}=0.$$

Equation (29) then becomes (writing $\mathcal{G}_\nu v^\nu=m$):

$$\mathcal{G}_\mu=mv_\mu-\mathfrak{M}_{\mu\nu}dv^\nu/d\tau,$$

which gives, when multiplied by $dv^\mu/d\tau$,

$$\mathcal{G}_\mu\frac{dv^\mu}{d\tau}=mv_\mu\frac{dv^\mu}{d\tau}-\mathfrak{M}_{\mu\nu}\frac{dv^\mu}{d\tau}\frac{dv^\nu}{d\tau}.$$

But because $v^\mu v_\mu=1$, the first term on the right-hand side vanishes; while because of the antisymmetry of $\mathfrak{M}_{\mu\nu}$ the second term also vanishes. Thus we obtain $(d/d\tau)(\mathcal{G}^\mu v_\mu)=0$; and $\mathcal{G}^\nu v_\nu=\text{constant}$. In fact $\mathcal{G}^\nu v_\nu$ plays just the role of a rest mass. Thus we have, from (29),

$$\mathcal{G}_\mu=mv_\mu-\mathfrak{M}_{\mu\nu}\frac{dv^\nu}{d\tau}. \qquad (30)$$

This is one of Weysenhoff's set of equations.

To obtain the other set of Weysenhoff's equations, we differentiate Eq. (30) with regard to τ noting that $d\mathcal{G}_\mu/d\tau=0$. This yields

$$m\frac{dv_\mu}{d\tau}-\frac{d\mathfrak{M}_{\mu\nu}}{d\tau}\frac{dv^\nu}{d\tau}-\mathfrak{M}_{\mu\nu}\frac{d^2v^\nu}{d\tau^2}=0.$$

By applying (28) and $\mathcal{G}_\mu(dv^\mu/d\tau)=0$, we obtain

$$\frac{d\mathfrak{M}_{\mu\nu}}{d\tau}\frac{dv^\nu}{d\tau}=0;$$

and are left with

$$m\frac{dv_\mu}{d\tau}=\mathfrak{M}_{\mu\nu}\frac{d^2v^\nu}{d\tau^2}. \qquad (31)$$

The physical meaning of these relations can be further clarified by the introduction of a spin vector[12] s_μ defined by the relation

$$s_\mu=\mathfrak{M}_{\mu\nu}{}'v^\nu=\tfrac{1}{2}\epsilon_{\mu\nu\alpha\beta}v^\mu\mathfrak{M}^{\alpha\beta}.$$

The vector s_μ is a space-like vector, for we have evidently

$$s_\mu v^\mu=0.$$

In the rest frame we get

$$s_i=\tfrac{1}{2}\epsilon_{ijk}\mathfrak{M}^{jk}\quad(\epsilon_{ijk}=\epsilon_{ijk0}),$$

which implies that the spin is the space dual of the angular momentum in the rest frame.

Reciprocally we can write $\mathfrak{M}_{\alpha\beta}$ in terms of **s** and **v**. The preceding relation gives evidently

$$\mathfrak{M}^{jk}=\epsilon_{ijk}s^i,$$

which can be written in the covariant form:

$$\mathfrak{M}_{\mu\nu}=\epsilon_{\mu\nu\alpha\beta}s^\alpha v^\beta. \qquad (32)$$

The above implies the identity

$$\epsilon_{\alpha\beta\mu\nu}\mathfrak{M}^{\mu\nu}=2(s_\alpha v_\beta-s_\beta v_\alpha). \qquad (33)$$

From Eq. (19), it is then possible to calculate the derivatives of s_μ. We find immediately;

$$\frac{ds_\mu}{d\tau}=\tfrac{1}{2}\epsilon_{\mu\nu\alpha\beta}\left(\mathfrak{M}^{\alpha\beta}\frac{dv^\nu}{d\tau}+\frac{d\mathfrak{M}^{\alpha\beta}}{d\tau}v^\nu\right)$$

$$=\tfrac{1}{2}\epsilon_{\mu\nu\alpha\beta}\left(\mathfrak{M}^{\alpha\beta}\frac{dv^\nu}{d\tau}+\mathcal{G}^\alpha v^\beta v^\nu-\mathcal{G}^\beta v^\alpha v^\nu\right)$$

$$=\tfrac{1}{2}\epsilon_{\mu\nu\alpha\beta}\mathfrak{M}^{\alpha\beta}\frac{dv^\nu}{d\tau}.$$

By utilizing the decomposition (32) of $\mathfrak{M}^{\alpha\beta}$, we obtain:

$$\frac{ds_\mu}{d\tau}=\frac{dv^\nu}{d\tau}(s_\mu v_\nu-s_\nu v^\mu)$$

$$=-\left(s_\nu\frac{dv^\nu}{d\tau}\right)v_\mu; \qquad (34)$$

where we have also used $v^\nu(dv_\nu/d\tau)=0$.

In terms of the spin components s_μ, Eq. (29) then becomes

$$\mathcal{G}_\mu=mv_\mu-\epsilon_{\mu\nu\alpha\beta}s^\alpha v^\beta\frac{dv^\nu}{d\tau},$$

which can be written as

$$\mathcal{G}_\mu=mv_\mu-p_\mu, \qquad (35)$$

if we introduce the four-vector

$$p_\mu=\epsilon_{\mu\nu\alpha\beta}\frac{dv^\nu}{d\tau}s^\alpha v^\beta \qquad (36)$$

---

[12] The need for the introduction of a spin 4-vector density, instead of a tensor density $\mathfrak{M}_{\mu\nu}$ has been stressed by De Broglie [*Théorie des Particules de spin* ½ (Gauthiers-Villars, Paris), p. 54]. It has also been introduced independently of us by F. Halbwachs, Compt. rend. **243**, 1022, and **243**, 1098 (1956).

orthogonal to s_μ and v_μ. The quantity p_μ evidently represents the usual energy momentum of rotation in the rest frame.

The Weysenhoff equations (31) can then be expressed in a simplified form. We obtain, after a simple calculation,

$$\frac{dp_\mu}{d\tau}=m\frac{dv_\mu}{d\tau}$$

$$=\epsilon_{\mu\nu\alpha\beta}\frac{d^2v^\nu}{d\tau^2}s^\alpha v^\beta+\epsilon_{\mu\nu\alpha\beta}\frac{dv^\nu}{d\tau}\frac{ds^\alpha}{d\tau}v^\beta. \tag{37}$$

Multiplying (37) by s_μ, we then obtain:

$$m\frac{dv_\mu}{d\tau}s^\mu=0, \tag{38}$$

which, when inserted into (34) yields the relation:

$$ds_\mu/d\tau=0. \tag{39}$$

This shows that the spin s_μ of the droplet is a constant, according to the Möller-Weysenhoff theory.

From (37) and (38), we then obtain

$$m\frac{dv_\mu}{d\tau}=\epsilon_{\mu\nu\alpha\beta}\frac{d^2v^\nu}{d\tau^2}s^\alpha v^\beta. \tag{40}$$

Equation (40) constitutes a set of second-order differential equations for the velocity. These equations imply that, unlike what happens with Newton's laws of motion, not only are the initial values of Y_i and $dY_i/d\tau$ arbitrary, but so also are those of $dv_i/d\tau=d^2Y_i/d\tau^2$. As a result, new motions are possible that are not contained within the framework of Newton's laws. These new motions must, however, be consistent with (30) and (35) from which they were derived by differentiation.

To investigate the solutions to (40), it will be adequate to consider what happens in a special frame, namely that in which the space components $\mathcal{G}_i$ of the momentum are zero. For because of the Lorentz invariance of the theory, another solution of these equations corresponding to a nonzero value of $\mathcal{G}_i$ can always be obtained by Lorentz transforming the solution corresponding to $\mathcal{G}_i=0$.

As the velocity of the center of mass corresponding to each frame Σ is given by the relation $dX_i/d\tau=\mathcal{G}_i/\mathcal{G}_0$, we see that in the frame Π_0 where $\mathcal{G}_i=0$ (inertial rest frame of Weysenhoff) the center of mass is at rest. In Π_0 the space components v_i are then the components of a space vector **v** which represents the velocity of the center of matter relative to the center of mass.

The velocity v_μ also has a time component $v_0=J_0/D$. Since $dJ_0/d\tau=dD/d\tau=0$, we obtain $dv_0/d\tau=0$.

We see also that in the special frame Π_0 the acceleration of the center of matter is a space-like vector $d\mathbf{v}/d\tau$ and the general relation $v_\mu(dv^\mu/d\tau)=0$ can be written $v_i(dv^i/d\tau=0$; which implies that the two vectors **v** and $d\mathbf{v}/d\tau$ are orthogonal.

In Π_0 the components of the four-vector p_μ are

$$p_i=mv_i,\quad p_0=mv_0-\mathcal{G}_0.$$

The relation $p_\mu s^\mu=0$ which follows from (36) can be written as $-mv_is^i+p_0s^0=0$. From $v_\mu s^\mu=0$ we then deduce $v_is^i=-v_0s_0$, so that we find

$$-mv_0s^0+p_0s^0=0,$$

or equivalently:

$$s^0(p_0-mv_0)=-s^0\mathcal{G}_0=0.$$

As $\mathcal{G}_0\neq0$ we see finally that $s^0=0$ in Π_0 which implies that s_μ is a space-like vector in that frame. Then $v^\mu s_\mu=0$ becomes $v_is^i=0$ and we get $s^\mu(dv_\mu/d\tau)=s_i(dv_i/d\tau)=0$ (because of $dv_0/d\tau=0$). This shows that the three vectors **v**, $d\mathbf{v}/d\tau$ and **s** form an orthogonal instantaneous system of axes which generalizes, to our case, the Darboux-Freinet moving system of axes. The integration of the laws of motion results immediately from these considerations. The space-like components of p_μ in the frame Π_0 can be written as

$$p_i=mv_i=\epsilon_{i0jk}\frac{dv^0}{d\tau}s^jv^k+\epsilon_{ij0k}\frac{dv^j}{d\tau}s^0v^k+\epsilon_{ijk0}\frac{dv^j}{d\tau}s^kv^0$$

$$=\epsilon_{ijk0}\frac{dv^j}{d\tau}s^kv^0,$$

since $dv^0/d\tau$ and s^0 are zero. In ordinary vector notation, this relation becomes

$$m\mathbf{v}=v^0\left(\frac{d\mathbf{v}}{d\tau}\times\mathbf{s}\right). \tag{41}$$

The above equation implies that the motion of the center of matter remains in a plane orthogonal to s_μ. As **s** and v_0 are constants in Π_0 (since $ds_\mu/d\tau=dv_0/d\tau=0$), this motion reduces to a circular uniform motion with an angular velocity $\omega=m/|s|$. If we multiply (41) vectorially by **s**, we obtain

$$m\mathbf{v}\times\mathbf{s}=v_0\left(\frac{d\mathbf{v}}{d\tau}\times\mathbf{s}\right)\times\mathbf{s}$$

$$=\left(\mathbf{s}\cdot\frac{d\mathbf{v}}{d\tau}\right)\mathbf{s}-s^2\frac{d\mathbf{v}}{d\tau}$$

$$=-s^2\frac{d\mathbf{v}}{d\tau},$$

since, by Eq. (37), **s** is orthogonal to $d\mathbf{v}/d\tau$. This shows that:

$$\frac{d\mathbf{v}}{d\tau}=\frac{m}{s^2}\mathbf{s}\times\mathbf{v}.$$

Weysenhoff's equations therefore completely determine the time rate of change of $\mathbf{v}$ (once the momentum, p_μ, is determined).

We conclude that in Π_0 Weysenhoff's equations imply that the center-of-matter density executes the above-mentioned uniform circular motion around a fixed center of mass. This gives a clear physical meaning to the motions described by the Weysenhoff equations.

Finally we remark that if the external forces are acting on the fluid droplet, their effect can be taken into account by adding them to the energy-momentum conservation equation (2), so that we have $\partial^\nu T_{\mu\nu}=F_\mu$, where F_μ is the applied force. External torques can be taken into account in a similar way. This, in fact, has already been done by Möller.

VI. PHYSICAL MEANING OF THE WEYSENHOFF MOTIONS

Thus far we have seen in Sec. IV that Weysenhoff's assumption $\mathfrak{M}_{\mu\nu}v^\nu=0$ leads to interesting new kinds of circular motion. Mathematically speaking, the new motions are possible because Eq. (40) permits arbitrary initial values of $dv^\mu/d\tau$ as well as of v^μ and Y^μ. But in Newtonian mechanics $dv^\mu/d\tau$ can differ from zero only if there is an applied force. We must now see what are the physical conditions in the fluid which could lead to nonzero values of $dv^\nu/d\tau$ even in the absence of an applied force.

We can obtain a better understanding of this problem by going to the Lorentz frame Σ_0 in which the space-like parts v_i of the velocity are zero (while v_0 is unity). Equation (29) then takes the form

$$\mathcal{G}_i=-\mathfrak{M}_{i\nu}dv^\nu/d\tau.$$

Thus a nonvanishing acceleration will imply that the momentum and velocity are not collinear, so that even when the mean matter current is zero, there is still some momentum.

To show how this situation could come about, consider a mass of fluid in its rest frame (so that the total current is zero). Now suppose that this fluid is, to begin with, in a symmetrical and uniform rotational motion about its center of matter density. Then, as is evident from the symmetrical distribution of the energy, the center of mass will coincide with the center of matter density and Weysenhoff's condition will be satisfied. However, there will be no acceleration since the mean momentum is zero because of the symmetry. Thus the fluid will simply continue to rotate about a fixed point.

If this fluid is viewed from another Lorentz frame in which the body of the fluid moves with a velocity $\mathbf{v}$, the part of the body in which rotational and translational velocities add will then be moving faster than the part in which they substract. Thus the energy density will be higher on the former side than on the latter; and as a result, the center of mass will move away from the center of matter density.[13]

In this way, we see qualitatively, the origin of the relation $d\mathbf{v}/d\tau=(m/s^2)\mathbf{s}\times\mathbf{v}$ since $d\mathbf{v}/dt$ is just proportional to the difference between these two centers.

However, it is clear that as long as the distribution of velocity in the rest frame is symmetrical, there will be no net acceleration. We may however, suppose a further disturbance in the rotating mass of fluid; for example, localized vortices which do not contribute to the net current, but which do contribute to the energy density. Such a vortex would have two effects on the net motion.

First, it would displace the center of mass away from the center of matter density. Secondly, it would contribute to the mean momentum, since it would place a high mass density in a region of high velocity. Thus the mean momentum could fail to be zero even when the mean current was zero.

Of course, to satisfy the Weysenhoff condition without reducing to the trivial case of rectilinear motion, it is necessary to bring the center of mass back to the center of matter density, without bringing the momentum back to zero. This could be done by supposing further vortices on the opposite side of the body which lead to an energy density that cancels the moments of the original vortices in the determination of the center of mass, without canceling their momentum completely. To show that this is possible, consider two vortices on opposite sides of a diameter of the body. Let r_1 be the distance of the first vortex from the center of matter density, r_2 that of the second. Let W_1 be the energy of the first vortex, W_2 that of the second. Then we choose $W_1r_1=W_2r_2$ in order to satisfy to the Weysenhoff condition. Now the momentum of the first vortex will be $\mathbf{p}_1=W_1\mathbf{u}_1$ where $\mathbf{u}_1$ is the local mean stream velocity around the vortex, while that of the second vortex will be $\mathbf{p}_2=W_2\mathbf{u}_2$. If the angular velocity ω were a constant throughout the body (so that it was rotating as if it were rigid), then we would have $\mathbf{u}_1=\omega\mathbf{r}_1$ and $\mathbf{u}_2=\omega\mathbf{r}_2$ so that $\mathbf{p}_1+\mathbf{p}_2$ would be $\omega(W_1r_1+W_2r_2)=0$. But suppose ω were a function of r. This would imply, of course, a nonrigid rotation (of a type which is evidently quite common in fluids). Then we would have

$$p_1+p_2=\omega(r_1)W_1r_1-\omega(r_2)W_2r_2=[\omega(r_1)-\omega(r_2)]W_1r_1,$$

which is in general not zero. We see then that the Weysenhoff condition would be satisfied by suitable distribution of motions in the fluid. Of course, it could also be satisfied with much more complex distributions, but the principle is essentially the same.

We can now easily see qualitatively the reason for the Weysenhoff motion; for the velocity of the center of mass is proportional to the total momentum. Thus,

[13] The center of matter density will also be displaced, (because it too is not a four-vector). But a simple calculation shows that it will suffer a smaller displacement than that of the center of mass.

in the frame where $v_i=0$, $\mathcal{G}_i$ is not zero, so that the center of mass will move and separate from the center of matter density. Since the fluid body tends to maintain a certain shape, this process cannot continue indefinitely without some change in the pattern of the fluid motion and internal tensions, which leads to acceleration of the center of matter density. Indeed, if the fluid satisfies the Weysenhoff condition (1) for all times, then the distribution of motion will be such as to lead to an acceleration of the center of matter density which satisfies the Weysenhoff equations. Thus, the Weysenhoff assumption implies certain restrictions on the general features of the internal motions of the fluid body.

VII. EXTENSION TO MORE GENERAL MOTIONS NOT SATISFYING THE WEYSENHOFF CONDITION

We have seen that the Weysenhoff condition (1) represents a certain state of the internal motion of the fluid. The most general state of motion evidently need not satisfy this condition. We shall now formulate the problem of how to treat this more general type of motion.

First of all, we no longer require that the time component of the angular momentum vanish in the rest frame. It then becomes convenient to split the angular momentum into two parts, one of which is purely space like, and the other which is purely time like. To do this, we first define the four-vectors:

$$t_\mu=\mathfrak{M}_{\mu\nu}v^\nu,$$

$$s_\mu=\mathfrak{M}_{\mu\nu}'v^\nu=\tfrac{1}{2}\epsilon_{\mu\nu\alpha\beta}v^\nu\mathfrak{M}^{\alpha\beta}, \tag{42}$$

where $\mathfrak{M}_{\mu\nu}'$ is the dual of $\mathfrak{M}_{\mu\nu}$.

We note that because of the antisymmetry of $\mathfrak{M}_{\mu\nu}$, we have

$$v_\mu t^\mu=0, \tag{43a}$$

and similarly

$$v_\mu s^\mu=0. \tag{43b}$$

Thus, t_μ and s_μ are both vectors whose time component vanish in the rest frame. Hence they have a total of six independent components. Since $\mathfrak{M}_{\mu\nu}$ also has six independent components, this suggests that it should be possible to express $\mathfrak{M}_{\mu\nu}$ completely in terms of s_μ and t_μ. This is indeed possible, and the expression is

$$\mathfrak{M}_{\mu\nu}=t_\mu v_\nu-t_\nu v_\mu+\tfrac{1}{2}\epsilon_{\mu\nu\alpha\beta}(s^\alpha v^\beta-s^\beta v^\alpha). \tag{44}$$

To verify this, one need merely to go to the rest frame (where $v_i=0$ and $v_0=1$ while t_0 and s_0 are zero). For since (44) is a tensor relation it will be true in every frame if it is true in any one frame. But in the rest frame we have, from (42),

$$\mathfrak{M}_{i0}=t_i.$$

Since $\frac{1}{2}\epsilon_{0\nu\alpha\beta}(s_\alpha v_\beta-s_\beta v_\alpha)=0$ in the rest frame, the time-like components of (44) are identical.

To deal with the space-like components, we take the dual of both sides of (29). This gives

$$\tfrac{1}{2}\epsilon_{\mu\nu\alpha\beta}\mathfrak{M}^{\alpha\beta}=\epsilon_{\mu\nu\alpha\beta}t^\alpha v^\beta+s_\mu v_\nu-s_\nu v_\mu.$$

In the rest frame, the time components of this relation reduce to:

$$\tfrac{1}{2}\epsilon_{0ijk}\mathfrak{M}^{jk}=-s_i$$

which is also an identity. Thus Eq. (44) is proved.

The Weysenhoff theory then corresponds to the choice $t_\mu=0$.

If we wish to make a more general choice, we are then faced with the problem of determining the equation of motion of the t_μ; for as we saw the original equations (26) on which the Weysenhoff theory is based are just sufficient to determine the equations of motion when t_μ is chosen equal to zero. To proceed further we shall therefore need some additional physical hypothesis. In terms of the model of the fluid droplet, such a hypothesis implies that in the rest frame, the center of mass and center of matter density are no longer the same. It is evident that by means of a suitable distribution of vortex motions, any conceivable relation between t_μ and s_μ is possible. In a later paper, we shall consider a specific model leading to the relationship, $t_\mu=\lambda s_\mu$, where λ is a pseudoscalar. It will be seen that this relationship leads to a particularly significant generalization of the Weysenhoff equations, which when quantized can be applied to the treatment of many different elementary particles as different states of the same rotating fluid mass.

VIII. COMPARISON WITH TREATMENTS OF OTHER AUTHORS

We shall now compare our treatment of this problem with that used by others.

First of all, an essential new step proposed here was the introduction of the concept of center of matter density. We recall that because this center is not a four-vector, it was necessary to evaluate it in a definite Lorentz frame; namely Π_0, the one in which the quantities $\mathcal{G}_i$ vanished.

As we have already pointed out in Sec. IV, however, there exists another natural center, namely the center of mass. Previous work on this problem has been based on defining the point Y_μ of Eqs. (25) and (26) as the "center of gravity" which is the center of mass evaluated in the rest frame.

However, as we have seen from Eq. (12), the velocity of the center of mass is proportional to the momentum. If we evaluate the angular momentum relative to the "center of gravity," Eq. (26) then reduces to $d\mathfrak{M}_{\mu\nu}/d\tau=0$. Thus, all components of the angular momentum remain constant; the center of gravity moves at a constant rate; and Weysenhoff's equations reduce to the trivial case of uniform rectilinear motion. Hence, the whole treatment loses its interest, and nothing qualitatively new is learned about the motion from the Weysenhoff equations.

In order to give more physical relevance to the Weysenhoff equations, two types of proposals have been made.

(a) Möller[2] has shown that these equations could describe the motion of a certain set of purely mathematically defined "pseudo centers of gravity," whose connection with any aspect of the motion of the body has not been defined,

(b) It has been shown that under suitable conditions points fixed in the body will undergo the Weysenhoff motions.

Possibility (a) is evidently not satisfactory, since it describes no real physical properties of the motion. Possibility (b) also is not entirely satisfactory, especially for a fluid where elements undergo complex motions and where it is arbitrary to choose a particular element to describe the behavior of the mass of fluid.

Our proposal of a center of matter density provides a natural reference point to describe the behavior of the fluid; for the matter current determines the general motion of the fluid, while the center of mass is a point which moves at a constant rate. By studying the motion of the center-of-matter density relative to that of the center of mass, we obtain a general idea of how the fluid motion differs from uniform rectilinear motion, without the need for going into a detailed treatment of all the complex motions inside the fluid body.[14]

We also give a clear interpretation of the Weysenhoff condition $\mathfrak{M}_{\mu\nu}v^{\nu}=0$ for it means that in the Lorentz frame in which the spatial components of the mean velocity are zero, there is no time component to the internal angular momentum, so that in this frame the center of mass and the center of matter density are the same. As we have seen in the previous sections, this is a possible state of motion of the fluid droplet, and one which could readily be set up. As we have also seen, however, more general states of motion are possible. Thus our physical interpretation of these equations opens up the possibility for studying a broader range of motions. Indeed, as we shall show in a later paper, some of these new possibilities correspond to a set of classical equations which, when quantized, lead to a generalization of the Dirac equation. In these generalized equations, there is a new set of quantum numbers very similar to those (such as isotopic spin and strangeness) which have been used recently for classifying the various types of elementary particles. Thus, the new quantum numbers can be interpreted as representing states of rotation and internal excitation of a relativistic liquid droplet.

This general theory of rotating relativistic droplets is also interesting from another point of view. It provides a clear physical model for the "molecules" which could constitute relativistic fluids with spin, that is, fluids characterized at the macroscopic level by a continuous distribution of internal angular momentum. By adding to $\mathcal{G}_{\mu}$ and $\mathfrak{M}_{\mu\nu}$ suitable tensions $\Theta_{\mu\nu}$ representing interactions between the rotating droplets which constitutes such fluids (so that the total energy-momentum tensor of such a fluid can be represented by $D\mathcal{G}_{\mu}v_{\mu}+\Theta_{\mu\nu}$ where the total current is Dv_{μ} and the internal angular momentum $D\mathfrak{M}_{\mu\nu}$), we can formulate new types of relativistic hydrodynamics. The theory of these types is now being developed. Indeed, it has already been shown that they provide a model for the hydrodynamical representation of the Dirac and Kemmer wave equations,[15] thus furnishing a physical basis for the causal interpretation of relativistic wave equations.

ACKNOWLEDGMENTS

Finally, we wish to express our gratitude to Professor De Broglie, Professor Takabayasi, and to Mr. F. Halbwachs for many interesting discussions and valuable suggestions. Professor Takabayasi in particular has greatly contributed to the clarification of the ideas in this paper.

[14] These details could for example be treated by considering moments of the first energy-momentum and current densities higher than the first. But since such moments are not conserved in general, their effects tend to be lost by random mixing processes analogous to collisions in the Boltzmann equation in statistical mechanics. Thus, the conserved moments not only satisfy equations that are independent of the higher moments, but also describe the major part of the "bulk" properties of the fluid averaged over some period of time.

[15] Halbwachs, Lochak, and Vigier, Compt. rend. **241**, 692 (1955), and Takehiko Takabayasi, Phys. Rev. **102**, 292 (1956).

Causal Superluminal Interpretation of the Einstein-Podolsky-Rosen Paradox.

N. CUFARO PETRONI

Istituto di Fisica dell'Università - Bari, Italia
Istituto Nazionale di Fisica Nucleare - Sezione di Bari, Italia

J. P. VIGIER

Institut Henri Poincaré - Paris, France

(ricevuto il 27 Giugno 1979)

In three recent papers ([1]) a possible causal interpretation of the EPR paradox has been suggested in terms of the superluminal propagation of de Broglie's ([2]) and Bohm's ([3]) quantum potential in the causal interpretation of quantum mechanics. Indeed non-local interactions now seem unavoidable in quantum theory if aspect's forthcoming experiment confirms (as believed by the authors) the experimental predictions of quantum mechanics and disproves the validity of Bell's inequalities.

The aim of the present letter is double. We first want to present a quantitative description of this model in some detail for the simple case of two identical, correlated, scalar particles and thus interpret in this context the first form of the EPR paradox (*i.e.* the simultaneous measurement of their positions and momenta) discussed within the frame of the causal interpretation, by Bohm and Hiley ([4]). We then want to analyse briefly the evident connection (and discrepancies) between our point of view and the well-known superluminal tachyonic interactions introduced in the literature by Sudarshan, Feinberg, Recami *et al.* ([5]). We show in particular that superluminal, phaselike, phononlike, collective motions of the quantum potential in Dirac's «ether» do not induce the well-known causal paradoxes ([5]) of tachyon theory.

Our starting point is just the two-particle generalization in configuration space of our one-particle model ([1]). Indeed let us assume two identical scalar particles labelled 1

([1]) J. P. VIGIER: *Lett. Nuovo Cimento*, **24**, 258, 265 (1979); N. CUFARO PETRONI and J. P. VIGIE: *Lett. Nuovo Cimento*, **25**, 151 (1979).
([2]) L DE BROGLIE: *La physique quantique restera-t-elle indeterministe?* (Paris, 1953).
([3]) D. BOHM: *Phys. Rev.*, **85**, 166, 180 (1952).
([4]) D. BOHM and B. HILEY: in *Quantum Mechanics a Half Century Later*, edited by J. LEITE-LOPES and M PATY (1975).
([5]) O. M. P. BILANIUK, V. K. DESHPANDE and E. C. G. SUDARSHAN: *Am. J. Phys.*, **30**, 718 (1962); G. FEINBERG: *Phys. Rev.*, **159**, 1089 (1967); E. RECAMI and R. MIGNANI: *Riv. Nuovo Cimento*, **4**, 209, 398 (1974); E. RECAMI: *Found. Phys.*, **8**, 329 (1978).

and 2 imbedded in Dirac's stochastic « ether » ([6]). The pair's motions along any world line, in configuration space-time, build a fluid in this space-time. These motions are not independent (since the presence of particle 1 disturbs the « ether » *i.e.* the motion of particle 2 and vice-versa) and one assumes that we are dealing (as in the one-particle case) with stochastic jumps at the velocity of light (in physical space-time) which pass the pair 1, 2 from one drift line of flow (in configuration space) to another. Physically this amounts (in the hydrodynamical model of Bohm and Vigier) to the superposition in space-time of two interacting fluids 1 and 2 which undergo lightlike internal stochastic motions, particle-antiparticle transitions and possible number-preserving transfers from one fluid to another ... so that we have a conserved scalar fluid particle density in configuration space.

Mathematically this model can thus be described by in an eight-dimensional configuration space where a pair position is defined by an eight-component vector X^i ($i = 1, ..., 8$) where

(1) $$\{X^i\}_{i=1,...,8} = \{x_1^\mu ; x_2^\nu\}_{\mu,\nu=0,...,3}$$

with x_1^μ, x_2^ν four-vectors of the position of each body. The metric is defined by

(2) $$g_{ij} = \begin{pmatrix} 1 & 0 & 0 & 0 & 0 & 0 & 0 & 0 \\ 0 & -1 & 0 & 0 & 0 & 0 & 0 & 0 \\ 0 & 0 & -1 & 0 & 0 & 0 & 0 & 0 \\ 0 & 0 & 0 & -1 & 0 & 0 & 0 & 0 \\ 0 & 0 & 0 & 0 & 1 & 0 & 0 & 0 \\ 0 & 0 & 0 & 0 & 0 & -1 & 0 & 0 \\ 0 & 0 & 0 & 0 & 0 & 0 & -1 & 0 \\ 0 & 0 & 0 & 0 & 0 & 0 & 0 & -1 \end{pmatrix}$$

so that

(3) $$X^2 = X_i X^i = g_{ij} X^i X^j = (x_1)^2 + (x_2)^2 .$$

If $x_1^\mu(\tau_1)$, $x_2^\nu(\tau_2)$ are the trajectories for the two particles, the trajectory in configuration space will be an $X^i(\tau_1, \tau_2)$. As a consequence of Nelson's equations ([1]) we can now generalize the differential operators defined by GUERRA and RUGGIERO ([7]) for the single-particle case to a system of two identical particles

(4) $$\begin{cases} D = \dfrac{\partial}{\partial \tau_1} + \dfrac{\partial}{\partial \tau_2} + b^i \partial_i , \qquad \delta D = \delta b_i \partial^i - \dfrac{\hbar}{2m} \Box , \\[2ex] \partial_i = \dfrac{\partial}{\partial X^i} , \qquad \Box = \partial_i \partial^i = \Box_1 + \Box_2 , \qquad i = 1, ..., 8, \\[2ex] b_i = DX_i ; \qquad \delta b_i = \delta D X_i . \end{cases}$$

Now a direct extension of Guerra and Ruggiero ([7]) formulae gives the following de-

([6]) P. M. A. DIRAC: *Nature (London)*, **168**, 906 (1951).
([7]) F. GUERRA and P. RUGGIERO: *Lett. Nuovo Cimento*, **23**, 529 (1978).

pendence of δb_i on a density $\varrho(X^i, \tau_1, \tau_2)$:

(5)
$$\delta b_i = -\frac{\hbar}{m}\,\partial_i \log \varrho^{\frac{1}{2}}\,,$$

where for the density we have as continuity equations

(6)
$$\frac{\partial\varrho}{\partial\tau_1} = -\,\partial_{1\mu}(\varrho b_1^\mu) \quad\text{and}\quad \frac{\partial\varrho}{\partial\tau_2} = -\,\partial_{2\mu}(\varrho b_2^\mu)\,,$$

with

(7)
$$b_1^\mu = Dx_1^\mu\,,\qquad b_2^\mu = Dx_2^\mu\,,\qquad \partial_1^\mu = \frac{\partial}{\partial x_{1\mu}}\,,\qquad \partial_2^\mu = \frac{\partial}{\partial x_{2\mu}}\,.$$

In our model, as a generalization of the assumption (7) that ϱ is independent of the proper time in the one-body case, we make the physical hypothesis that the total number of particles (*i.e.* pair in the real space-time) is conserved, and thus we write

(8)
$$\frac{\partial\varrho}{\partial\tau_1} + \frac{\partial\varrho}{\partial\tau_2} = 0\,,$$

so that our continuity equations in configuration space is

(9)
$$\partial_i(\varrho b^i) = 0\,.$$

We assume as before (7) that our fluid motion is irrational, so that

(10)
$$b^i = \frac{1}{m}\,\partial^i\Phi\,,$$

where $\Phi(X^i, \tau_1, \tau_2)$ is a phase function, and, if we look for a steady state (*i.e.* proper time independent) equation,

(11)
$$\Phi(X^i, \tau_1, \tau_2) = \frac{mc^2}{2}(\tau_1 + \tau_2) + S(X^i)\,.$$

Now it is clear that (as generally assumed and later demonstrated by CUFARO PETRONI and VIGIER (1,8)) Newton's equations for the two free particles can be written in the compact form

(12)
$$(DD - \delta D\,\delta D)X^i = 0\,.$$

Starting from (9), (12) and using (5), (10), (11) we obtain an Hamiltonian-Jacobi-type equation ($R = \varrho^{\frac{1}{2}}$) for our two-body system *i.e.*:

(13)
$$\left(\partial_i\partial^i - \frac{\partial_i S\,\partial^i S}{\hbar} - 2\,\frac{m^2c^2}{\hbar^2}\right)R = 0$$

(8) N. CUFARO PETRONI and J. P. VIGIER: *A Markov process at the velocity of light: the Klein-Gordon statistic*, preprint Inst. H. Poincaré, Paris (June 1979).

which yields for the continuity equation the form

$$2\partial_i R\,\partial^i S + R\,\partial_i\partial^i S = 0\,. \tag{14}$$

Finally if we consider (13) as the real part and (14) as the imaginary part the total equation for $\psi = R\exp[iS/\hbar]$ is

$$\left(\Box - 2\,\frac{m^2c^2}{\hbar^2}\right)\psi = 0\,. \tag{15}$$

From relation (15) one evidently deduces, in the nonrelativistic limit, the usual two-particle Schrödinger equation which (writing $\psi(\boldsymbol{x}_1, \boldsymbol{x}_2, t) = R(\boldsymbol{x}_1, \boldsymbol{x}_2, t)\exp[iS/\hbar]$) splits into real and imaginary parts *i.e.*:

$$\frac{\partial P}{\partial t} + \boldsymbol{\nabla}_1\left(P\,\frac{\boldsymbol{\nabla}_1 S}{m}\right) + \boldsymbol{\nabla}_2\left(P\,\frac{\boldsymbol{\nabla}_2 S}{m}\right) = 0 \tag{16}$$

with $P = R^2 = \psi^*\psi$ and

$$\frac{\partial S}{\partial t} + \frac{(\boldsymbol{\nabla}_1 S)^2}{m} + \frac{(\boldsymbol{\nabla}_2 S)^2}{m} + Q = 0\,, \tag{17}$$

with $Q = -(\hbar^2/2m)[(\nabla_1^2 R/R) + (\nabla_2^2 R/R)]$. Clearly relation (16) represents the conservation of the probability $P = \psi^*\psi$ in configuration space $(\boldsymbol{x}_1, \boldsymbol{x}_2)$ while relation (17) as discussed by BOHM and HILEY (4) corresponds to a Hamilton-Jacobi equation for two particles which interact through a nonlocal quantum potential Q with which they have interpreted the first form of the EPR paradox in its original position-momentum formulation. In the causal interpretation of this situation one adds of course that our particle momenta are described in real space by $V_1 = \boldsymbol{\nabla}_1 S/m$ on $V_2 = \boldsymbol{\nabla}_2 S/m$ as the mapping of configuration space into real space suggests (9).

According to plan we conclude this letter with a brief discussion of the physical implications of nonlocality, since this question is now strongly reproposed by recent developments of the analysis of measurement process for correlated systems (10). The first implication is the possibility of time inversions of such events under specific Lorentz transformations. As one knows the question of the time exchange of two causally correlated events has already been discussed (for tachyons) by several authors on the basis of the reinterpretation principle (5) and rests on the remark that a Lorentz transformation which exchanges time co-ordinates of two spacelike events also exchanges energy signs and hence (on the basis of the particle-antiparticle symmetry (11)) also exchanges the cause-effect role: so that the cause always precedes the effect. Finally we can preserve the right time succession of causes and effects if we abandon the independence from the observer of what is cause and what is effect. (For a detailed discussion see ref. (5).)

---

(9) J. ANDRADE E SILVA: *La théorie des systémes de particles dans l'interprétation causale de la mécanique ondulatoire* (Paris, 1960).

(10) A. GARUCCIO and F. SELLERI: *Action at distance in quantum mechanics*, in *Communication at Einstein's Centenary Commemoration* (Paris, June 1979); N. CUFARO PETRONI, A. GARUCCIO, F. SELLERI and J. P. VIGIER: *Sur la contradiction d Einstein-Bell entre les théories de la mésure quantique et la théorie locale de la rélativité restreinte*, preprint Inst. H. Poincaré (June 1979).

(11) R. P. FEYNMAN: *Phys. Rev.*, **76**, 749, 769 (1949).

More care must be used to solve the second implication *i.e.* the so-called « causal anomalies » which can be condensed in the following paradox (see fig. 1):

Let us consider two relatively moving observers O_1, O_2 with respective rest frame S, S'. At the event ε_1, O_1 sends (in its relative future) a superluminal signal to O_2 which absorbs it at ε_2; after some time, at the event ε_3, O_2 sends (in its relative future) another superluminal signal to O_1 which absorbs it at ε_4. It is easy to verify ([5]) that we can always arrange this experiment so that ε_4 precedes ε_1 ... so that we can use superluminal signals in order to modify the absolute past of O_1!

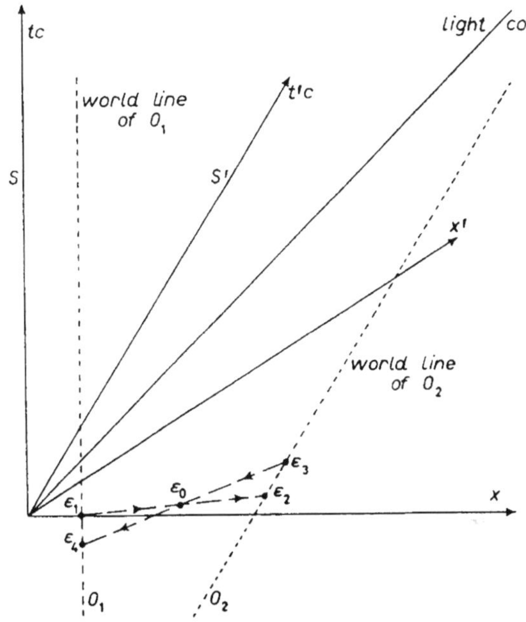

Fig. 1. – Causal anomaly.

Of course possible solutions of this causal paradox have already been proposed ([5]) for the case of signals carried by tachyons: but it is significant to note that the problem does not exist in our model where superluminal signals are not tachyons, since their propagation is now carried by a collective motion of the extended particles of Dirac's vacuum ([6]). This collective phaselike motion behaves like an heat flux ([1]) having, as it is known ([12]) a superluminal diffusion velocity. In these models the causality is first preserved in the sense that, although heat diffusion velocity is infinite, the carrier particles always remain within the light cone ([1]). However, as in the preceding example, the possibility apparently remains in principle to send signals which can modify the absolute past of any physical system. To avoid this causal anomaly, one must forbid the closed paths of fig. 1 which are evidently responsible of all causal paradoxes. This is true, as we shall show, if superluminal signals are real collective motions carried by extended vacuum particles. In that case we can require, indeed, that each particle has an intrinsic absolute flux of time (its own « proper » time) so that, with respect to this time and for this particle, no causal effect can precede its cause. Since our vacuum

([12]) R. Hakim: *Lett. Nuovo Cimento*, **25**, 108 (1979).

particles are extended ([1,13]) the superluminal signal must always « cross » the world-tube of these « carrier » particles in the positive sense of their own time flux (as in fig. 2). So that the propagation of a signal from ε_1 to ε_2 is always possible provided that $\tau_1 < \tau_2$.

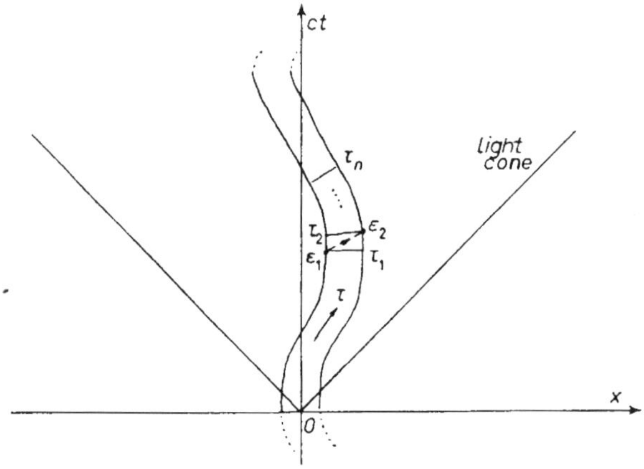

Fig. 2. – World tube of a vacuum particle with proper time τ defined as proper time of the centre of matter ([13]). A signal between $\varepsilon_1, \varepsilon_2$ always travels in the positive τ direction.

Of course for superluminal signals the time succession of ε_1 and ε_2 can be reversed for another Lorentz frame, but we can now show that this feature is irrelevant in order to avoid causal anomalies. Indeed the condition $\tau_1 < \tau_2$ (locally verified for each signal which crosses a world tube) is sufficient to forbid a path like that of fig. 1 because in the event ε_0 there are at least two criss crossing superluminal signals, so that at least one of these two signals cannot satisfy the condition $\tau_1 < \tau_2$ ... provided that vacuum particles always move with infraluminal velocity. An analog analysis can be made in the four-dimensional case if we consider superluminal signals to be « acoustical » waves with associated quantum potential propagating in the vacuum in all space directions. It is interesting to note that this elimination of causal paradoxes is only possible in a subquantum model built on a Dirac's vacuum and cannot be applied to theories where superluminal signals are carried by tachyonic particles, and to theories of the Costa de Beauregard ([14]) type where the causal connection between two spacelike events is always possible in principle through time travel into the absolute past of any physical system.

(13) D. Bohm and J. P. Vigier: *Phys. Rev.*, **109**, 1882 (1958); D. Gutkowski, M. Moles and J. P. Vigier: *Nuovo Cimento B*, **39**, 193 (1977); F. Halbwachs: *Théorie relativiste des fluides γ spin* (Paris, 1960).
(14) O. Costa de Beauregard: *Phys. Lett. A*, **67**, 171 (1978); *Ann. Fond. de Broglie*, **2**, 231 (1977)

Reprinted from *Lettere al Nuovo Cimento*, Vol. 31, No. 12, pp. 415-420, Copyright (1981) with permission from the Società Italiana di Fisica.

Action-at-a-Distance and Causality in the Stochastic Interpretation of Quantum Mechanics.

N. Cufaro Petroni

Istituto di Fisica dell'Università - Bari, Italia
Istituto Nazionale di Fisica Nucleare - Sezione di Bari, Italia

Ph. Droz-Vincent

Collège de France et Université Paris VII
Laboratoire de Physique Théorique et Mathématique
Tour 33-43, 1er étage, 2 Place Jussieu, 75231 Paris Cedex 05, France

J. P. Vigier

Equipe de Recherche Associée au CNRS N. 533, Institut Henri Poincaré
11 rue P. et M. Curie, 75231 Paris Cedex 05, France

(ricevuto il 15 Aprile 1981)

Since the results of the first stage of Aspect's experiment (without the switches) ([1]) have 1) strongly confirmed quantum-mechanical predictions, 2) disproved Bell's inequalities (and thus eliminated the possibility of local hidden variables), it is now probable that the second stage (with switches) will also confirm the existence of a nonlocal correlation between distant polarizers (practically separated by 12 meters) measuring the relative polarization of photon pairs emitted in the calcium $4p^2\,{}^1S_0$-$4s4p\,{}^1P_1$-$4s^2\,{}^1S_0$ transitions. It is thus to be expected that the same result will appear in any concrete realization of the EPR gedanken-experiment ([2]) proposed by Bartell ([3]) to test (with Furry microscopes) the initial version of the EPR paradox on correlated p^μ and q^μ values in pairs of noninteracting scalar particles.

In brief we are now confronted with the very probable result that the forthcoming experiment of Aspect and Rapisarda et al. ([1]) *will confirm in the near future the nonlocality predicted by quantum mechanics.*

This is an important event, since many people still believe in the antagonistic character of nonlocality and causality. This is understandable, since one knows that

---

([1]) A. Aspect, Ph. Grangier and Q. Roger: *Experimental test of realistic local theories via Bell s theorem*, Orsay preprint (1981); A. Aspect: *Phys. Lett. A*, **67**, 117 (1975); *Prog. Sci. Cult.*, **1**, 439 (1976); *Phys. Rev. D*, **14**, 1944 (1978).
([2]) A. Einstein, B. Podolsky and N. Rosen: *Phys. Rev.*, **47**, 777 (1935).
([3]) L. S. Bartell: *Phys. Rev. D*, **22**, 1352 (1980).

unless one imposes particular restrictions on possible superluminal interactions one faces causal paradoxes tied with possible retroaction in time (4).

The aim of the present letter is to discuss the relation of these two concepts in the particular case of two identical noninteracting quantum particles in order to interpret causally the corresponding EPR situation analysed (in the nonrelativistic limit) by BOHM and HILEY (5).

We first define what we mean with the word « causality » by three properties:

a) the system of our two particles can be solved in the forward (or backward) time direction in the sense of the Cauchy problem;

b) the paths of all material particles must be timelike;

c) the formalism must be invariant under the Poincaré group $P = T \otimes \mathscr{L}\uparrow$.

As shown by one of us (PDV) (6), one can have action at a distance between two identical particles and preserve causality in the following case: we start with the two free Hamiltonians $H_{01} = p_1^2/2 = m_1^2/2$, and $H = p_2^2/2 = m_2^2/2$, completed with additive interaction terms V_1 and V_2 which are nonlocal potentials (*). Note that

I) we shall call V_1, V_2 potentials for convenience, although they have not the dimensions of energy: they must have the dimensions of squared masses;

II) the Hamiltonians are not directly related with energy but related with half the squared masses.

We thus get

$$H_1 = H_{01} + V_1\,, \qquad H_2 = H_{02} + V_2 \tag{1}$$

now defined in the sixteen-dimensional phase space $q_1^\mu, q_2^\mu, p_1^\mu, p_2^\mu$. One sees immediately that the potentials cannot be chosen arbitrarly, since the existence of world-lines requires for identical particles, the vanishing of Poisson's brackets $\{H_1, H_2\}$ (6). The phase space has 16 dimensions and the standard brackets are assumed among q_1^μ, q_2^μ and unconstrained p_1^μ, p_2^μ.

We now perform the following separation of internal and external variables:

$$\begin{cases} P^\mu = p_1^\mu + p_2^\mu\,, & y^\mu = \frac{1}{2}(p_1^\mu - p_2^\mu)\,, \\ Q^\mu = \frac{1}{2}(q_1^\mu + q_2^\mu)\,, & z^\mu = q_1^\mu - q_2^\mu\,, \end{cases} \tag{2}$$

so that, in the case that $V_1 = V_2 = V$, and $m_1 = m_2 = m$, we have

$$\begin{cases} H_1 + H_2 = 4P^2 + y^2 + 2V\,, \\ H_1 - H_2 = yP\,. \end{cases} \tag{3}$$

---

(4) C. MØLLER: *The Theory of Relativity* (Oxford, 1962), p. 52.

(5) D. BOHM and B. HILEY: in *Quantum Mechanics a Half Century Later*, edited by J. LEITE-LOPES and M. PATY (1975).

(6) PH. DROZ-VINCENT: *Ann. Inst. Henry Poincaré*, **27**, 407 (1977); *Phys. Rev. D*, **19**, 702 (1979) and references quoted therein.

(*) Relativistic action-at-a-distance dynamics is generally nonlocal. However, it has been exhaustively shown that also conventional local field theories (electromagnetism and gravitation) can be cast into this scheme: L. BEL, A. SANS and J. M. SANCHEZ: *Phys. Rev. D*, **7**, 1099 (1973); L. BEL and J. MARTIN: *Phys. Rev.*, **8**, 4347 (1973).

The condition for the existence of causal timelike world-lines then reduces to the relation

$$\{yP, V\} = 0 . \tag{4}$$

If we define the projector $\Pi^{\mu}{}_{\nu} = \delta^{\mu}{}_{\nu} - P^{\mu}P_{\nu}/P^2$ and $\tilde{z}^{\mu} = \Pi^{\mu}{}_{\nu}z^{\nu}$, $\tilde{y}^{\mu} = \Pi^{\mu}{}_{\nu}y^{\nu}$, relation (4) implies that V depends on $\tilde{z}^2$, P^2, $\tilde{y}^2$, $\tilde{z}\tilde{y}$, yP, but does not depend on zP which is, in the rest frame of the system, the relative time co-ordinate up to a factor $|P|$. Moreover, one finds $\{P^{\mu}, H_1\} = \{P^{\mu}, H_2\} = 0$, so that the centre-of-mass momentum P^{μ} is constant and one can slice space-time with 3 planes orthogonal to P^{μ} and connect the two particles by spacelike lines in these hyperplanes.

We now come to the description of two quantum noninteracting particles. For a system of two classical relativistic particles interacting at distance the evolution, in our multitemporal formalism (6), is described by two parameters: τ_1, τ_2, *i.e.* the proper times of the two particles. The movement is generated in the phase space $T(M_4) \times T(M_4)$ in a symplectic way by the covariant Hamiltonians H_1 and H_2 analysed in the first part of this letter. Of course we can build the canonical transformation theory in this covariant framework (7). The transformation which solves the motion equation is generated by Jacobi's principal function S, but it is simpler to consider the covariant Hamiltonian-Jacobi characteristic function $W = S - (m^2/2)(\tau_1 + \tau_2)$ which is determined by the Hamiltonian-Jacobi *system*:

$$H_1\left(q_1^{\mu}, q_2^{\mu}; \frac{\partial W}{\partial q_1^{\mu}}, \frac{\partial W}{\partial q_2^{\mu}}\right) = \frac{m^2}{2}, \qquad H_2\left(q_1^{\mu}, q_2^{\mu}; \frac{\partial W}{\partial q_1^{\mu}}, \frac{\partial W}{\partial q_2^{\mu}}\right) = \frac{m^2}{2}. \tag{5}$$

One remarks here (in accordance with the well-known no-interaction theorem (8) that the canonical variables q_1^{μ}, q_2^{μ} are not coincident with the positions x_1^{μ}, x_2^{μ} except when the interaction vanishes.

By straightforward quantization of this multitemporal canonical formalism we obtain, for a system of two *free particles*, the Klein-Gordon system (for $\hbar = c = 1$)

$$-\Box_1\psi(x_1, x_2) = m^2\psi(x_1, x_2) \quad \text{and} \quad -\Box_2\psi(x_1, x_2) = m^2\psi(x_1, x_2) , \tag{6}$$

where ψ is a two-point–dependent function. Out of (6) we can extract the usual main equation

$$(\Box_1 + \Box_2)\psi(x_1, x_2) = 2m^2\psi(x_1, x_2) \tag{7}$$

completed by the so-called « subsidiary » condition:

$$(\Box_1 - \Box_2)\psi = 0 . \tag{8}$$

We now introduce in a relativistic way the concept of quantum potential (9). Following the original de Broglie's method, we set $\psi = \exp[R + iW]$, where R, W are

---

(7) D. Hirondel: Thesis, Paris (1977).
(8) D. G. Currie: *J. Math. Phys.* (*N. Y.*), **4**, 1470 (1963); *Phys. Rev.*, **142**, 817 (1966).
(9) L. de Broglie: *Une interprétation causale et non linéaire de la mécanique ondulatoire* (Paris, 1972). D. Bohm and J. P. Vigier: *Phys. Rev.*, **96**, 208 (1954); **109**, 1882 (1958).

real functions. Separating eq. (7) into the real and the imaginary part, we get for the real part

$$(9)\qquad \begin{cases} \frac{1}{2}(\partial_{1\mu} W \partial_1^\mu W) + U_1 = \frac{1}{2}m^2\,, \\ \frac{1}{2}(\partial_{2\mu} W \partial_2^\mu W) + U_2 = \frac{1}{2}m^2\,, \end{cases}$$

where we have

$$(10)\qquad \begin{cases} U_1 = -\frac{1}{2}(\Box_1 R + \partial_2^\mu R \partial_{1\mu} R)\,, \\ U_2 = -\frac{1}{2}(\Box_2 R + \partial_1^\mu R \partial_{2\mu} R)\,. \end{cases}$$

In spite of an obvious analogy (*) the system (9) cannot be immediately identified with eq. (5). To be more specific, we will consider the case of a ψ eigenstate of $P^\mu = i(\partial_1^\mu + \partial_2^\mu)$:

$$\psi = \exp\left[i\left(K_\mu \frac{x_1^\mu + x_2^\mu}{2}\right)\right]\varphi(z_\mu)\;;$$

where K_μ is a constant timelike vector: so we have

$$(11)\qquad (\partial_1^\mu + \partial_2^\mu)R = 0\,, \qquad (\partial_1^\mu + \partial_2^\mu)W = K_\mu\,.$$

Moreover, since the difference of eq. (9) gives

$$(12)\qquad K^\mu \frac{\partial}{\partial z^\mu} R = 0\,,$$

we see that R only depends on $z^\mu = x_1^\mu - x_2^\mu$ and more precisely only through its spatial part with respect to K^μ, namely $z_\perp^\mu = z^\mu - (z_\nu K^\nu)K^\mu/K^2$. In this case from (11) we have $U_1 = U_2 = U = f(z_\perp^\mu)$. But, as seen before, U has not a suitable expression because it depends only on $z_\perp$ and it cannot satisfy the condition $\{yP, V\} = 0$. In fact this process gives U as a function of z^μ and K^μ and not of z^μ, P^μ.

Making the substitution

$$(13)\qquad z_\perp^\mu \to \tilde{z}^\mu$$

in U, we get finally $V = f(\tilde{z})$ which depends on P^μ in a correct way, so that we can interpret it as a relativistic potential. Equations

$$(14)\qquad \frac{1}{2}(\partial_{1\mu} W \partial_1^\mu W) + V(\tilde{z}) = \frac{m^2}{2}\,, \qquad \frac{1}{2}(\partial_{2\mu} W \partial_2^\mu W) + V(\tilde{z}) = \frac{m^2}{2}\,,$$

(*) Of course eqs. (5) are written in terms of q_1^μ, q_2^μ, while eq. (9) involves x_1^μ, x_2^μ; but for the original *free* system the position variables are canonical, so that we can write without problems $q_1^\mu = x_1^\mu$, $q_2^\mu = x_2^\mu$ which makes the analogy between (5) and (9) manifest.

are now coincident with (5) if $q^\mu = x^\mu$, *i.e.*

$$(15)\qquad \Pi_1 = \frac{P_1^2}{2} + V\,, \qquad \Pi_2 = \frac{P_2^2}{2} + V\,.$$

We remark here that the variables $x_1^\mu = q_1^\mu$ and $x_2^\mu = q_2^\mu$ are canonical for the free quantum system as well as for the classical interacting system. Moreover, they are also position variables for the quantum free system, but they do not represent the positions in the classical interacting system except in the particular rest frame system where we recover the Hamilton-Jacobi equations for a classical system in interaction through the potential V.

At this stage of our work, as was the case for the old de Broglie's derivation, we have only exhibited a mathematical analogy between a system of two quantum free particles and a system of « fictitious », but causally interacting particles. We are going now to recall and summarize the physical interpretation of this fictitious system (in the framework of the stochastic interpretation of quantum mechanics) in two points.

A) We can give a physical basis to our quantum potential only if we consider the ψ-field of a quantum particle not as a pure mathematical tool but as a real wave field on a subquantal medium ([10]). Indeed it is well known, since Dirac's pioneer work ([11]), that Einstein's relativity theory (and Michelson's experiment) are perfectly compatible with an underlying relativistic stochastic aether model, so that quantum statistic will reflect the real random fluctuations of a particle embedded in this aether ([12]). More precisely the quantum potential, introduced at the beginning of this paper on the basis of a pure formal analogy, is now interpreted as a real interaction among the particles and the subquantal fluid polarized by the presence of the particles ([13]). The quantum potential now represents a true stochastic potential. In this sense we can also understand how, starting from classical free particles, we have obtained, through quantization, two classical interacting (at a distance) particles. In fact the quantization procedure, which brought (5) into (6), is equivalent, in our aether interpretation, to add to our original free system (described by eq. (5)) the action of the subquantal medium so that finally the « free » quantum system (6) is equivalent to a system of classical interacting (via Dirac's aether) particles described by (14).

B) One has shown that the existence of the quantum aether allows one to deduce ([12]) the relativistic quantum equations for single free particles ([12,14]) and for systems of two particles ([13]) as describing the stochastic motion of classical particles in interaction with the aether, if the random jumps are made at the velocity of light ([12]).

We conclude with the remark that the causality implied in our model is absolute in the sense that the measuring processes themselves (and the observers) satisfy the same causal laws and are real physical processes with antecedents in time. The measuring process (observer plus apparatus plus observed particles) is a set of particles which are part of an overall causal process. In this scheme the intervention of a measuring process contains no supranatural « free will » or « observer consciousness », since quantum

---

([10]) J. P. Vigier: *Lett. Nuovo Cimento*, **29**, 467 (1980).
([11]) P. A. M. Dirac: *Nature (London)*, **168**, 906 (1951).
([12]) W. Lehr and J. Park: *J. Math. Phys.* (*N. Y.*), **18**, 1235 (1977); F. Guerra and P. Ruggiero: *Lett. Nuovo Cimento*, **23**, 529 (1978); J. P. Vigier: *Lett. Nuovo Cimento*, **24**, 265 (1979).
([13]) N. Cufaro Petroni and J. P. Vigier: *Lett. Nuovo Cimento*, **26**, 149 (1979).
([14]) N. Cufaro Petroni and J. P. Vigier: *Phys. Lett. A*, **73**, 289 (1979); **81**, 12 (1981).

measuring devices act as spectral analysers which split into subpackets the real de Broglie's waves associated with particles (which behave as planes flying at Mach 1 within their own sound waves): the particle entering into one of them according to its random causal motion (15). In that scheme there is no « free will » signal production and thus no possible causal paradoxes (12): nothing exists beyond the motion and interactions of material particles in a random stochastic aether.

* * *

The authors would like to thank Prof. SELLERI for helpful discussions on the implications of nonlocality in quantum mechanics.

---

(15) M. CINI, M. DE MARIA, G. MATTIOLI and F. NICOLO: *Found. Phys.*, **9**, 479 (1979).

Reprinted from *Foundations of Physics*, Vol. 12, No. 12, pp. 1057-1083, Copyright (1982) with permission from Kluwer Academic/Plenum Publishers.

De Broglie's Wave Particle Duality in the Stochastic Interpretation of Quantum Mechanics: A Testable Physical Assumption

Ph. Gueret[1] **and J.-P. Vigier**[2]

Received April 21, 1982

If one starts from de Broglie's basic relativistic assumptions, i.e., that all particles have an intrinsic real internal vibration in their rest frame, i.e., $h\nu_0 = m_0 c^2$; that when they are at any one point in space-time the phase of this vibration cannot depend on the choice of the reference frame, then, one can show (following Mackinnon[1]*) that there exists a nondispersive wave packet of de Broglie's waves which can be assimilated to the nonlinear soliton wave U_0 introduced by him in his double solution model of wave mechanics.*[2] *Since de Broglie's linear pilot waves can be considered to be real waves propagating as collective motions on a covariant subquantum chaotic "aether,"*[3] *these new soliton waves can be considered as describing the particle's immediate neighborhood, i.e., the aether's reaction to the particle's motion in the stochastic interpretation of quantum mechanics. The existence of such a physical aether (which provides a perfectly causal interpretation of the action-a-distance implied by the Einstein–Podolsky–Rosen experiments) can now be proved by establishing the reality of de Broglie's waves in realizable experiments.*

1. INTRODUCTION

As one knows, recent discussions on Aspect's[4] and Rapisarda's[5] experiments to test nonlocal superluminal quantum correlations in Bohm's version[6] of the Einstein–Podolsky–Rosen experiments[7] have led to the development of two conflicting interpretations[8,9] of the quantum

[1] Institut de Math. Pures et Appliquées de l'Université P. et M. Curie, 4, Place Jussieu, 75230 Paris Cedex 05.

[2] Équipe de Recherche Associée au C.N.R.S. no. 533, Institut Henri Poincaré, 11, rue P. et M. Curie, 75231 Paris Cedex 05.

nonlocality suggested by preliminary results.[4] Both interpretations appear as natural extensions of the antagonistic views of Bohr and Einstein in their 1927–1935 controversy, i.e., lead to renewed confrontation between the a-causal Copenhagen Interpretation of Quantum Mechanics (CIQM) and the causal Stochastic Interpretation of Quantum Mechanics (SIQM). Indeed, in CIQM one has proposed real advanced potentials[10] and in SIQM phase like collective causal actions at a distance propagating superluminaly within Dirac's stochastic subquantal aether model.[11]

Despite evident similarities (both interpretations utilize the same wave equations and yield, except for a few crucial cases,[12,13,14] the same statistical predictions), CIQM and SIQM differ on three essential points.

1. In CIQM the quantum states (waves) are associated with individual systems and represent an ultimate statistical knowledge. Microphenomena are particles *or* waves, never the two simultaneously.

In SIQM the quantum states correspond to real physical fields (waves) associated with both individual particles and ensembles of identically prepared systems. Microobjects are thus particles *and* waves simultaneously. Indeed, it has been shown:

(a) that these waves represent (for spins $J=0$,[15] $J=1/2$,[16] and $J=1$[17]) collective motions on the top of a chaotic medium which induces random stochastic jumps at the velocity of light.

(b) that the associated statistics is correctly represented by the $\Psi\Psi^*$ distribution of these pilot Ψ waves.[9]

2. In CIQM a measurement on a system implies an unanalyzable discontinuous instantaneous (for all observers) collapse of state (wave packet reduction) so that no microphenomena is a phenomena until it is an observed phenomena.[8]

In SIQM these states (particles plus associated real waves) always evolve causally in time. There is thus no wave packet reduction (i.e., no wave collapse) but they are modified (split) by interactions with any real physical macroscopic apparatus: the particle entering into one of the apparatus's measurable eigenwave states.

3. In CIQM the uncertainty principle restricts the simultaneous measurability of noncommuting observables on individual systems.

In SIQM the Heisenberg uncertainty principle does not restrict simultaneous measurability of noncommuting observables on individual objects (since the corresponding extended localized particles follows world lines in space-time) but represent dispersion relations on their measured values resulting from their simultaneous dual (i.e., wave plus particle) character and the associated subquantal stochastic motion.

The aim of the present paper is to present a new concrete model of point 1 of SIQM in the scalar particle case, i.e., a new explicit, detailed mathematical relativistic description of de Broglie's wave particle duality in the $J = 0$ case.

To clarify its physical meaning, we must briefly return to the origin of wave mechanics. As one knows, the assumption 1 of SIQM (i.e., the real existence of de Broglie's waves) was first submitted in its simplified pilot wave version by de Broglie himself at the 1927 Solvay Congress.

Following its enlargment with subquantal stochastic motions by Bohm and Vigier,[9] it was later given the form of the "double solution theory" by de Broglie et al.,[2] with de Broglie's waves propagating on a real subquantal vacuum model.[9] Its essential characteristics are the following: in SIQM a particle is represented:

(a) by a real physical wave (pilot wave henceforth denoted P wave) $\Psi = R\exp(iS/\hbar)$ which propagates in space and satisfies the linear wave equations of quantum mechanics, i.e., in the $J = 0$ case

$$(\Box - m_0^2c^2/\hbar^2)\Psi = 0 \tag{1}$$

As one knows, these P waves:

(1) are built with an ensemble of de Broglie's plane phase waves (henceforth denoted B waves): $\Psi(x, t) = a\exp[2\pi i\nu(t - x/V)]$ where ν is the observed frequency, $V = c^2/v$ the phase velocity (v denoting the particle's velocity) which also satisfies relation (1).

(2) define "drift" world lines of flow tangent to the unitary four vector $u_\mu = \partial_\mu S/M_0^2$ (so that $u_\mu u^\mu = -c^2$) where $M_0^2 = m_0^2 - (\hbar^2/c^2)(\Box R/R)$ represents de Broglie's and Bohm's quantum potential which appears in the relativistic Jacobi equation $\partial_\mu S\partial^\mu S + M_0^2 = 0$ which represents the real part of (1).

(3). carry a conserved density $\rho = \sqrt{-g}\,(M_0/m)R^2$ along the preceding drift lines of flow, so that $\dot{\rho} = d\rho/d\tau = 0$.

(4). necessarily disperse according to the well-known relation $\omega = \pm(c^2|k_B|^2 = m_0^2c^4\hbar^{-2})^{1/2}$ with $|k_B| = 2\pi\lambda_B^{-1}$, k_B representing the usual wave vector.[5]

(5) can be physically considered (as shown by Lehr and Park,[9] Guerra–Ruggiero[18] and Vigier[19]) as a Brown–Markov stochastic wave propagating on a random covariant thermostat. The corresponding diffusion coefficient $D = \hbar/2m$ results[19] from the necessity of preserving the phase locking of all its oscillating components in the subquantal random jumps at the velocity of light, and the P wave satisfies[19] the relativistic form given by Eckart[20] to the basic principles of thermodynamics.

(b) by a singular, extended, nonlinear and nondispersive wave packet $U_0 = H \exp(iS/\hbar)$ superposed on the P wave which characterizes the particle aspect of matter. As one knows, for de Broglie, this moving localized region contains an energy and four-momentum iE/c and $p_B = \hbar k_B$ equal to the preceding P waves drift motion's four-momentum. Moreover it has been shown[21,22] that this soliton like wave packet U_0 (which we henceforth call S wave) follow the drift lines of the P waves if their phases coincide, i.e., if the S wave beats in phase with the P wave. The S wave is an evident possible representation of Einstein's "Nadelstrahlung" concept.[33]

(c) by an associated probability wave, i.e., the Ψ wave of quantum mechanics. As first suggested by Vigier in de Broglie's model,[24] this associated stochastic probability wave (which describes the particle's random distribution on the P wave) can be written $\Psi = Cv$, C being a constant normalizing factor. Indeed, as shown by Bohm and Vigier,[9] the introduction in this model of random subquantal Markov fluctuations, which move in a random way the S wave form one P wave's line of flow to another, leads (for U_0) to the quantum mechanical probability $\Psi\Psi^*$.

Despite its appealing character, the essential weakness of de Broglie's preceding model of wave particle duality resides in the failure of de Broglie[2] and his followers to present an explicit form of the nonlinear equation which would yield an S wave soliton moving along the P wave drift lines of flow. As one knows,[25] there exist many such possible nonlinear terms (including the sine Gordon equation) which open the way to a generalization of (1) that contains solitons. How is one to choose between them in the absence of a new physical idea? As stated in our abstract, the essential point of this paper is to demonstrate that this choice is indeed possible if one develops an idea introduced by Mackinnon in the discussion of this problem.[1]

Indeed in Sections 2 and 3 we shall show that if one starts from de Broglie's basic relativistic assumptions, i.e.,

- *that all particles have an intrinsic real internal vibration in their rest frame;*

- *that when they are at any point in space-time, the phase of this vibration cannot depend on the choice of the reference frame, then there exists a unique nondispersive packet of de Broglie's B waves which can be assimilated to the soliton S wave, and that this choice determines the form of the corresponding nonlinear equation.*

In Section 2, we shall show how de Broglie's phase equality principle implies that this soliton S wave moves along the P wave's drift lines, so that they can be considered as describing physically the particle's immediate neighborhood, i.e., the aether's reaction to the particle's motion in the SIQM.

In conclusion we shall re-analyse a proposal made with Garuccio and Rapisarda[14] to show that, far from being an abstract nontestable proposal, this model of de Broglie's wave particle duality leads to concrete explicit experiments which can help us make a crucial choice between CIQM and SIQM.

2. DE BROGLIE'S INTERNAL FREQUENCY AND SOLITON WAVE

To construct a nondispersive wave packet of de Broglie's B waves, we shall briefly rediscuss de Broglie's thesis[26] and its development by Mackinnon.[1] As all physicist now knows, de Broglie's essential starting point cannot be reduced to the famous assumption of the dual aspect of matter (i.e., the wave particle duality) but rests essentially

1. on the idea that, since any free particle at rest with respect to an observer has mass m_0 and energy $E = m_0 c^2$ and its energy is equal to one quantum $h\nu_0$, it can be compared to a real physical oscillator with an internal rest frequency

$$\nu_0 \times m_0 c^2/h \tag{2}$$

2. on the assumption that all particles are accompanied in real space-time be real plane monochromatic phase waves (the preceding B waves), which can be written

$$\Psi_B = a \exp 2\pi i(\nu t - x/\lambda + \theta) \tag{3}$$

with well determined amplitudes a and phase constant θ. As first remarked by de Broglie,[33] these B waves cannot serve to locate the particle. As long as the particle remain at rest, the particle's internal frequency and the frequency of the de Broglie waves remain the same.

Three coments can be made at this stage:

1. Point 1 raises immediately the question of the nature of the oscillations described by (3).

Clearly de Broglie himself has always believed that (2) described real physical internal oscillations, and thus departed from the pointparticle model supported by a majority of relativists and quantum physicists.

This idea has engineered a long set of researches starting for example with Yukawa's bilocal particle model[27] and Bohm and Vigier's liquid droplet mode.[9] The essential point is that, independently of the internal motions which yield a classical model of spin,[28] it has generally been demonstrated by Souriau et al.[29] that any extended particle model yields an internal rotation of the particle's center of matter density around its center of mass with the exact frequency of de Broglie's relation. Of course, such extended particle models have received (until now) no direct experimental support. They open nevertheless interesting paths of research since.

- they offer the possibility to interpret the particle's newly discovered quantum numbers (T, Y, C, B, L, etc...) in terms of internal oscillations.[30]
- they can contain in their interior (as suggested by Dirac[31] and Vigier[15]) nonlocal hidden variables which have been utilized to support the superluminal propagation of the quantum potential[15] and lead to a perfectly causal action-at-a-distance interpretation of nonlocal correlations suggested by preliminary experiments[4] on the Einstein–Podolsky–Rosen paradox.[7]

2. As stressed by de Broglie himself, the Ψ_B (B waves) can be interpreted as real superluminal phase waves propagating in space-time on the top of a continuous field of physical oscillators. Since, following Dirac's pioneer work,[31] one has now discovered random covariant possible distributions of extended particles, one can consider these B waves as propagating (according to relation (1)) on such a material subquantal oscillating level of matter. This is all the more tempting since the possibility of superluminal phase wave propagations (with large amplitudes) has been recently considered on real physical plasmas.[32]

Both these views justify immediately the next step of de Broglie's discoveries.

First, if one assumes that particles have nonzero mass (photon included) the vacuum now acts as a dispersive medium which propagates different frequencies with different velocities. Indeed, the relation $h\nu = m_0 c^2/\sqrt{1-\beta^2}$ ($\beta = v/c$) implies a dispersion of space with $n = (1 - \nu^2/\nu^2)^{1/2}$.

Second, this dispersive character justifies de Broglie's identification of particle velocity with wave group velocity. Indeed, if we transform the Ψ_B wave in the particle's rest frame Σ^0, i.e., $\Psi_B = a\exp(2\pi i\nu_0 t_0)$ into a frame Σ moving with a velocity with regard to this particle, one knows that the wave function becomes

$$\Psi = a\exp\left\{2\pi i\nu_0\left(\frac{t-\beta x/c}{\sqrt{1-\beta^2}}\right)\right\} \tag{4}$$

t is the time in Σ. Relation (1) then now represents a Ψ_B wave of frequency $\nu = \nu_0\sqrt{1-\beta^2}$, of wavelength $\lambda = h/p$ (p representing the momentum in Σ), and of the phase velocity $V = \nu\lambda = c^2/v$. From the point of view of Σ, $\nu_0 \to \nu_0\sqrt{1-\beta^2}$, so that the oscillation appears to slow down w.r.t. Σ^0. A combination of such waves yields a dispersive wave packet with a group velocity $v = dv/d(1/\lambda)$, a result which led de Broglie to assume that such packets act like a "pilot wave" of the particle's behavior. Indeed, as stressed by Mackinnon, de Broglie's guiding principle just represents the fact that the internal vibration and the de Broglie P wave remain in phase where the particle is located.

3. The third remark is that, if one considers particles as real oscillators and P wave as real wave, *then one should consider the phases* (and the phase constant θ_k of relation (3) *as real physical quantities.* In other words in particles *and* waves are real vibrations when the particle is at any point in space-time the phase of this vibration cannot depend on the choice of the reference frame. To quote Mackinnon,[1] "If the particle has a vibration which has a phase then the phase at any point in space-time cannot by any reasonable process of logic depend for its value on the choice of reference frame used to assign coordinates to that point." De Broglie himself, who did not mention this in his early work, later furnished[33] strong indirect evidence in favor of phase reality. For him (as for Einstein) a laser is just a coherent wave packet carrying many photons which have a well determined phase constant. Moreover, since a boson wave can carry an arbitrary number n of photons in space-time, i.e., the so-called occupation number, it can be written in the form (3), and in second quantization theory n and θ are canonically conjugated quantities, so that one can write $(\theta)_{0p} = \partial/\partial n$ and $(n)_{0p} = -i\partial/\partial\theta$. Writing then $(n_k)_{0p}\,\phi(\theta_k) = -i\partial\Phi(\theta_k)/\partial\theta_k)/\partial\theta_k = n_k\phi(\theta_k)$, one finds (up to a normalization constant) $\phi(\theta_k) = \exp(in_k\theta_k)$, and since θ_k is a variable with a 2π period, ϕ_k can only be uniform if n_k is an integer number, so that for any such number the phase constants is a physically measurable quantity in Dirac's sense, i.e., corresponds to a physical reality. The fact that $\delta n\,\delta\theta \geqslant 1$ shows that, for $n \to \infty$, θ can be measured with an arbitrary precision. The existence of θ is also confirmed by interference measurements on single particles,[51] despite the fact that its value is then undetermined since the observed pattern only depends on the difference of march of correlated paths.

3. CONSTRUCTION OF THE SOLITON WAVE

We now come to the essential point of this paper i.e., the idea that instead of deducing the soliton's form from an arbitrarily given nonlinear equation, we shall show, following Mackinnon, that the very existence of a physical phase implies

- an explicit form for the S wave
- an explicit form of the nonlinear Klein–Gordon equation.

To show this, let assume that in a Minkowski diagram space-time is described by the axes $(O; x, ct)$ of a stationary observer who watches a particle of rest mass m_0, moving with a constant velocity u in the $(-x)$-direction. Its associated frame $(O; x_0, ct_0)$ is symetrical w.r.t. the second bisectrix of the axes $(O; x, ct)$ and—Ox_0 makes with—Ox an angle α with $tg\,\alpha = u/c$. (See Fig. 1.)

If the particle contains an intrinsic oscillation of frequency ν_0, an observer tied to it will see it in the same state after equal time $(1/c)OA$, $(1/c)AB$, etc... equal to the rest period $T_0 = 1/\nu_0 = h/m_0c^2$.

The parallel lines to Ox_0 passing through A, B, etc... are "equal phase" lines for the observer moving with the particle and the points a', O, a,... represent their intersections with the space axis of the observer at rest at time O so that $Oa = \lambda$ represents de Broglie's wavelength seen by the observer at rest.

Let us now consider a second observer moving at uniform velocity v in the x direction. Its associated frame $(O; x', ct')$ is symetrical w.r.t. the first bisectrix of the axes $((O; x, ct)$ and Ox' makes an angle δ with Ox such that $tg\,\delta = v/c$. (See Fig. 2.) From the point of view of this second observer the de Broglie's wave length is changed and appears to the rest observer with the value $\lambda_1 = Od$.

The velocity of the moving observers w.r.t. $(O; x, ct)$ can vary from $-c$ to $+c$ and this determines the boundaries between with all wavelengths can vary in their turn. The observer's frame which moves with velocity $+c$ is mapped in the rest frame with the first bisectrix of $(O; x, ct)$ and the

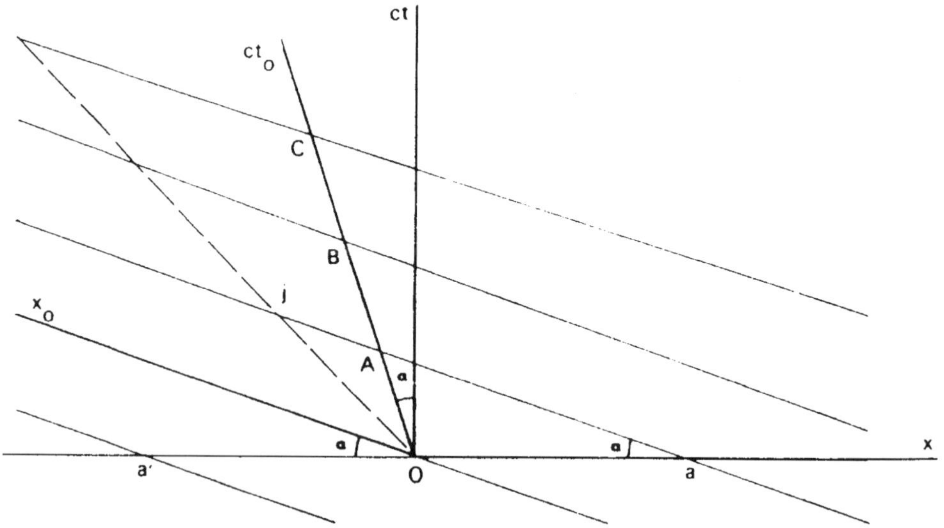

Fig. 1. Minkowski diagram representing a particle moving in the $-x$ direction with velocity u. Its associated wavelength is given by Oa.

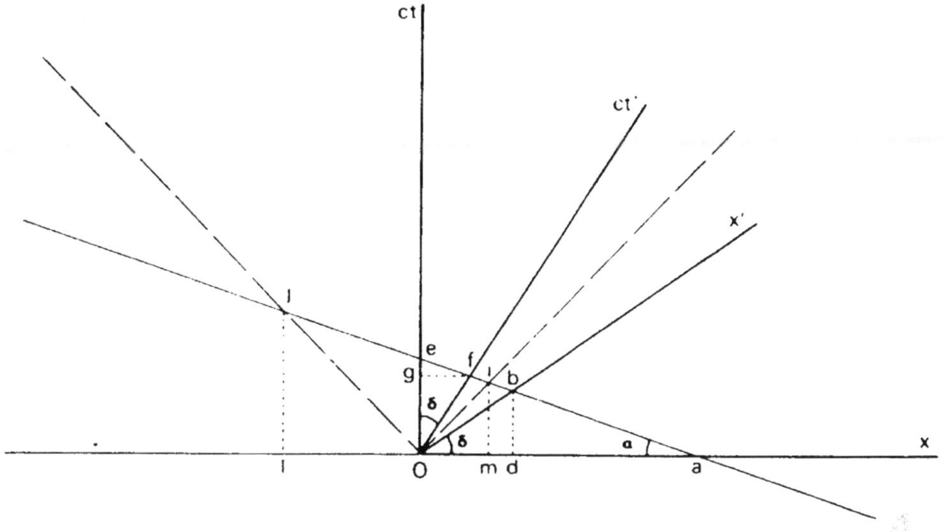

Fig. 2. Minkowski diagram representing the same moving particle and the motion of an observer in the $+x$ direction with velocity v. Oa still denotes the particle wavelength in the rest frame, Ob denotes its wavelength for the moving observer, and Od the same wavelength seen from the rest frame.

observer's frame which moves at velocity $-c$ is mapped in the rest frame with the second bisectrix of $(O; x, ct)$. In this way the de Broglie wavelengths measured, respectively, by Oi and Oj in the frames which have the limiting velocities $+c$ and $-c$ appear, respectively, as Om and Ol for an observer at rest.

From the point of view of an observer at rest, there now appears a wave packet associated with the particle. It is built from a superposition of B waves moving in the $(-x)$ direction with wavelengths greater than Ol and from B waves moving in the x direction with wavelengths greater than Om.

Let us now calculate the limiting value Ol and Om. In fig. 3, one can write

$$Oj^2 = jl^2 + Ol^2 = 2Ol^2, \qquad Oi^2 = jm^2 + Om^2 = 2Om^2 \tag{5}$$

The triangles jAO and iAO have two equal sides so $jA = AO = Ai$. Moreover we have $Oj^2 = 2AO^2[1 - \cos(90 + 2\alpha)]$ and $Oi^2 = 2AO^2[1 - \cos(90 - 2\alpha)]$. Replacing the Oj and Oi by their values deduced from (5) we obtain $Ol^2 = AO^2(1 + \sin 2\alpha)$ and $Om^2 = AO^2(1 - \sin 2\alpha)$. Since $\beta = u/c = tg\,\alpha$ we get $\sin 2\alpha = 2\beta/(1 + \beta^2)$ so that $O1^2 = AO^2(1 + \beta)^2/(1 + \beta^2)$ and $Om^2 = AO^2(1 - \beta)^2/(1 + \beta^2)$. Noting that $AO/c = h/m_0 c$ we then get finally

$$Ol = (h/m_0 c)[(1 + \beta)/(1 - \beta)]^{1/2}, \qquad Om = (h/m_0 c)[(1 - \beta)/(1 + \beta)]^{1/2} \tag{6}$$

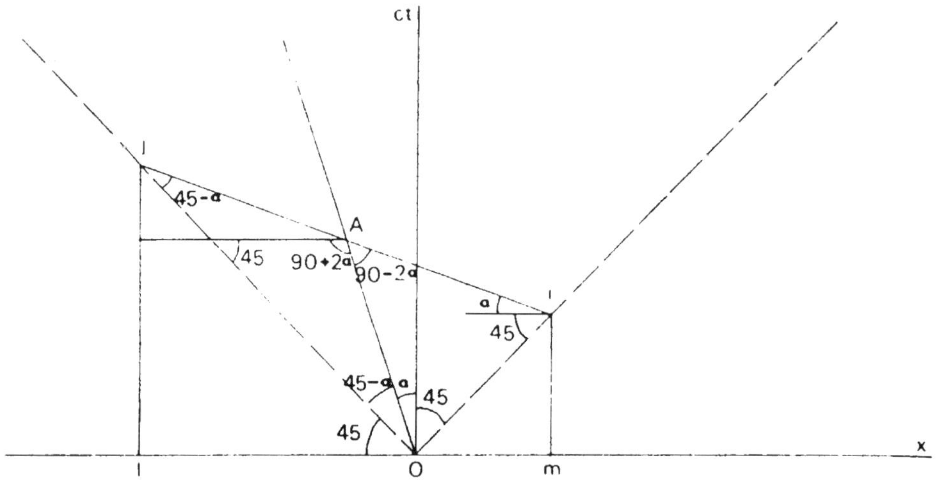

Fig. 3. Minkowski diagram representing the triangles which yield the limit wavelengths Ol and Om.

We obtain in this way a wave packet (S wave) for which the wave vector recovers the domain $\Delta k = 2\pi(Ol^{-1} + Om^{-1})$ i.e., $\Delta k = m_0 c/\hbar\sqrt{1-\beta^2}$. The central wave vector k_0 from Δk can then be written $k_0 = |2\pi(Om^{-1} - Ol^{-1})|/2$ i.e., $k_0 = m_0 u/\hbar\sqrt{1-\beta^2}$. If $m = m_0/\sqrt{1-\beta^2}$ is the apparent mass for the observer at rest, Δk and k_0 can be finally written $\Delta k = mc/\hbar$ and $k_0 = mu/\hbar$.

Let us now show that the S wave does not spread with time. If we come back to fig. 2, we have seen that

$Oa = \lambda =$ de Broglie wavelength for the observer at rest.

$Ob = \lambda' =$ de Broglie wavelength for a moving observer for which its apparent value is $Od = \lambda_1$ for the observer at rest.

We obtain in the same way:

$Oe = cT =$ de Broglie wave period for the observer at rest.

$Of = cT' =$ de Broglie wave period for a moving observer for which the apparent value is $Og = cT_1$ for the observer at rest.

We thus obtain through a simple geometric reasoning the relations

$$fg/Og = bd/Od = tg\,\delta = v/c \text{ i.e., } fg/cT_1 = bd/(\lambda - \lambda_1) = u/c$$

$$gl/fg = bd/ad = tg\,\alpha = u/c = \beta \text{ i.e., } c(T - T_1)/fg = bd/(\lambda - \lambda_1) = u/c$$

$$Oe/Od = tg\,\alpha = u/c = \beta \text{ i.e., } cT/\lambda = u/c.$$

By eliminating of fg and bd in this system, and putting $\omega(k) = 2\pi/T$ with $\omega(k_0) = 2\pi/T_0$, $k = 2\pi/\lambda$, $k_0 = 2\pi/\lambda_0$, we get $\omega(k) - \omega(k_0) = u(k - k_0)$. Since

u, k_0, and $\omega(k_0)$ are constants, it follows that $(d/dk)[\omega(k)] = u$ for all ω and $(d^2/dk^2)[\omega(k)] = 0$. This is the necessary condition to ensure that the S wave preserves its shape for all time.

We thus see that the S wave preserves its shape under Lorentz transformations and appear as a relativistic effect.

If we then utilize an inverse Fourier transform, we can express the S wave's shape (up to a multiplicative constant) in space-time in the form

$$F(x, t) = \frac{\sin \Delta k(x - ut)}{\Delta k(x - ut)} \exp\{i[\omega(k_0)t - k_0 x]\} \tag{7}$$

One notices that this form of the S wave is obtained without approximation, that it implies its nondispersive character in time and that it is directly deduced from the basic assumptions of wave mechanics. The spatial distance between the two first zeros of the S wave is $h/mc = (h/m_0 c)\sqrt{1-\beta^2}$ so that it is submitted to the usual relativistic contraction of length with velocity.

In the system tied to the particle, $u = 0$, and the S wave's amplitude is proportional to $(\sin kx)/kx$ with $k = m_0 c/h$ so that the spatial distance between the two first zeros of the S wave is just equal to Compton's wavelength $h/m_0 c$, i.e., $2.43 \times 10^{-13} m$ for an electron and $1.32 \times 10^{-15} m$ for a proton.

One notes that the representations of $F(x, t)_{t=0}$ given in fig. 4 can be multiplied by an arbitrary constant factor. This has a physical meaning since it has (as we shall see in the fourth section) consequences on the ratios F/R, $\nabla F/\nabla R$, etc... whivh are multiplied by this factor. We have found no physical criteria to determine its numerical value and must leave the question open.

Let us now deduce the form of the nonlinear equation which acepts F as a soliton solution. Writing $F(x, t) = G(x, t)\, H(x, t)$ with $G(x, t)$ and $H(x, t)$, respectively, for $\exp\{i[\omega(k_0)t - k_0 x]\}$ and $\sin \Delta k(x - ut)/\Delta k(x - ut)$ calculating $\Box F = F_{xx} - F_{tt}/c^2$ one checks that the imaginary part in the multiplier of G cancels, and that we obtain[1] the modified Klein–Gordon equation

$$\Box F - (m_0^2 c^2/\hbar^2)F = (1 - \beta^2)\, G H_{xx} \tag{8}$$

Moreover one has $\theta(x, t) = \Delta k(x - ut)$ from which $H(x, t) = H[\theta(x, t)]$ and one finds $H_{xx} = \Delta k^2 H_{\theta\theta} = (m^2 c^2/\hbar^2) H_{\theta\theta}$ so that H_{xx} depends on θ but not in the precise form in which θ depends on x and t. Relation (8) can be written more generally in the form

$$\Box F - (m_0^2 c^2/\hbar^2)F = (m_0^2 c^2/\hbar^2)\, G H_{\theta\theta} \tag{9}$$

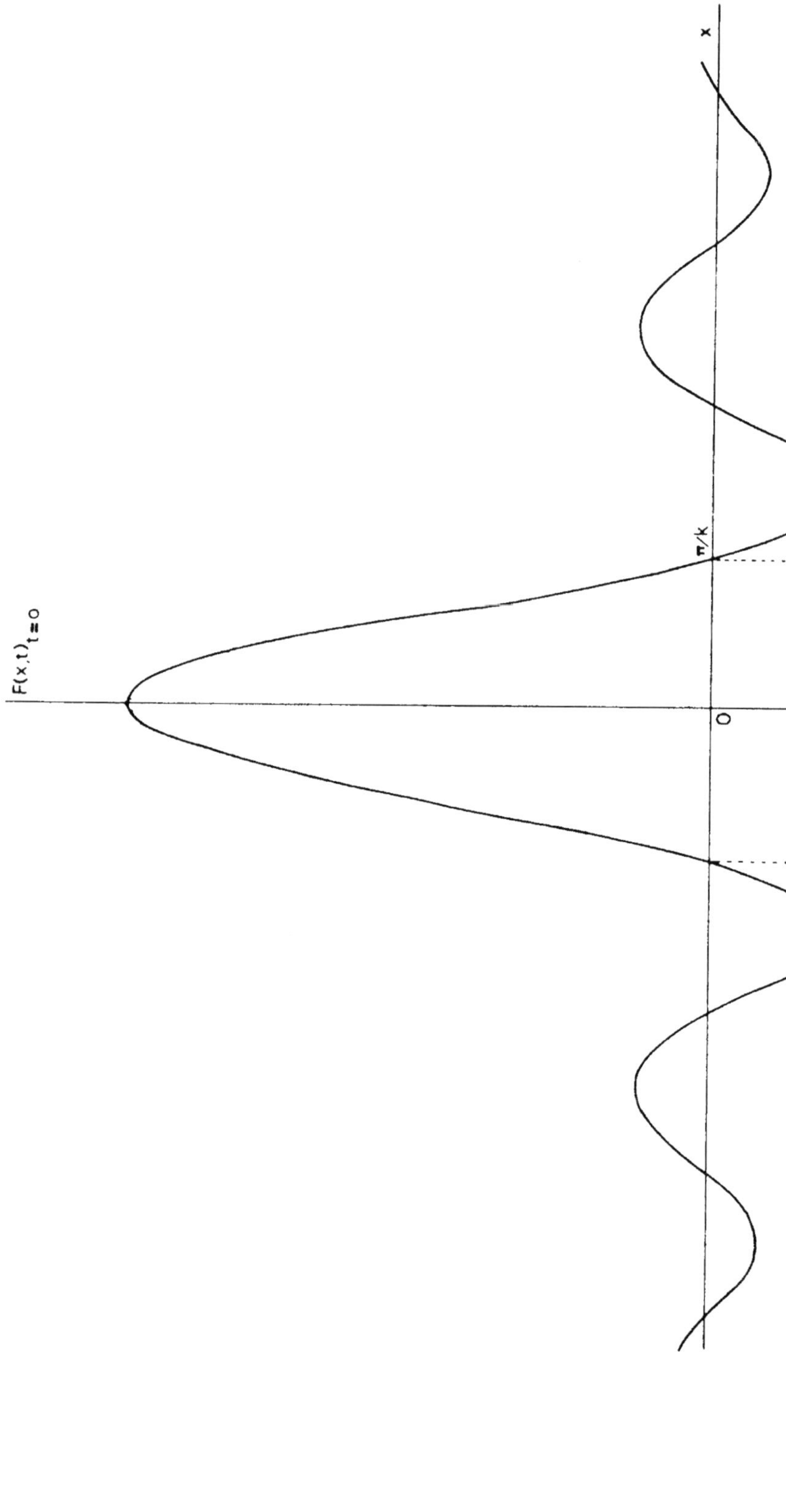

Fig. 4. Representation of the form of the soliton S wave amplitude with an arbitrary scale.

or still more generally as

$$\partial_\mu \partial_\mu U - (m_0^2 c^2/\hbar^2) U = (m_0^2 x^2/\hbar^2)\, G H_{\theta\theta} \tag{10}$$

with $U = U_0 + \Phi$ in which

- U_0 is the S wave defined by $U_0(x, t) = aF(x, t) = aH(x, t)\exp(iS/\hbar)$, a is a constant factor and $S = m(c^2 t^2 - ux) = Et - \mathbf{px}$.
- Φ is a P wave of some phase as U_0: $\Phi(x, t) = R(x, t)\exp(iS/\hbar)$ which satisfies the Klein–Gordon condition $\partial_\mu \partial_\mu \Phi = (m_0^2 c^2/\hbar^2)\Phi$.

The covariant character of the first member of (10) is evident. The covariance of the second nonlinear member results from the fact that $H_{\theta\theta}$ corresponds to $\Box_S = \partial^2/\partial y_\mu\, \partial y^\mu$ the dy^μ representing the space like distance orthogonal to the drift lines $\nabla_\mu S$ on the surfaces $S =$ constant.

One recorvers in this way de Broglie's double solution model.[2] As it has been shown by Mackinnon,[34] in the four-dimensionnal space time, the S wave traveling in the x direction takes the form

$$\Psi = (\sin kr/kr)\exp[i(\omega t - k_0 x)] \tag{11}$$

with $k = m_0 c/\hbar$ and

$$r = \left[\frac{(x - ut)^2}{1 - \beta^2} + y^2 + z^2\right]^{1/2} \tag{12}$$

and Ψ is a solution of the equation $\Box\Psi = 0$.

The solution (11) is well-known to represent the superposition of two spherically symetrical waves, one converging and one diverging, both having phase velocity c.[35] If the waves are electromagnetic waves, Ψ behaves as a phase locked cavity similar to those analyzed by Jennison[36] and which have inertial properties of classical particles.

In the nonrelativistic approximation $\sin \Delta k(x - ut)/\Delta k(x - ut)$ so that the S wave does not appear at that stage. As for the P wave equation, it becomes, as shown by Kemble,[37] a stationary Schrödinger equation.

In equation (8) one can express G and H_{xx} as a function of F. Indeed since $H = (\operatorname{sgn} H)\,|F|$ and $G = (HG)/H = F/(\operatorname{sgn} H)\,|F|$ with sgn H for sign of H, one has $GH_{xx} = (F/|F|)\,|F|_{xx}$.

As a consequence, if one accepts the physical reality of the B waves the wave equation of de Broglie's double solution theory is

$$u_{xx} - u_{tt}/c^2 - (m_0^2 c^2/\hbar^2)(1 + |u|_{xx}/|u|)u = 0 \tag{13}$$

which is a nonlinear Klein–Gordon equation in which U_0 describes a soliton, or

$$[\Box - (m_0^2c^2/h^2)]\Psi = (m_0^2c^2/h^2)\sqrt{\Psi/\Psi^*}\,\Box_S(\sqrt{\Psi\Psi^*}) \qquad (14)$$

where $\Box_S$ is a Dalembertian on the surface $S = \text{constant}$.

4. DEMONSTRATION OF DE BROGLIE'S "GUIDING PRINCIPLE" IN THE SIQM

Following our plan, we shall now study the associated motions of the preceding S waves and P waves, i.e., analyse the nature and interpretation of their relative physical behavior in the Stochastic interpretation of quantum mechanics.

As one knows,[9] the deduction of relation (1) results in this interpretation from a derivation of Nelson's equation in the Guerra–Ruggiero notations

$$m(D_c V - D_s U) = F^+ \qquad (15)$$

i.e., from a relativistic stochastic model which assimilates the P waves to fluids which satisfy three new physical structures:

(i) The fluid elements (and the particles) which follow the lines of flow of the fluid with irregular fluctuations are built from extended elements in the sense discussed by Bohm[9] and Souriau.[29]

(ii) The stochastic fluctuations occur at the velocity of light.

(iii) The fluid is a mixture of extended particles (and antiparticles): the latter being mathematically equivalent to negative energy particles moving backward in time.

The existence of such fluctuations (which induce in the particle a Markov type of Brownian motion) has been shown[9] to lead any initial distribution of the particles "piloted" by the fluid into a limiting equilibrium distribution constant $\Psi^*\Psi$ proportional to the fluid's average conserved drift density $\rho(x_\mu, \tau)$. This means that the fluctuations of our Madelung fluid induce on our particles stochastic jumps at the velocity of light (from one line of flow to another) and that such jumps can be decomposed into the regular drift motion v_d plus an apparent space like random part u_s with $v_d^\mu = dx^\mu/d\tau$, τ representing the proper time along the drift lines so that $v_d v_d = -c^2$.

Assuming then that the drift motions are irrotational (so that the drift velocity is proportional to ∇S) one can show that the demonstration of (15) and (1) rests on the definition of average velocities, and accelerations defined in a four-dimensional volume $d\omega$ limited on the side by the fluid's regular lines of flow and, at both extremities, by two space like constant phase surfaces[9] S_1 and S_3. If the domain is small enough such surfaces are separated by an interval $2\Delta\tau$: an interval $\pm\Delta\tau$ separating S_1 and S_3 from a median section S_2. Of course $|\Delta\tau| \geqslant \overline{\Delta\tau}$ (fig. 5).

As a consequence of the assumed stochastic equilibrium we can treat on the same footing the fluid behavior and an ensemble of similary prepared particles characterized by the density $\rho(x_\mu, \tau)$ in configuration space where x represents a point in four-dimensional space-time. Indeed, if one assumes (following Einstein's famous treatment of Brownian motion) that the average velocities and mean acceleration (which correspond mathematically to second order proper time derivatives) should be defined physically only by the motions of fluid elements surrounding i.e., enclosed within the four-dimensional volume element limited by S_1 and S_2 to define mean quantities. Then one obtains in Guerra–Ruggiero's notation[9]

$$(D_c D_c - \delta D_s \, \delta D_s) x_\mu = 0 \tag{16}$$

where $x_\mu(\tau)$ corresponds to the particle's position in Minkowski's space-time: $D_c = \partial/\partial\tau + b^\mu \partial_\mu$ and $\delta D_s = \delta b^\mu \partial_\mu - (h/2m_0)\Box$ representing, respectively, the drift derivative (w.r.t. the variable τ) and the stochastic derivative

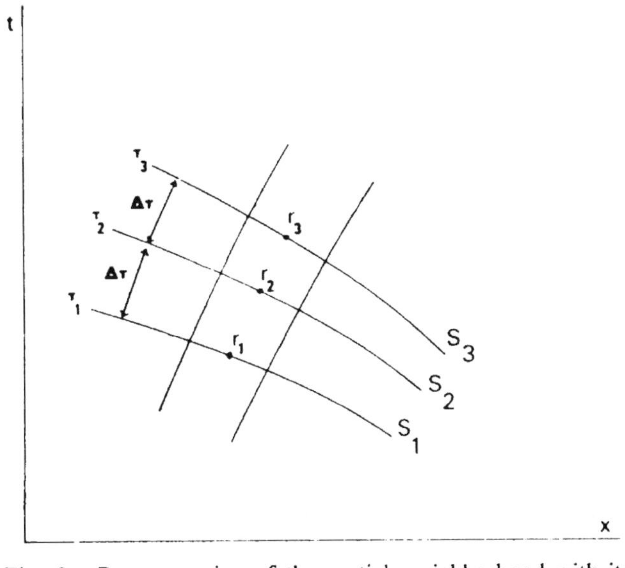

Fig. 5. Representation of the particle neighborhood with its drift lines.

$b^\mu = D_c x_\mu$ is the drift four velocity ($b^\mu b_\mu = -c^2$) and $\delta b^\mu = \delta D_s x^\mu$ the stochastic velocity. If one then writes: $\Psi[x_\mu(\tau)] = \exp\{[P[x_\mu(\tau)] + iS[x_\mu(\tau)]]/\hbar\}$ one sees that (1) can be split into two independent relations i.e.,

$$\partial_\mu [\exp(2P/\hbar)\sqrt{-g}\, g^{\mu\nu}\partial_\nu S] = 0 \tag{17}$$

$$g^{\mu\nu}\partial_\mu S \partial_\mu S + m_0^2 c^2 - \hbar \Box P - \partial_\mu P \partial_\mu P = 0 \tag{18}$$

which correspond, respectively, to a continuous equation (on the scalar density $\rho = \exp(2R/\hbar)$ and a generalized Jacobi equation in which the "variable mass" is $M_0 = [m_0^2 - (\hbar/c^2)\Box P - (\partial_\mu P \partial_\mu P)/c^2]^{1/2}$.

The drift four velocity and the stochastic velocity being written

$$M_0 b_\mu = \partial_\mu S \qquad \text{and} \qquad \delta b_\mu = (\hbar/2m_0)\,\partial_0 \log \rho = (1/m_0)\,\partial_\mu P$$

one sees that our oscillator-particles cannot be considered as free particles. We obtain for the unitary current: $b_\mu = (\partial_\mu S(/M_0$ and for the scalar density: $\rho = \sqrt{-g}\exp(2P/\hbar)M_0/m_0$ with $\dot\rho = \partial_\mu(\rho b^\mu) = 0$ the quantum stochastic force (per unit of M_0) being written $K_\mu = -c^2\,\partial_\mu \log M_0$.

As one knows, the continuity equation is equivalent to a forward and backward Fokker–Planck equation:

$$\partial\rho/\partial\tau + \nabla v_\pm \rho \mp D\Box\rho = 0 \tag{19}$$

where $D = \hbar/2m$ to maintain the phase coherence in the quantum jumps at the velocity of light. Since we have:

$$\langle(\Delta x)^2\rangle_\rho \langle(\Delta u_x)^2\rangle_\rho \geqslant D^2 = \hbar^2/4m^2 \tag{20}$$

where the subscript means average over ρ in $\delta\omega$, we obtain Heisenberg's relation $\Delta x\, \Delta m_0 v_x^d = \Delta x\, \Delta p_x \geqslant \hbar/2$ as stochastic dispersion relation.

One sees from (17) and (18) that the preceding derivation yields time reversible linear equations to that $\Psi(x,t) = \Psi(x,-t)$. This is an evident result of the neglect in our model of possible frictional or dissipative forces (added to F^+ in (16)) which would generalize to "Dirac vacuum" the terms usualy introduced in Brownian motion theory. For example the addition to F^+ of nonlinear of the form $K = -2mD\beta\nabla \log \Psi_+$ destroys Brownian time reversibility of Einstein's initial nonrelativistic linear treatment in the classical relation[38]

$$\mp 2m_0 D\,\partial\Psi_\pm/\partial t = 2mD^2\,\nabla^2\Psi_\pm + 2m_0 D\beta \log \Psi_+\,\Psi_\pm \tag{21}$$

In the case of our model we are then justified to treat the nonlinear term $(m_0^2c^2/h^2)(\Psi/\Psi^*)^{1/2}\Box_s(\Psi\Psi^*)^{1/2}$ of (16) as a representation of time reversible frictional force term which are important in the U_0 region but can be neglected outside a sphere of radius λ_0 in the P wave region.

We now study motion of S waves into associated P waves. In the four-dimensional space-time, let us write in (11)

$$\{\sin kr/kr\}_0 = \left| \begin{array}{ll} \sin kr/kr & \text{if } \ |r| \leqslant \pi/k \\ 0 & \text{if } \ |r| > \pi/k \end{array} \right. \tag{22}$$

we have for such a "free" truncated S wave

$$\{\sin kr/kr\}_0 \exp i(\omega t - k_0 x) = \exp[(P_0 + iS)/\hbar] \tag{23}$$

with $\Box\{\exp[(P_0 + iS)/\hbar]\} = 0$, i.e.,

- a real part: $\partial_\mu S\, \partial^\mu S + m_0^2 c^2 = 0$
- an imaginary part: $\partial^\mu \partial_\mu S + 2\partial_\mu P_0\, \partial^\mu S = 0$

so that the center c_0 of the free soliton behaves exactly like a classical point like relativistic free particle surrounded in its rest frame by a stable spherical wave i.e., follows the geodetics of the external gravitational field.

From a stochastic point of view we can now describe the S wave in SIQM as a combination of outgoing and ingoing B waves satisfying $\Box U_0 = 0$ (i.e., corresponding to zero mass) centered on $c_0 = r_2$ in fig. 5. This is equivalent physically to a set of particles moving out or into c_0 at the velocity of light: a picture which fits nicely with out assumption of random jumps at the velocity of light.

We then analyze the behavior of a general wave $U = U_0 + V = \exp[(P_0 + P + iS)/\hbar]$ with $\exp(P_0/\hbar) = \{\sin kr/kr\}_0$ and $V = \exp[(P + iS)/\hbar)]$ with $R_2 = VV^* = \exp(2P/\hbar)$, in the rest frame Σ^0 defined by $b_i = 0$, $i = 1, 2, 3$. If this regular P wave has around c_0 the same phase S as the associated S wave, one realizes that everything goes as if the particle described by U_0 is located at the center c_0 of a singularity of the amplitude of a P wave which satisfies outside a small region of radius $r_0 = \hbar/2m$ to the usual linear relation (1) which can be analyzed in the hydrodynamical representation given by Halbwachs[21] and Vigier.[3] One sees moreover that the total wave U begins to differ from the P wave only inside a sphere of radius r_0 ($r_0 \gg r$). In this sphere (in Σ^0) one thus observes a singularity of the density in the hydrodynamical representation and one assumes that within a sphere of radius $r < r_0$ (in Σ_0) the phase S' of V, its gradient, and those of R are practically uniform. As we shall now see, this justifies de Broglie's guiding principle provided these variations can be neglected at the scale of r_0.

Indeed, let us assume that inside $r < r_0$ the density $\rho = \exp[2(P_0 + P)/\hbar]$

growth rapidly with $1/r$ (so that $\rho/(\partial\rho/\partial r) = 0$) and that at the scale r_0 we know from Section 3 that the motion of the total density's singularity does not vary with time so that all values of the singularity of P_0 (for $r < r_0$) move "as a block" with the same four velocity u_μ. In order to establish the wave behavior of u one can average all quantities of the wave equation between two spherical surfaces of radii r_1 and r_0 ($r_1 < r_0$) centered on c_0.

Indeed, the essential point is that according to SIQM, the total wave behavior of u is obtained by averaging over the infinitesimal four volume $d\omega_1$ (see fig. 5) now characterized by S_1 and S_3, and two cross-sections of radii r_0 and r_1. In such a domain we can consider $\langle\partial_\mu S\rangle$, $\langle\partial_\mu P\rangle$, $\langle P\rangle$, $\langle P\rangle$, and $\langle S\rangle$ as being practically constant for $r < r_0$. The total nonlinear wave equation (14) now yields for $U = U_0 + V$:

- a real part which can be written

$$\partial_\mu S\,\partial_\mu S + (\Box P + \partial^\mu P\,\partial_\mu P) + m_0^2c^2 + \{\Box P_0 + \partial_\mu P_0\,\partial^\mu P_0\} + 2\partial_\mu P\,\partial^\mu P_0$$
$$= (m_0^2c^2/h^2)(\partial^2/\partial y_\mu\,\partial y^\mu)(P_0 + P) = (m_0^2c^2/h^2)\,\Box_S(P_0 + P) \qquad (24)$$

- an imaginary part which yields

$$\Box S + 2\partial^\mu S\,\partial_\mu P + 2\partial^\mu S\,\partial_\mu P_0 = 0 \qquad (25)$$

To average over $d\omega_1$ we can go to the instantaneous rest frame Σ^0 ($\nabla_i S = 0$) where the P wave V is locally a plane wave and U_0 is spherically symmetric. We thus have locally $\langle\Box_S P\rangle = 0$ and $2\langle\partial_\mu P\rangle\partial^\mu P_0 = 0$, and since the $\{\ \}$ now cancels the second member we get in a general frame from (24) in Σ^0

$$\partial_\mu S\,\partial^\mu S + \Box P + \partial^\mu P\,\partial_\mu P + m_0^2c^2 = 0 \qquad (26)$$

which shows that c_0 must follow the P wave's drift lines of flow. Moreover the averaging of (25) yields

$$\Box S + 2\partial^\mu S\,\partial_\mu P = 0 \qquad (27)$$

since $\langle\partial_\mu S\rangle\,\partial^\mu P_0 = 0$ in Σ^0. This implies the conservation of U_0 with the drift's tubes of flow. Relations (26) and (27) clearly demonstrate de Broglie's "guiding" of "pilot" assumption. The preceding reasoning is evidently inspired by the famous argument through which Einstein *et al*[39,40] showed that if one considered the gravitational field of test particles as a singularity of the gravitational field, these singularities would necessarily follow geodetics path of the surrounding continuous gravitational field ... which thus necessarily "pilots" them according to the assumptions of relativistic mechanics. First suggested by one of us (J.P.V.)[3] the analogy is more than formal and can be turned into the following independent demonstration of de

Broglie's "guiding principle." Indeed, writing the P waves equation in the general relativistic form

$$(1/\sqrt{-g})\,\partial_\mu\, g^{\mu\nu}\partial_\nu \sqrt{-g}\,\partial_\nu \Psi(r,t) = (m_0^2 c^2/h^2) \tag{28}$$

where $g_{\mu\nu}$ represents the external gravitational field, one can show[21]

(a) that the drift lines are geodetics of the conformal metric $g'_{\mu\nu} = (M_0^2/m_0^2)\, g_{\mu\nu}$. The scalar stochastic quantum potential can thus be mapped into the gravitational potential of this new external conformal metric... which be considered as describing the disturbance of space-time resulting from the real existence of the P wave.

(b) that the frequency $M_0^2 c^2/\sqrt{1-\beta^2}$ represents exactly the frequency $m_0^2 c^2/\hbar$ modified by the action of the conformal gravitational potential.

Of course, the same arguments apply separately to the S wave which satisfies $\partial_\mu S\, \partial_\mu S - M_0^2 c^2 = 0$ since in that case the drift lines are geodetics of the metric $g_{\mu\nu}$.

Point (a) is important since one knows[40] that in general relativity the only way to combine "external" and "internal" gravitation fields is along time like geodetic surfaces... *as is exactly the case with hypertubes tangent to S when S is the phase of both U_0 and V.*

CONCLUSION

We conclude this paper with a brief estimate of the present theoretical and experimented stage of the Bohr–Einstein controversy—in which de Broglie always sided with Einstein.

As one knows, the root of the confrontation lies in Einstein's and de Broglie's refusal to accept any ultimate limit to our understanding and analysis of probabilities, and their belief in the causal character of *all* phenomena—observers' behavior and observations included. Indeed, when Einstein and de Broglie (with "ghost waves" and "double solution") first suggested that microobjects were both particles *and* waves they wanted to interpret the observed statistics of quantum theory in terms of subquantal *local* hidden variables. The reason for their choice is now clear: they were both realists and determinists in the relativistic sense of the word. This explains, for example, the motivations behind Einstein's attack against the CIQM as being both *incomplete* and (as we now know since John Bell's 1964 discovery) *nonlocal.* Their rejection of nonlocality was evidently tied to their belief that causality implied locality.

This belief, evidently widely shared in the physics community, explains the commotion caused by Bell's discovery that any type of local hidden parameters yields predictions in contradiction with quantum mechanical predictions in Bohm's version of EPR experiments. For the first time, it seemed that relativistic causality could be contradicted by experiment—so that Einstein's world vision would have to be abandoned or drastically transformed. This also explains the growing number of experimental efforts to test quantum predictions in correlated measuring processes separated by space-like separations.[4,5]

However, as seen for example in reference,[11] this general opinion is too drastic. Even (as is now practically certain) nonlocal superluminal correlations between space like separated measuring devices are definitely established, this does not imply Bohr's ultimate victory, but only that both Einstein's and de Broglie's followers as well as the present supporters of CIQM will have to modify their theories to some extent.

For example, even before Aspect's and Rapisarda's experiments probably confirm the nonlocal predictions of quantum mechanics, CIQM follows are already in the process of:

(a) modifying their conceptions of measurement in quantum theory
(b) introducing retroaction into time[10]
(c) developing the nonlocal character of conscious measurements[8]—at the expense of the existence of space and time.

For the same reasons, CIQM opponents are trying to destroy Bell's inequalities[42] (a hopeless attempt in the author's opinion) or have been led to abandon point particle locality, i.e., to stress the nonlocal character of the quantum potential. This later choice which rests either on nonlinear subquantum fluctuations[43] and/or the nonlocal extended structure of the elements which construct Dirac's subquantal vacuum (or "aether") had led to the *new result*[11] *that superluminal actions-at-a-distance can be perfectly causal provided they satisfy supplementary relativistic conditions—and that is precisely the case for the quantum potential in configuration space.* Evidently this is only possible if one accepts de Broglie's basic assumption that a particle is surrounded by a real *P* wave field carrying real stochastic quantum forces.

In other terms, an experimental proof of the existence of causal nonlocality is tied in SIQM to proofs of the physical truth of de Broglie's conception of wave particle duality: the nonlocal quantum potential being interpreted in the EPR paradoxon through perturbations of the subquantal aether induced by correlated pairs of particles which disturb the vacuum around them.

Since Selleri's first proposal[44] various possible *Gedanken* experimental tests of the existence of *P* waves have been proposed, one evidently reaches here a new stage of the Bohr–Einstein controversy, since it is the first time (to our knowledge) that the nonlocal form of SIQM and CIQM predict different testable experimental results. As well as for EPR experiments, the proposed set-ups turn out to be limit experiments with individual quantas. To break Bell's inequalities (which are satisfied in all classical and almost every quantum situation), one has to measure 100 couples of photons per second.[4]

To establish the possible existence of P waves, one has to reach very low photon intensities where separation of CIQM and SIQM raises difficult theoretical and experimental problems. To comment on this, we will reanalyze the most recent proposal (made with Garuccio and Rapisarda)[14] to detect P waves. This set-up is described in Fig. 6.

A single very weakened pulsed source (i.e., an impulsed l.e.d.) produces, one by one, independent wave packets which contain single photons at the same frequency.

These successive wave packets containing 1 photon only, denoted $IW_n = (n = 1, 2, \ldots)$, are split by a semitransparent beam splitter (the semitransparent mirror M_1 with a transmission coefficient 1/2) and one knows from CIQM that the photons pass, one by one, either in the reflected packets RW_n or in the transmitted packets TW_n with a probability 1/2; there is a 100% anticorrelation between reflected and transmitted photons, as has been recently verified experimentally by Mandel *et al.*[48].

Following a suggestion of Selleri,[45] we then introduce along the path of TW_n an organic laser gain tube (L.G.T.) which multiplies the photon number by two (on the average) and preserves the different phases of TW_n.[3] This is perfectly feasible. As one knows, gain tubes amplify externally introduced photons beams and preserve their phases with sharp temporal correlation: so that one can expect this to hold in the one photon limit also. Such amplifiers have been recently used (as will be discussed presently) in a different context by Martinolli[46] in Prof. Gozzini's laboratory in Pisa to test for the two photons incoming case the outgoing distribution generated by a semitransparent mirror.

Of course, the use of this multiplying gain tube implies a preliminary test i.e., one should test if in the set-up of fig. 6, where incoming isolated photon beams are split by M_1, there is significant anticoincidences. If not then the photon empty TW de Broglie wave packet would excite, without photons, the LGT and, contrary to CIQM predictions induce a real physical effect.

[3] The question of the phase conservation between the phases of separated incoming wave packets and the corresponding outgoing photon pairs, theoretically valid because of the laser character of LGT should be tested separately by verifying that the set-up of Fig. 6 maintains an interference pattern at ordinary light intensity.

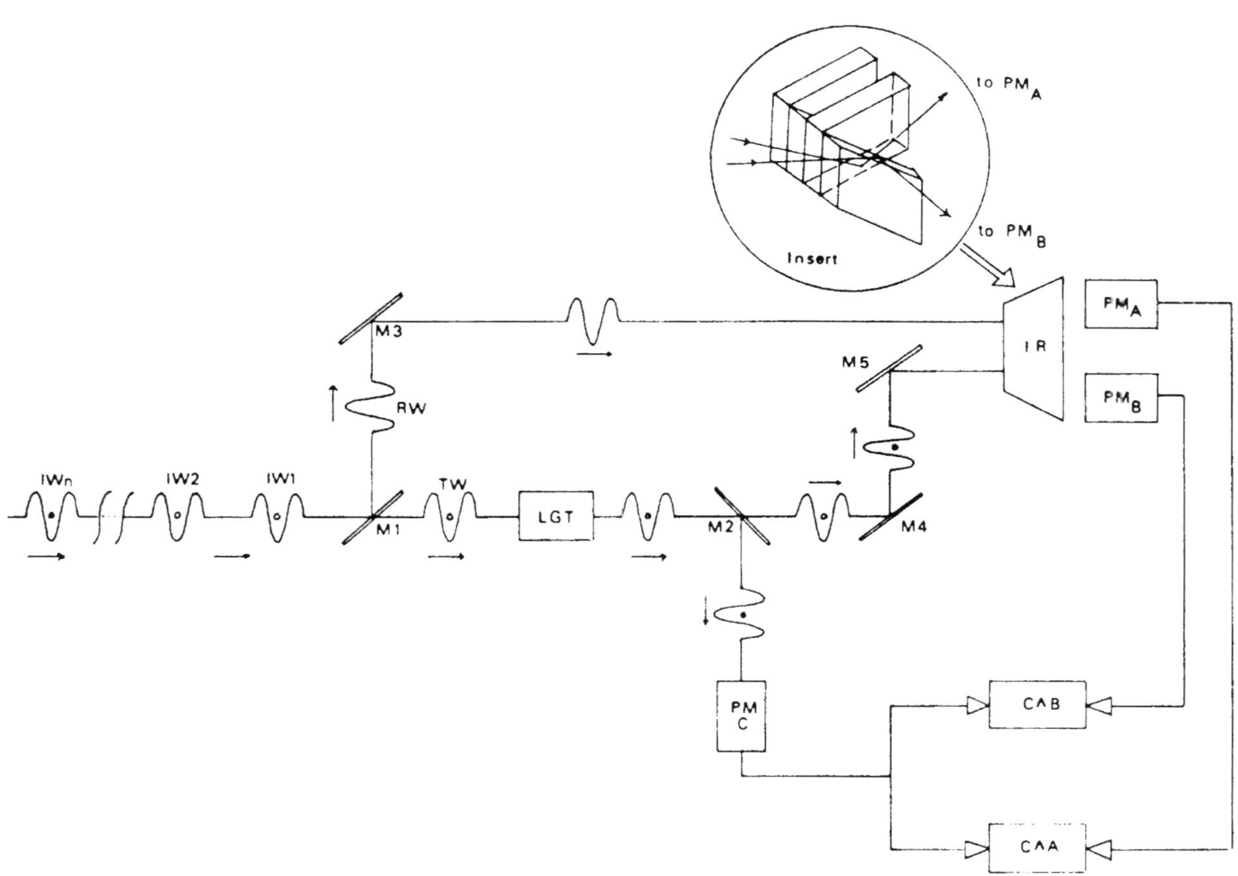

Fig. 6. Experimental set-up to prove the existence of de Broglie's waves. Incoming wave packets IW_n, containing one photon only, are split by the semitransparent mirror M_1 with transmission coefficient 1/2. If the laser gain tube is excited, two outcoming photons are split by M_2 and coincidences measured between PM_A, PM_B, and PM_C. The insert represents Mandel's device with $\lambda/2$ plate to separate maximums and minimums in the interference pattern.

If anticoincidences between PM_1 and PM_2 persist, then one can move a step further since only a photon in TW can induce two photon wave packets in LGT, which are split in their turn by a beam splitter M_2 identical to M_1.

If further tests confirm this, then one sees that the LGT outgoing photons have a 1/3 probability of producing coincidences between the photomultiplier PM, and the interference region (IR), where a detector devised by Mandel and Pfleegor[47] is built with a stack of thin glass plates (see insert of Fig. 6 each of which has a thickness corresponding to a half fringe width. The plates are cut and attanged so that any photons failing on the odd plates are fed to one photomultiplier PM_A, while photons falling on the even plate are fed to the other PM_B. So the LGT outgoing photons have a 1/3 probability of produring coincidences between PM_1 and one of two photomultipliers PM_A or PM_B.

The experiment is now to analyze the rate of observed coincidences between PM_C, PM_A, and PM_B. Two perfectly conflicting results are now theoretically possible.

A. According to CIQM no interferences should be observed, i.e., one rate of observed coincidences should be equal to 1. Indeed, if 2 photons appear at PM_C, no photons should appear at PM_A or PM_B. Moreover, if one photon appears at PM_C and one in PM_A or PM_B. then since one now *knows* that no photons have taken the $M_1 - M_3 - IR$ path, then no interference should appear. Indeed, the passage of a photon in the LGT path has collapsed the RW packet.

B. According to Maxwell's theory of light (or the SIQM)[9] interference fringes should appear with the maximum contrast, i.e., the rate of observed coincidences should be different from 1, since one has $I(RW) = I(TW'') = 1/2$. Indeed, even if the incoming photons of IW have entered the LGT, then a real physical de Broglie wave is moving with RW along the path $M_1 - M_3 - IR$. This induces a stable interference pattern since the path difference between $M_1 - M_3 - IR$ and $M_1 - \mathrm{LGT} - M_2 - M_4 - M_5 - IR$ is constant, in the mean.

At first sight, this set-up presents evident advantages. Indeed, if LGT (according to CIQM) needs real incoming photons to be excited, then no photon in our coincidence scheme can travel along the $M_1 - M_3 - IR$ path, so that the detection of interferences (according to SIQM or Maxwell's theory) would really constitute a crucial contradiction between CIQM and reality. Things however are more complex in a real measurement. As remarked by a referee: "A laser gain tube works through stimulated emission. The fundamental quantum mechanical rule here is that the pobability of a photon appearing in a given state must be proportional to

$1 + n$ where n is the number of photons already in that final state. In the one-incoming-photon case. n will be equal to 1. This implies that the stimulated emission rate cannot be greater than the spontaneous emission rate, at best it can be qual to it. So, if one photon comes in the best that can happen is three coming out, two in phase, the original one and one produced by stimulated emission, and one produced by spontaneous emission (hence not generally in phase with the incoming photon). However the spontaneous emission ones will also be appearing throughout the period when there is no incoming photon. It seems that they must therefore mask any expected interference pattern produced by the pairs which are in phase. This problem does not show up in the many-incoming-photons case since there the ratio of n to 1 will be very large, so there is no need to have a rate of spontaneous emission which is significant in relation to the rate of stimulated emission. But I do not see how in the one-incoming-photon case there can be any way round the difficulty: a whole series of such processes stimulated by a single initial photon will still always produce 1/3 of outcoming photons not in phase with the original photon, and this emission rate will be maintained throughout the time there is no photon in in."

Two remarks can be made to overcome this objection. The first remark is that if one accepts the proceding theory that the rate of spontaneous noise can be calculated by assuming one stimulating photon per mode continually present, then between the noise spontaneous photon (phase decorrelated) and the two phase correlated photons according to Scorle's result, there is no time correlation in excess of the Hanbury–Brown–Twiss effect: the three photons can thus be concentrated within the gain tube's transit time i.e., one should try to reduce their size to the minimum and separate as much as possible the time intervals of the incoming (one photon carrying) wave packets.

The second remark is that, following Prof. Gozzini,[49] if one uses the three photon spectroscopy technique (based on Cagnac proposal[50] to build a LGT, this particular LGT produces two phase locked photons with the same (incoming) frequency (one of them retarded in time w.r.t. the incoming one) and one spontaneous photon *with a different frequency*. As suggested to us by Prof. Gozzini, this is very interesting in our case since the "noise" (spontaneous) photon can now be eliminated form measurement by an interfering filter or a Fabry–Pérot device. Two cases thus only remain when one measures the $C - IR$ coincidences:

1. The initial photon reaches IR and the delayed one C, so that one observes interferences fringes in IR, in Maxwell's or SIQM theories.

2. The delayed photons reach IR and the initial one C, in which case one will obtained in IR a uniform spot without interferences fringes.

The probability of (1) and (2) being equal, we can thus conclude that the creation of delayed photons in LGT preserves a signal/noise ratio equal to 1, i.e., does not destroy the proposed test. As one sees it is the coincidence set-up of *IR* and *PMC* which suppresses observability of the permanent emission of phase decorrelated photons when there is no photon coming in. We therefore conclude on the experimental "falsifiable" character of this model of de Broglie's wave particle duality—so that its fate no rest, as it should, in the hand of experimental physicists.

ACKNOWLEDGMENTS

The authors want to thank Prof. van der Merwe for the possibility to publish this work in *Foundations of Physics* in honor of Prof. de Broglie on his 90th birthday. One of us (J.-P.V.) is happy for this opportunity to thank publicly Prof. de Broglie for many years of inspiration, support and common work. Great physicists fight great battles; without men like Albert Einstein and Louis de Broglie modern physics would be different. Their vision, courage, and leadership have carried the epistemological discussion on the Bohr–Einstein controversy to its present experimental level. Whatever its final outcome, they have stuck to their minority convictions with fortitude, an admirable example of the way scientific debate should be conducted. The authors are also grateful to many colleagues, including Profs. D. Bohm, F. Selleri, V. Rapisarda, A. Gozzini and Drs. J. Bell, A. Shimony, A. Garuccio, and N. Cufaro-Petroni for helpful criticisms and suggestions. During the completion of this work, they have been shocked by Prof. V. Rapisarda's tragic death. They share with his family, many friends and the Italian physics community a sense of irretrievable loss.

REFERENCES

1. L. Mackinnon, *Found. Phys.* **11**, 907 (1981).
2. L. de Broglie, *Une tentative d'interprétation causale et non-linéaire de la Mécanique Ondulatoire* (Gauthier-Villars, Paris, 1956).
3. P. A. M. Dirac, *Nature* (London), **168**, 906 (1951). For a recent review, see J. P. Vigier, *Lett. al Nuovo Cimento* **29**, 467 (1980).
4. A. Aspect, *Phys. Rev. D* **14**, 1944 (1978); *Phys. Rev. Lett.* **47**, 480 (1980).
5. L. Pappalardo and V. A. Rapisarda, *Lett. Nuovo Cimento* **29**, 221 (1980). A. Garuccio and V. A. Rapisarda, *Nuovo Cimento, A* **65**, 269 (1981).
6. D. Bohm, *Quantum Theory* (Prentice-Hall, Englewood Cliffs, N.J., 1951).
7. A. Einstein, B. Podolsky, and N. Rosen, *Phys. Rev.* **47**, 777 (1935).
8. N. Bohr. *Phys. Rev.* **48**, 696 (1935).
9. D. Bohm and J.-P. Vigier, *Phys. Rev.* **96**, 208 (1954); L. de Broglie, *Thermodynamique de*

la particule isolée (Gauthier-Villars, Paris, 1964); W. Lehr and J. Park, *J. Math. Phys.* **18**, 1235 (1977); E. Nelson, *Phys. Rev.* **150**, 1079 (1966); J. P. Vigier, *Lett. al Nuovo Cimento* **24**, 258, 265 (1979).

10. O. Costa de Beauregard, *Nuovo Cimento* **42**, B, 41 (1977); *Nuovo Cimento* **51**, B, 267 (1979).
11. N. Cufaro-Petroni, Ph. Droz-Vincent, and J. P. Vigier, *Lett. al Nuovo Cimento* **31**, 415 (1981).
12. A. Garuccio, K. R. Popper, and J.-P. Vigier, *Phys. Lett.* **86**, A, 397 (1981).
13. E. E. Fitchard, *Found. Phys.* **9**, 525 (1979).
14. A. Garuccio, V. A. Rapisarda, and J. P. Vigier, "New Experimental Set-up for the Detection of the de Broglie Waves," *Phys. Lett.* (1982).
15. J. P. Vigier, *Lett. Nuovo Cimento* **24**, 258, 265 (1979).
16. L. de la Pena Auerbach, *J. Math. Phys.* **12**, 453 (1971); N. Cufaro-Petroni and J.-P. Vigier, *Phys. Lett.* **81**, A, 12 (1981).
17. N. Cufaro-Petroni and J.-P. Vigier, *Phys. Lett.* **73A, 4**, 289 (1979).
18. F. Guerra and P. Ruggiero, *Lett. al Nuovo Cimento* **23**, 529 (1978).
19. J.-P. Vigier, *Lett. Nuovo Cimento* **29**, 467 (1980).
20. C. Eckart, *Phys. Rev.* **58**, 919 (1950).
21. F. Halbwachs, *Théorie Relativiste des Fluides à Spin* (Gauthier-Villars, Paris, 1960).
22. Ph. Guèret and J. P. Vigier, *Une équation non linéaire de Klein-Gordon en Mécanique Ondulatoire, possèdant une solution non dispersive du type soliton*, IHP Preprint, March 1982.
23. A. Einstein, "Physik und Realität," *Journ. of the Franklin Inst.* **221**, 313 (1936).
24. L. de Broglie, *La Physique restera-t-elle indéterministe*? (Gauthier-Villars, Paris, 1953).
25. A. C. Scott, F. Y. F. Chu, and D. Mc Laughlin, *Proc. of the IEEE* **61**, 10 (1972).
26. L. de Broglie, *Ann. Phys.* (Paris) **3**, 22 (1925).
27. H. Yukawa, *Proc. of the International Conference on Elem. Particles* (Kyoto, 1965).
28. C. Fenech, M. Moles, and J.-P. Vigier, *Lett. al Nuovo Cimento* **24**, 56 (1979).
29. F. Halbwachs, J.-M. Souriau, and J.-P. Vigier, *J. Phys. Radium* **22**, 26 (1981).
30. Ph. Guèret, M. Moles, P. Merat, and J.-P. Vigier, *Lett. Math. Phys.* **3**, 47 (1979); N. Cufaro-Petroni, Z. Maric, Dj. Zivanovic, and J.-P. Vigier, *J. Phys. A, Math. Gen.* **14**, 501–508 (1981); N. Cufaro-Petroni, Z. Maric, Dj. Zivanovic, and J.-P. Vigier, *Lett. al Nuovo Cimento* **29**, 17, 565 (1980).
31. P. A. M. Dirac, *Proc. Roy. Soc.* **167**, A, 448 (1933).
32. P. C. Clemmow and R. D. Harding, *J. Plasma Phys.* **23**, 71 (1980).
33. L. de Broglie, *Cahiers de Phys.* **147**, 1 (1962).
34. L. Mackinnon, *Lett. al Nuovo Cimento* **32**, 10 (1981).
35. L. de Broglie, *C. R. Acad. Sci.* **180**, 498 (1925).
36. R. C. Jennison, *J. Phys. A, Math. Gen.* **11**, 1525 (1978).
37. E. C. Kemble, *Fundamental Principles of Quantum Mechanics* (McGraw-Hill, New York, 1937).
38. L. de la Pena and A. M. Cetto, *Found. Phys.* **5**, 355 (1975).
39. A. Einstein and J. Grommer, *Sitz. Preuss. Acad. Wiss.* **1** (1927).
40. A. Einstein and L. Infeld, *Ann. Math.* **41**, 455 (1940).
41. B. d'Espagnat, *La Physique et le réel.*
42. L. de Broglie, G. Lochak, J. A. Beswick, and J. Vassalo-Pereira, *Found. Phys.* **6**, 1, 3 (1976).
43. D. Bohm and J. Hiley, *Found. Phys.* **11**, 529 (1981).
44. F. Selleri, "Can an Actual Existence be Guaranted on Quantum Waves?" *Nuov.-Cim.* (to be published).

45. F. Selleri, *Lett. al Nuovo Cimento*, **9**, 8 (1979).
46. R. Martinolli (Reporter A. Gozzini), *Un esperimento in Ottica a Intensità Molto Basse* (Tesi di Laurea, Università di Pisa (October 1980).
47. L. Mandel and R. L. Pfleegor, *Phys. Rev.* **159**, 1084 (1967); *Journ. Opt. Soc. Amer.* **58**, 946 (1968).
48. L. Mandel and K. Dajenais, *Phys. Rev. Lett.* **A18**, 2217 (1978).
49. A. Gozzini, *Programma di Ricerca* (Università di Pisa Preprint, 1982).
50. G. Grynberg, F. Biraben, M. Bassini, and B. Cagnac, *Phys. Rev. Lett.* **37**, 284 (1976).
51. L. Mackinnon, *Found. Phys.* **11**, 907 (1981).

This page is deliberately left blank.

Reprinted from *Lettere al Nuovo Cimento*, Vol. 35, No. 8, pp. 256-259, Copyright (1982)
with permission from the Società Italiana di Fisica.

Nonlinear Klein-Gordon Equation Carrying a Nondispersive Solitonlike Singularity.

PH. GUERET

Institut de Mathèmatique Pures et Appliquées de l'Université P. et M. Curie
4 place Jussieu, 75230 Paris Cedex 05.

J. P. VIGIER

Institut Henri Poincaré, Laboratoire de Physique Théorique
II rue P. et M. Curie, 75231 Paris Cedex 05.

(ricevuto il 16 Luglio 1982)

The aim of this letter is to show that, if one accepts the physical reality of the de Broglie's waves associated to particles and the relativistic covariance of the phase equality of wave and particle at the point where it is localized, one can construct a nondispersive wave packet tied to the particle, which constitutes a possible model of the singular wave introduced by DE BROGLIE in his causal interpretation of wave mechanics. This soliton-like wave, first constructed by MACKINNON, is a solution of a nonlinear Klein-Gordon equation.

The basic idea of wave mechanics is that each particle with rest mass m_0 is endowed at rest with a real physical oscillation $\nu_0 = m_0 c^2/h$ (1) and that its uniform rectilinear motion with velocity v in the x-direction of space-time is associated with real phase plane waves (de Broglie's B-waves) whose propagation is described by the function

$$\Psi(x, t) = a \exp\left[2\pi i\nu\left(t - \frac{x}{V}\right)\right] \qquad (a = \text{const}), \tag{1}$$

where $\nu = \nu_0/\sqrt{I - \beta^2}$. $V = c^2/v$ represents its associated phase velocity. The corresponding wave-length is given by the well-known expression $\lambda = h/p$, where $p = |\boldsymbol{p}|$ represents the relativistic impulsion of the said particle. It is this wave-length which one can measure in diffraction experiments (2).

If one now superimposes a set of such B-waves, one can construct (3,4) a wave packet (henceforth denoted as P-wave) which presents the following characteristics:

---

(1) L. DE BROGLIE: *Ann. Phys. (Paris)*, **3**, 22 (1925).
(2) L. MACKINNON: *Found. Phys.*, **11**, 907 (1981).
(3) D. BOHM: *Quantum Theory* (Englewood Cliffs, N.J., 1951).
(4) E. C. KEMBLE: *Fundamental Principles of Quantum Mechanics* (New York, N. Y., 1937).

Its group velocity $\mathscr{U}$ is equal to the velocty v of the particle.

It spreads out without limit in a short time.

It differs from a Schrödinger wave packet on its phase [4,5] in spite of the fact that the latter is also endowed with the two first properties.

As the B-waves, the P-wave satisfies the Klein-Gordon equation

$$\Box\Psi + (m_0^2c^2/\hbar^2)\Psi = 0\,. \tag{2}$$

As one knows, de Broglie has then introduced a supplementary assumption which has played a leading part in the causal interpretation of wave mechanics [6], *i.e.* the « phase connection principle » which states that, in all points in which the particle is located, the phase of its internal oscillation coincides with the phase of the B-wave. This assumption implies that this equality must be a covariant one, identical for all observers. Mackinnon has accordingly utilized this important property to construct a new nondispersive wave packet [7] (henceforth denoted as S-wave). Indeed, if one considers a particle moving with a constant uniform velocity u in the x-direction of a fixed laboratory frame, one can superimpose the B-waves detected by a set of observers also moving in the x-direction with velocities included between $-c$ and $+c$ with respect to the laboratory frame. As a result, one obtains, following Mackinnon, the nondispersive wave packet

$$F(x,t) = \frac{\sin\Delta k(x-ut)}{\Delta k(x-ut)}\exp\left[i[\omega(k_0)t - k_0x]\right], \tag{3}$$

where $\Delta k = mc/\hbar = m_0c/\hbar\sqrt{1-\beta^2}$, $k_0 = mu/\hbar$, $\omega(k_0) = k_0V = mc^2/\hbar$.

This S-wave packet conserves its shape under a Lorentz transform and appears as essentially relativistic in its origin in spite of the fact that it does not satisfy relation (2).

In the particle rest frame ($u = 0$), the amplitude of the S-wave packet becomes $(\sin kx)/kx$ with $k = m_0c/\hbar$, where x is the distance from the centre of the wave packet. The first zero of this function arises so that the foot of its central part is equal to the Compton wave-length h/m_0c.

Let us now determine the wave equation which accepts solution (3) as solitary wave. Writing

$$F(x,t) = G(x,t)H(x,t)$$

with

$$H(x,t) = \frac{\sin\Delta k(x-ut)}{\Delta k(x-ut)}\,, \quad G(x,t) = \exp\left[i[\omega(k_0)t - k_0x]\right],$$

if one calculates $\Box F = F_{xx} - F_{tt}/c^2$, one checks that the imaginary part cancels and that a possible wave equation, given by Mackinnon [7], is the modified Klein-Gordon's equation

$$\Box F - (m_0^2c^2/\hbar^2)F = (1-\beta^2)GH_{xx}\,. \tag{4}$$

---

(5) E. Mackinnon: *Am. J. Phys.*, **44**, 1047 (1976).

(6) L. de Broglie: *Tentative d'interprétation causale et non linéaire de la Mécanique Ondulatoire* (Paris, 1956).

(7) L. Mackinnon: *Found. Phys.*, **8**, 157 (1978).

Moreover, if one writes $\xi(x,t) = \Delta k(x - ut)$, so that $H(x,t) = H[\xi(x,t)]$, one finds that $H_{xx} = (\Delta k)^2 H_{\xi\xi}$, so that H_{xx} depends on ξ but not in the precise form in which ξ depends on x and t. Equation (4) can thus be written more generally as

$$\Box F - (m_0^2 c^2/\hbar^2) F = (m_0^2 c^2/\hbar^2) G H_{\xi\xi} \,, \tag{5}$$

or still more generally

$$\partial_\mu \partial_\mu U - (m_0^2 c^2/\hbar^2) U = (m_0^2 c^2/\hbar^2) G H_{\xi\xi} \tag{6}$$

with $U = U_0 + \Phi$.

U_0 is the S-wave defined by $U_0(x,t) = aF(x,t) = a\big((\sin\xi)/\xi\big)\exp[iS/\hbar]$, where a is an arbitrary multiplicative constant and $S = m(c^2 t - ux) = Et - \boldsymbol{p}\cdot\boldsymbol{x}$.

Φ is a P-wave with the same phase S as U_0, *i.e.* $\Phi(x,t) = R(x,t)\exp[iS/\hbar]$ which satisfies the Klein-Gordon condition $\partial_\mu\partial_\mu\Phi = (m_0^2 c^2/\hbar^2)$.

The covariant feature of the first member of (5) is obvious. The second member is also covariant since the derivation which has given $H_{\xi\xi}$ corresponds to $\partial^2/\partial y_\mu \partial y^\mu$, where dy_μ denotes the spacelike normal distance taken from the drift lines parallel to $d_\mu S$ along the surfaces S = const.

Thus one finds again de Broglie's model of the double solution theory (6) with

the splitting (8)

$$\partial_\mu(R^2 \partial_\mu S) = 0 \,, \tag{C}$$

$$\partial_\mu S \partial_\mu S + m_0 c^2 - h^2(\Box R/R) = 0 \,. \tag{J}$$

the guiding theorem (6,8). The fitting of the S-wave on the P-wave corresponds to Einstein's famous result (9,10) which shows that the singularities of the gravitational field (test particles) necessarily follow, in the general relativity theory, the geodetics of the external gravitational field.

At the nonrelativistic approximation $(\sin\xi)/\xi \to 1$, so that the S-wave does not appear at this level. As for the P-wave equation, it reduces to a time-independent Schrödinger equation, as shown by KEMBLE (4).

In a recent paper (11), R. HORODECKI has shown the necessity to introduce the three types of waves considered in this letter, but without giving an explicit form for the S-wave.

As MACKINNON has shown (12), in the quadridimensional space-time the nondispersive S-wave packet for a particle of rest mass m_0 travelling in the x-direction with velocity u relative to the laboratory system takes the form

$$F(r,x,t) = \big((\sin kr)/kr\big)\exp\big[i[\omega(k_0)t - k_0 tx]\big] \,, \tag{7}$$

---

(8) F. HALBWACHS: *Theorie relativiste des fluides à spin* (Paris, 1960).
(9) A. EINSTEIN and J. GROMMER: *Sitz. Preuss. Akad. Wiss.*, **1** (1927).
(10) A. EINSTEIN and L. INFELD: *Ann. Math.*, **41**, 455 (1940).
(11) R. HORODECKI: *Phys. Lett.*, **87**, no. 3, 95 (1981).
(12) L. MACKINNON: *Lett. Nuovo Cimento*, **31**, 37 (1981); **32**, 311 (1981).

where $r^2 = ((x - ut)^2/(1 - \beta^2)) + y^2 + z^2$ and $k = m_0 c/\hbar$. By simple substitution, one sees that this S-wave is solution of

$$\square F = 0 . \tag{8}$$

It is one form of the standard spherically symmetrical solution of (8) after it has been subjected to a Lorentz transformation. In a more general form, this result was obtained in the case of electromagnetic waves by DE BROGLIE himself [13], who has shown that this solution represents the superposition of two spherically symmetrical waves, one converging and one diverging, both having the phase velocity c. More recently, JENNISON [14] has shown, still in the electromagnetic case, that this combination constitutes a phase-locked cavity having the inertial properties of particles.

Now let us return to relation (4). If we express G and H_{xx} as functions of F: $H = |F| \operatorname{sgn} H$ and $G = F/|F| \operatorname{sgn} H$, where $\operatorname{sgn} H$ denotes sign of H, we have $GH_{xx} = F|F|_{xx}/|F|$. In this way, if one accepts the physical reality of B-wave phase, a nonlinear equation for the de Broglie S-wave, in the double-solution theory is, in units $\hbar = c = 1$,

$$u_{xx} - u_{tt} = m_0^2 u \left(1 + \frac{l^2 |x|_{xx}}{|u|}\right), \tag{9}$$

i.e. a nonlinear Klein-Gordon equation for which U_0 is a soliton-like solution (l^2 is a homogeneity constant).

However, the solitonlike solution does not serve to characterize this nonlinear equation. Indeed let us consider a solution $u(x, t) = v \exp[is]$ defined by its modulus $|u| = v(x, t)$ and its phase $s(x, t)$, eq. (9) splits into the real and imaginary parts:

$$s_x v_x - s_t v_t = 0 , \tag{10}$$

$$v_{xx} - v_{tt} - v(s_x^2 - s_t^2) = m_0(v + l^2 v_{xx}) = 0 . \tag{11}$$

Equation (11) for the modulus is a linear one, for which $v = |(\sin \xi)/\xi|$ is only a peculiar solution.

Let us remark to conclude, that if we put $\Psi = |(\sin \xi)/\xi| \exp[is]$, we can write eq. (9) in the form

$$(\square - m_0^2)\Psi = m_0^2 \sqrt{\Psi/\Psi^*}\, \square_s \sqrt{\Psi^* \Psi} , \tag{12}$$

where $\square_s$ denotes a Laplace-Beltrami operator taken on the curves $s = $ const.

---

(13) L. DE BROGLIE: *C. R. Acad. Sci.*, **180**, 498 (1925).
(14) R. C. JENNISON: *J. Phys. A: Math., Nucl. Gen.*, **11**, 1525 (1978).

Reprinted from *Lettere al Nuovo Cimento*, Vol. 38, No. 4, pp. 125-128, Copyright (1983) with permission from the Società Italiana di Fisica.

Relativistic Wave Equations with Quantum Potential Nonlinearity.

PH. GUERET

Institut de Mathematique Pures et Appliquées de l'Université P. et M. Curie
4 place Jussieu, 75230 Paris Cedex 05

J. P. VIGIER

Institut Henri Poincaré, Laboratoire de Physique Théorique
II rue P. et M. Curie, 75231 Paris Cedex 05

(ricevuto il 27 Giugno 1983)

PACS. 11.10. - Field theory.

It has been shown (¹) that, by considering de Broglie's waves from the point of view of all possible observers, it is possible to form a nondispersive de Broglie wave packet for a free particle. More specifically, let it be supposed that a particle, of rest mass m_0, is travelling freely with uniform velocity v in the $+\,x$-direction. In such a case, the nondispersive wave packet is given in two-dimensional space-time by

$$u(x,t) = A\,\frac{\sin \Delta k\xi}{\Delta k\xi}\exp[iS]\,, \tag{1}$$

where $S = \omega(k_0)t - k_0 x$, $\Delta k = mc/\hbar$, $k_0 = mv/\hbar$, $\omega(k_0) = mc^2/\hbar^2$, $\xi = x - vt$. $u(x,t)$ is a solution of the nonlinear Klein-Gordon equation

$$u_{xx} - \frac{1}{c^2}u_{tt} = \frac{m_0^2c^2}{\hbar^2}\left(1 + \frac{l^2|u|_{xx}}{|u|}\right)u \tag{2}$$

(l^2 is a homogeneity constant). This result has been shown (²) to be consistent with de Broglie's original hypothesis of his double-solution interpretation of wave mechanics (³).

In this letter some extensions of eq. (2) are given. First, let it be considered a non-disperive wave packet of arbitrary shape, exactly as in the case of a nonlinear quantum

(¹) L. MACKINNON: *Found. Phys.*, **8**, 157 (1978).
(²) PH. GUÉRET and J. P. VIGIER: *Lett. Nuovo Cimento*, **35**, 256 (1982).
(³) L. DE BROGLIE: *Nonlinear Wave Mechanics* (Amsterdam, 1960).

Then, in eq. (2) the nonlinear term $(|u|_{xx}/|u|)u$ becomes $R_{\xi\xi}\exp[iS]$, with

$$R_{\xi\xi}\exp[iS] = \frac{\hbar^2}{m_0^2c^2}\frac{\Box|u|}{|u|}u \tag{4}$$

after trivial manipulations. So, eq. (2) exhibits a nonlinear quantum potential term when it is written as

$$\Box u - \frac{m_0^2c^2}{\hbar^2}u = \frac{\Box\sqrt{\varrho}}{\sqrt{\varrho}}u\,, \tag{5}$$

where $\sqrt{\varrho} = (\bar{u}u)^{\frac{1}{2}} = |u|$. This new equation satisfies the superposition principle in conformity with the Guerra and Pusterla requirements [5] and describes the propagation of two-dimensional space-time kinks and solitons of arbitrary shape. However, only the wave packet (I) is made of waves which will have the same phase at the same place where the particle is, in a covariant way.

In writing

$$\Box = \partial_\mu\partial^\mu = \frac{\partial^2}{\partial x^2} - \frac{1}{c^2}\frac{\partial^2}{\partial t^2},$$

it can be shown that eq. (5) springs from the Lagrangian

$$\mathscr{L} = \frac{1}{4}\left[\frac{u}{\bar{u}}(\partial_\mu\bar{u})(\partial^\mu\bar{u}) + \frac{\bar{u}}{u}(\partial_\mu u)(\partial^\mu u)\right] - \frac{1}{2}(\partial_\mu\bar{u})(\partial^\mu u) - \frac{m_0^2c^2}{\hbar^2}\bar{u}u\,. \tag{6}$$

From this Lagrangian the same current as in the linear case can be derived, *i.e.*

$$J_\mu = -\frac{i\hbar}{2mc}[\bar{u}(\partial_\mu u) - u(\partial_\mu\bar{u})] \tag{7}$$

and the energy-momentum tensor

$$T_{\mu\nu} = \frac{1}{2}\left[\frac{u}{\bar{u}}(\partial_\mu\bar{u})(\partial_\nu\bar{u}) + \frac{\bar{u}}{u}(\partial_\mu u)(\partial_\nu u) - (\partial_\mu u)(\partial_\nu\bar{u}) - (\partial_\mu\bar{u})(\partial_\nu u)\right] - \delta_{\mu\nu}\mathscr{L}\,. \tag{8}$$

In the four-dimensional space-time, instead of (3), the nondispersive wave packets associated to a particle, of rest mass m_0, travelling in the $+x$-direction are defined by

$$u(r,t) = R(r)\exp[iS]\,, \tag{9}$$

---

(4) R. W. Hasse: *Z. Phys. B*, **37**, 83 (1980).

(5) F. Guerra and M. Pusterla: *Lett. Nuovo Cimento*, **34**, 351 (1982).

where

$$r = |\boldsymbol{r}| = \left(\frac{\xi^2}{1-\beta^2} + y^2 + z^2\right)^{\frac{1}{2}}.$$

They are solutions of

$$\Box u - \frac{m_0^2 c^2}{\hbar^2} u = \left(R_{rr} + 2\frac{R_r}{r}\right)\exp[iS]. \tag{10}$$

But a simple calculation shows immediately that for u given by (9)

$$\frac{\Box\sqrt{\varrho}}{\sqrt{\varrho}} u = (\Box R)\exp[iS] = \left(R_{rr} + 2\frac{R_r}{r}\right)\exp[iS] \tag{11}$$

so that eqs. (10) and (5) are equivalent in four dimensions.

Among the solutions (9) there are solutions for which

$$\Box u = 0 \quad \text{and} \quad m_0^2 = \frac{\hbar^2}{c^2}\frac{\Box\sqrt{\varrho}}{\sqrt{\varrho}} \tag{12}$$

arise simultaneously, *i.e.* with $\Delta k_0 = m_0 c/\hbar$, the Eulerian differential equation

$$R_{rr} + 2\frac{R_r}{r} + (\Delta k_0)^2 R = 0, \tag{13}$$

which admits the general solution

$$u(r, t) = \left(A\frac{\sin \Delta k_0 r}{\Delta k_0 r} + B\frac{\cos \Delta k_0 r}{\Delta k_0 r}\right)\exp[iS], \tag{14}$$

where A and B are constants. If one now chooses to reject the cosine term by making $B = 0$ on the grounds of the awkward infinity at $r = 0$, one is left with

$$u(r, t) = A\frac{\sin \Delta k_0 r}{\Delta k_0 r}\exp[iS], \tag{15}$$

this precisely defines the nondispersive wave packet built by MACKINNON [6] as generalization of (I) to four-dimensional space-time. This solution is well known to represent the superposition of two spherically symmetrical waves, one converging and one diverging [3]. In the electromagnetic case, it behaves as a phase locked cavity similar to those analysed by JENNISON [7] and which have the inertial properties of classical particles.

Now, let it be supposed that a particle, of rest mass m_0 with charge e, is travelling freely with uniform velocity v in the $+x$-direction in a region where a uniform electromagnetic potential A^μ exists. In such a case, eq. (5) becomes

$$\left(\partial_\mu + \frac{ie}{\hbar c}A_\mu\right)\left(\partial^\mu + \frac{ie}{\hbar c}A^\mu\right)u - \frac{m_0^2 c^2}{\hbar^2}u = \frac{\Box\sqrt{\varrho}}{\sqrt{\varrho}}u, \tag{16}$$

---

(6) L. MACKINNON: *Lett. Nuovo Cimento*, **31**, 37 (1981).
(7) R. C. JENNISON: *J. Phys. A*, **11**, 1525 (1978).

it can be derived from the Lagrangian

$$\mathscr{L} = \frac{1}{4}\left[\frac{u}{\bar{u}}(\partial_\mu \bar{u})\partial^\mu \bar{u})) + \frac{\bar{u}}{u}(\partial_\mu u)(\partial^\mu u)\right] - \\ - \frac{1}{2}(\partial_\mu \bar{u})(\partial^\mu u) + \frac{ic}{\hbar c} A^\mu[\bar{u}(\partial_\mu u) - u(\partial_\mu \bar{u})] - \frac{c^2}{\hbar^2 c^2} A_\mu A^\mu \bar{u}u - \frac{m_0^2 c^2}{\hbar^2}\bar{u}u + \frac{1}{4}F_{\mu\nu}F^{\mu\nu} \tag{17}$$

in conformity with the Guerra and Pusterla statements [5].

Three main implications result from physical conjectures introduced to build this mathematical model:

1) As is pointed out in the first paragraph, the model is consistent with the double-solution interpretation of wave mechanics. On account of the superposition principle, to each solution (15) can be added, with the same phase S, a solution Ψ of the linear Klein-Gordon equation and it can be shown that the guiding principle is available [8].

2) According to the stochastic interpretation of quantum mechanics, the linear waves can be considered as Brown-Markov stochastic waves on a random covariant thermostat (Dirac aether) [9,10,11]. The continuity equation $\partial_\mu J^\mu = 0$ is equivalent to a forward and backward Fokker-Planck equation, *i.e.* in Guerra-Ruggiero notation

$$\frac{\partial \varrho}{\partial \tau} + \nabla v_\pm \varrho \pm D \Box \varrho = 0 \,. \tag{18}$$

The corresponding diffusion coefficient $D = \hbar/2m$ resulting [11] from the necessity of preserving the phase coherence in the quantum jumps at the velocity of light.

The nonlinear quantum potential term can be associated to a time reversible frictional force (important in the particle region, but negligible elsewhere) in a way generalizing to Dirac's vacuum Einstein-Smoluchowski treatment of Brownian motion.

3) The fact that an interference pattern is not dependent on the observer may be explained by de Broglie's waves, but cannot be explained by Schrödinger's waves [12]. For this reason, if one accepts the physical reality of de Broglie's waves, one must consider relativistic waves, even in the nonrelativistic limit.

---

(8) PH. GUÉRET and J. P. VIGIER: *Found. Phys.*, **12**, 1057 (1982).
(9) W. LEHR and J. PARK: *J. Math. Phys.*, **18**, 1235 (1977).
(10) F. GUERRA and P. RUGGIERO: *Lett. Nuovo Cimento*, **23**, 529 (1978).
(11) J. P. VIGIER: *Lett. Nuovo Cimento*, **29**, 467 (1980).
(12) L. MACKINNON: *Found. Phys.*, **11**, 907 (1981).

Reprinted from *Essays in Honour of David Bohm*, Eds. B.J. Hiley and F.D. Peat, pp. 169-204,

9

Causal particle trajectories and the interpretation of quantum mechanics

J.-P. Vigier, C. Dewdney, P. R. Holland and **A. Kyprianidis** *Institut Henri Poincaré.*

1 Introduction

The fundamental disagreement between Bohr and Einstein at the 1927 Solvay conferences concerned not only the interpretation of quantum mechanics but also general philosophical orientations as to the nature of physical theory. Although these two aspects of the debate can never be fully separated, it is clear that, since quantum mechanics is after all a theory about the behaviour of matter, specific claims of the various interpretations can be more or less adequate in the face of experimental evidence, and even shown to be false in certain cases[1,2].

In relation to this debate perhaps the greatest significance and contribution of Bohm's causal interpretation of quantum mechanics[3] is that it not only exposes the arbitrary philosophical assumptions underlying the claims of the Copenhagen interpretation but also brings into relief the essentially new content of quantum mechanics, which is reflected in different ways in Bohr's interpretation. Indeed the claim that the quantum formalism itself requires us not only to abandon the quest for explanation of quantum phenomena but also the concepts of causality, continuity and the objective reality of individual micro-objects, is shown to be false. However the existence of the single counter-interpretation proposed by Bohm constitutes sufficient grounds for rejecting the absolute and final necessity of complementary description and indeterminacy, along with the inherent unanalysable and closed nature of quantum phenomena.

This in itself was a major contribution, but further than this, since the possibility of alternative interpretations is not ruled out, specific models may be proposed which allow a space-time description of individual micro-events and the possibility of a deeper understanding, perhaps leading to an approach which transcends current perceptions.

Although the causal interpretation has in effect been in existence since the very beginnings of quantum mechanics, in the form of the pilot-wave model proposed by de Broglie, it has not been widely adopted in the physics community, perhaps for reasons more ideological and metaphysical than physical, and many people remain ignorant of it.

In this contribution we wish to reconsider Bohm's interpretation in the light of recent developments and to return to the question of the interpretation of quantum mechanics. In particular we examine the adequacy of the Copenhagen interpretation (CIQM), the causal stochastic interpretation (SIQM – an extension of Bohm's original approach) and the statistical interpretation in accounting for quantum interference phenomena and quantum statistics. We further demonstrate that the assumption of the existence of particle trajectories entails the elimination of negative probabilities from quantum mechanics.

Such phenomena are at the heart of quantum mechanics and interference experiments were crucial in the early stages of the Bohr–Einstein debate, in which the discussion was centred on the two-slit experiment. In fact they have become of central concern once again since the recent neutron interferometry experiments present more strikingly the same puzzling behaviour, and offer wider possibilities to examine the adequacy of the various interpretations. However, let us consider first the three interpretations of the two-slit experiment.

2 Interpretation and the two-slit experiment

(i) Bohr and the Copenhagen interpretation

Although several versions of this approach to quantum mechanics exist, the most consistent and coherent version was formulated by Bohr.[4] For Bohr science is only possible through unambiguous communication of results. A necessary condition for this is that a clear distinction be possible between subject and object (system and instrument). The concepts and language of classical physics automatically entail such a distinction and we have to communicate within its framework. If in any situation the subject/object distinction can be made in alternative ways, then the descriptions arising are to be termed complementary rather than simply contradictory. Bohr regarded complementarity as a general relationship evident in all areas of knowledge. The similarity between diverse domains regarding complementary description was not based on a more or less vague analogy, but on a thorough investigation of the conditions for the proper use of our conceptual means of expression. Complementarity, then, is not derived from quantum mechanics; it simply has a well-defined application in this area of knowledge where the existence and indivisibility of the quantum of action implies the unanalysability of the interaction between system and instrument. The placing of the division between system and instrument becomes arbitrary and unambiguous communication impossible. From Bohr's philosophical position the only possibility is to retreat to the classical description of the results of experiments. Their classical nature is taken as given, but then quantum phenomena become hermetically sealed. The fundamental unit for description in these terms is then the whole 'phenomenon' constituted by the system and experimental apparatus which together form an indivisible and unanalysable whole. Altering a part of the apparatus in order to define more closely the quantum process, by elucidating a conjugate quantity, simply produces a complementary phenomenon. In this view, 'There is no quantum world, there is only an abstract quantum physical description'[5].

Quantum mechanics only concerns the statistical prediction of the results of well-defined experiments and nothing more; it represents an ultimate limit to our knowledge. The wave function ψ is the most complete description of an individual state; it is merely a probability amplitude which states the odds on various results and is subject to instantaneous changes during measurement. If some preparation device (source, shutter, collimator) is designed to produce a wave packet, then all we can say is that the wave packet represents the fact that a single particle has a probability of appearing at a position x given by $|\psi(x)|^2$, if a measurement is made. Until such a time it is not legitimate even to conceive of a particle, let alone its properties.

In the specific case of the two-slit experiment (see Figure 9.1), what happens between source and screen when interference is observed cannot be described, even in principle. In fact the quantum system, detected at the plate, cannot even be said to have an existence in the usual sense. There is no possibility of defining the process giving rise to the interference pattern. Either we design an apparatus to observe interference, and hence the wave properties of matter, and forgo the

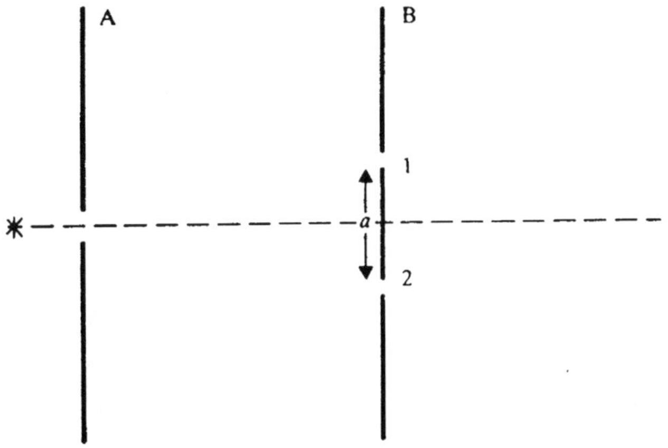

description in terms of space-time co-ordination, or we design an incompatible arrangement to determine more closely the space-time motion, particle properties, and forgo the possibility of observing interference. The two are complementary phenomena.

When Einstein[6] proposed a *gedanken* experiment which would enable the path of the particle to be determined by measuring the momentum it transfers to the slits, Bohr argued that if a screen is to be used in this way then its own momentum must be controlled with such a precision that by application of the uncertainty relations its position becomes uncertain by an amount sufficient to destroy the interference. It is a curious fact that in order to arrive at this conclusion Bohr must assume rectilinear particle trajectories between source and slits and slits and screen. The quantum object behaves classically whereas the macroscopic slit system behaves quantum mechanically. Indeed, to be consistent it must be said that the screen actually has no position; its existence has become 'fuzzy', not that its definite position is just unknown. Greenberger[7] has shown in detail how the interaction with such a 'fuzzy' object (in the neutron case) destroys the coherence of the overlapping wave functions.

In proscribing the possible in quantum physical description, Bohr has ruled out explanations in terms of determinate individual physical processes taking place in space and time. This is not the task of physics; the quantum theory is just an algorithm for predicting results and its theoretical entities need no interpretation. In this way, by epistemological re-definition, Bohr can avoid all the problems and paradoxes which arise when an attempt is made to provide the formalism and its rules with a physical interpretation in terms of the behaviour of matter. In Bohr's view the 'observer' plays no more special a role in quantum mechanics than in any other area of knowledge, and his or her consciousness of a given situation has no special effect. This is the core of Bohr's position and the unambiguous basis of the Copenhagen interpretation. Many other versions of the Copenhagen interpretation exist and these have led to extended discussions as a result of attempting to provide a physical, or psycho-physical, interpretation of the entities and laws of quantum mechanics, in terms of which the phenomena and the interphenomena may be described and explained. These should really be distinguished from Bohr's position, which does not constitute a physical interpretation in the usual sense.

In the following we separate Bohr's position from those versions of CIQM which attempt to interpret the formalism physically.

(ii) The statistical interpretation

As emphasized by Ballentine[8], the statistical interpretation is to be distinguished from the Copenhagen interpretation. He asserts that the wave function simply represents an ensemble of similarly prepared systems and does not provide a complete description of an individual system: 'In general, quantum theory predicts nothing which is relevant to a single measurement.'[8]

The interpretation of a wave packet is that, although each particle has always a definite position r, each position is realized with relative frequency $|\psi(r)|^2$ in an ensemble of similarly-prepared experiments. It follows that each particle has a well-defined trajectory, but its specification is beyond the statistical quantum theory; probabilities arising in the predictions of the theory are to be interpreted as in classical theory.

In the two-slit experiment this means that each particle in fact goes through one or other of the slits. Clearly the interference of particles is something new in quantum theory which this model must reproduce. If the particle goes through one or other of the slits, the two possibilities are in principle distinguishable; we should write a mixture instead of a pure state and the interference disappears. In order to explain the persistence of interference in this interpretation Ballentine refers to the work of Duane[9] in 1927, more recently revived by Landé[10]. The result is obtained by considering the possibilities for momentum transfer between the individual particle and the screen containing the slits. The matter distribution of the screen is Fourier analysed into a 'three-fold infinity of sinusoidal elementary lattices of spacings l_1, l_2, $l_3 \ldots$ and amplitude $A(l_i)$'. According to an extension of the Bohr-Sommerfeld quantum conditions, each such lattice is capable of changing its momentum in the direction of the periodicity only by amounts;

$$\Delta p_i = h/l_i$$

The intensity of an l component in the harmonic analysis is proportional to the statistical frequency of the corresponding momentum transfer. Thus each particle does not simply interact locally with the screen but non-locally with the matter distribution of the screen as a whole. Now a change in this matter distribution, e.g. closing a slit, results in an instantaneous change in the components of the harmonic analysis and thus in a corresponding change in the possible momentum transfers, resulting in a single-slit distribution of intensity. We are bound to ask what, in this analysis, determines which of all the possible momentum transfers actually occurs in the individual particle's passage. There is no answer and so individual events are inherently statistical.

It is not clear in this model why the matter distribution consisting the screen should be Fourier analysed but not the matter distribution which constitutes the particle. The screen is, after all, made of particles. The physical status of the Fourier components which exist with certain amplitudes is also unclear. Einstein originally denied that the wave function gives a complete description of an individual because he saw that this assumption contradicted the notion of locality[11]. If we assign the wave function only a meaning in a statistical ensemble and resort to the above arguments to explain interference and diffraction, then clearly non-locality is introduced, but in a way which is not intuitively clear.

If we reconsider Einstein's modification of the two-slit experiment in this model then we see that the meaning of the uncertainty in the position of the slits, resulting in the loss of interference, is to be interpreted differently. In each experiment the screen has a definite position (this position has a statistical dispersion in the ensemble, Δx) and so in each individual case the particle is forbidden to land in the positions of the minima of the pattern. (This incidently is a definite prediction for the outcome of an individual experiment, in contradiction to Ballentine's statement above.) However, because the position of the maxima and minima in each case is different, in the ensemble interference is lost – a different explanation to that of the Copenhagen interpretation but with the same results.

The statistical interpretation claims to be a minimal interpretation which removes the 'dead wood' of the Copenhagen interpretation. However, nothing is gained in the understanding of the quantum world and the mysteries remain complete.

(iii) The causal interpretation

This interpretation was originally proposed by de Broglie[12] and independently by Bohm[3]. It has recently been extended by Vigier[13] to include a sub-quantum Dirac ether as an underlying physical model.

In fact it is the only known interpretation of quantum mechanics in terms of which all quantum effects can be explained on the basis of causal continuous motions in space and time. The quantum mechanical description of an individual through the wave-function is held to be incomplete in the sense of Einstein. The description may be supplemented with real physical motions of particles, without ambiguity or contradiction, but in a manner which introduces severely non-classical features. Bohr's epistemological position is set aside and the task of physics is held to consist not only of the attempt to predict the statistical frequency of results in ensembles but also to provide explanations and descriptions of the individual processes between source and detection. No problems arise in the analysis of 'phenomena' into constituent parts, as the essential feature of their unity is now manifested by the quantum potential which arises from the non-locally correlated stochastic fluctuations of the underlying covariant ether (see Introduction). The quantum potential exhibits radically new properties. In the single-particle case its form depends on the state of the system as a whole, a feature which is the analogue of its non-local character in the many-body system[14].

Clearly if we consider that individual particles really exist in the interphenomena between source and screen and follow determinate trajectories, then the motion of each particle must be inextricably linked with the structure of its environment. Any change in the apparatus affects the whole ensemble of possible trajectories. This undivided connection is mediated by the quantum potential which arises as an extra potential term in the Hamilton-Jacobi-like equation, which may be derived by substituting $\psi = Re^{iS/\hbar}$ in the Schrödinger equation and separating the real and imaginary parts, as Bohm did in 1952. In addition to the Hamilton-Jacobi equation:

$$-\frac{\partial S}{\partial t} = \frac{(\nabla S)^2}{2m} + V - \frac{\hbar^2}{2m}\frac{\nabla^2 R}{R} \qquad [1]$$

one finds a continuity equation with $P = R^2$:

$$\frac{\partial P}{\partial t} + \nabla\cdot\left(P\frac{\nabla S}{m}\right) = 0 \qquad [2]$$

R and S are interpreted as the amplitude and phase of the real ψ field. The possible real average motions of a particle may be represented by trajectories derived from the relation that the particle momentum is given by:

$$\bar{p} = \nabla S \qquad [3]$$

In the many-body case, particle motions are correlated by the quantum potential in a non-local way, although in the scalar case this action at-a-distance does not give rise to any special problems. Even in the relativistic case, where non-locality may be thought to conflict with the requirements of relativistic causality, it can be shown that this connection is mediated superluminally, yet causally, and cannot lead to any results conflicting with the predictions of special relativity[15,16].

An exact calculation has been carried out in detail by Philippidis *et al.*[17] in the causal interpretation of the one-particle Schrödinger equation description of the two-slit experiment. Here we represent the form of the quantum potential and the associated trajectories in Figure 9.2 and Figure 9.3 respectively. The intensity distribution at the screen depends on the density of trajectories along with their occupation probability, and of course agrees in the Fraunhoffer limit with that expected from the usual considerations.

The precise form of each trajectory is sensitive to changes in variables describing the particle's environment. The distribution of trajectories demonstrates that each particle travelling in the apparatus

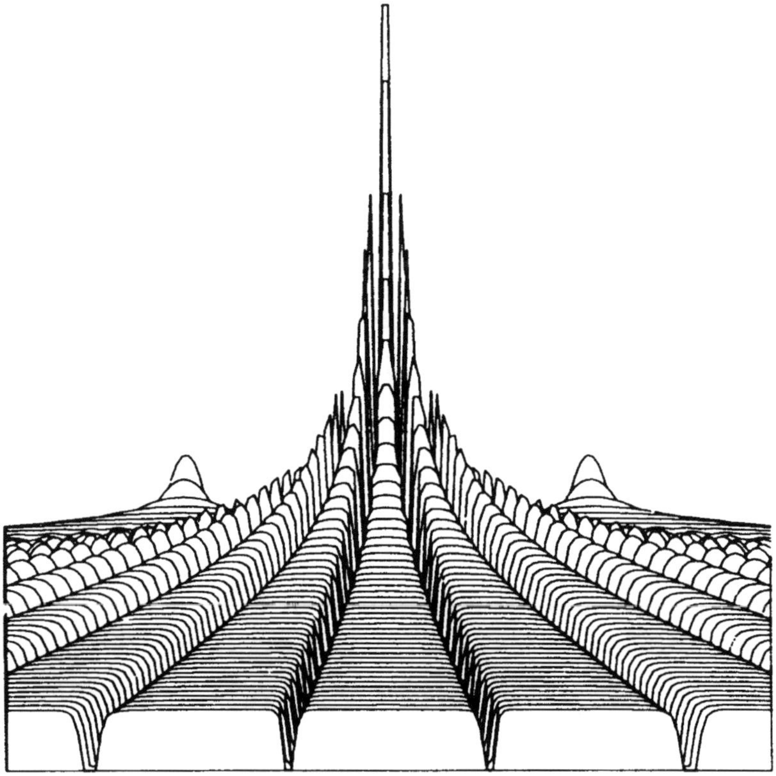

Figure 9.2 The quantum potential for two Gaussian slits viewed from a position on the axis beyond screen B.

'knows about' or responds to the global structure of its environment (e.g. the presence of two slits, not one) and so exhibits a wholeness completely foreign to mechanistic models in classical conceptions.

The quantum potential approach provides a way of understanding the feature of the quantum wholeness of phenomena emphasized by Bohr. Yet we are not required to relinquish the attempt to explain the interphenomena in terms of space-time co-ordination and causal connection simultaneously.

The unity of system and environment, so clearly demonstrated in the double-slit trajectories, is then revealed as the essentially new non-classical feature of quantum mechanics. Of course the single-particle description is an abstraction and this unity is really a reflection of the non-local character of the correlations that arise in the many-body case. The non-separability of quantum systems had been emphasized by Schrödinger[18] and Einstein, Podolsky and Rosen[11] in 1935. That such non-local correlations exist can no longer be doubted, as the

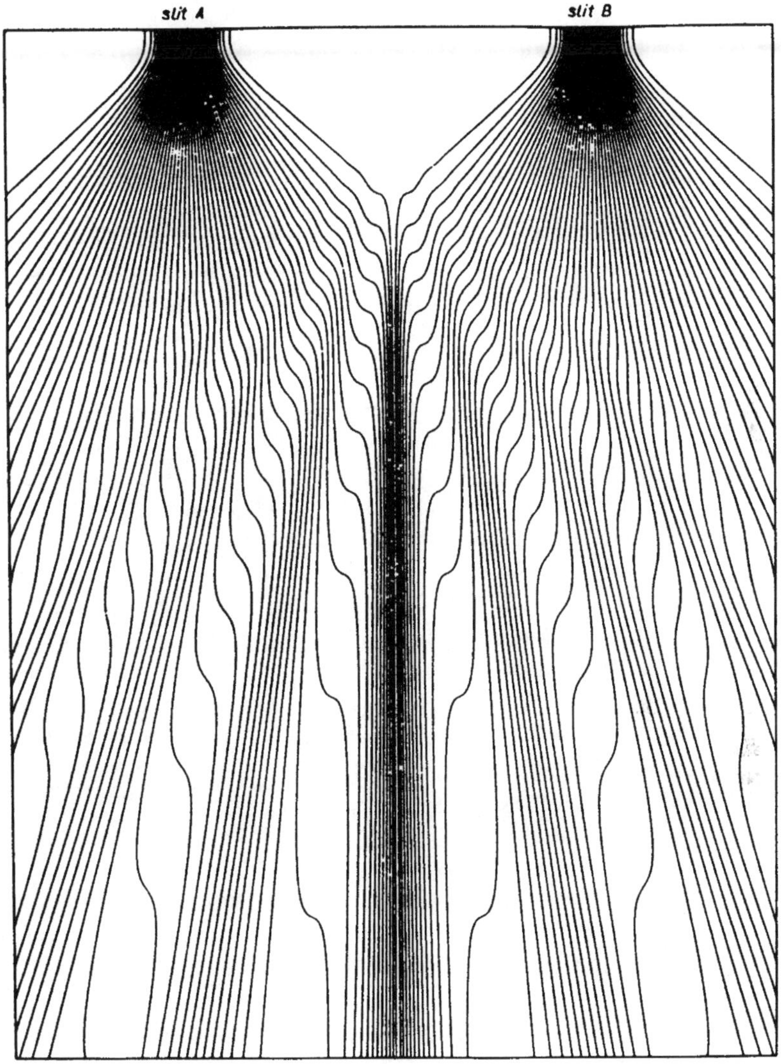

Figure 9.3 Trajectories for two Gaussian slits with a Gaussian distribution of initial positions at the slits.

results of Aspect's experiment demonstrate[19]. Indeed these experiments find a perfectly causal explanation through the quantum potential[20].

Reconsidering Einstein's modified two-slit experiment, the explanation of the loss of the interference pattern upon path determination in an ensemble of results is similar to that of the statistical interpretation, but now with precisely-definable individual particle trajectories determined by the quantum potential. None of the trajectories crosses the line of symmetry (a point confirmed by Prosser, who calculated lines of energy flow in the electromagnetic case[21]) and this is a new macroscopic prediction.

(iv) Wave-packet collapse

In the Bohr–Einstein debate, Bohr was able to defend the complementarity principle by showing that attempts to use detailed energy or momentum conservation in individual processes to determine particle trajectories more closely requires a change of the experimental arrangement which results in a loss of interference and the wave aspect. Bohr never referred to wave-packet collapse in these arguments. However, many of his supporters did and the concept is associated with his interpretation. Using the concept of wave-packet collapse to provide a physical interpretation of the quantum formalism is contrary to the spirit of Bohr's epistemological position. Nevertheless the argument is often put in the following way in CIQM. The introduction of a device capable of determining through which slit a particle passes induces a collapse of the wave function in the apparatus and a consequent loss of interference. Such a collapse follows from the assumption that the wave function provides a complete description of an individual system. In fact it is an hypothesis added to the quantum formalism and not an integral part of that formalism, although this is not the impression given in most texts. Arguments demonstrating the redundancy of this *ad hoc* postulate in quantum theory have existed for a long time.

Bohm, followed more recently by Cini[22], have argued that wave packet collapse does not correspond to any objective real physical change in the state of a system. By including the interaction with the measuring device in the quantum description, it can be shown that the interference between the components of a pure state is destroyed as a result of the interaction between the system and the macroscopic instrument, which is in effect irreversible. That is, there are no observable differences between the description of the composite in terms of the pure-state density matrix with vanishing interference terms and that in terms of the mixture density matrix (with wave-packet collapse). If we find out which component of the pure state is actualized in the process, the others may be disregarded as they can have no further influence on the behaviour of the system after the (irreversible) interaction with the macroscopic instrument has taken place. Wave-packet collapse is now simply a matter of convenience. The physical interpretation of individual quantum processes in terms of particle trajectories excludes the necessity of introducing wave packet collapse.

Consider the general case of particle interference with two wave functions ψ_{I} and ψ_{II} (this could be a two-slit experiment with ψ_{I} from slit 1 and ψ_{II} from slit 2 or an interferometer with ψ_{I} in path 1 and ψ_{II} in path 2). Also let the wave function of an apparatus introduced only in path II be φ_i initially and φ_f finally, then we have:

$$\Psi_i = \varphi_i\psi_{\mathrm{I}} + \varphi_i\psi_{\mathrm{II}} \rightarrow \Psi_f = \varphi_i\psi_{\mathrm{I}} + \varphi_f\psi_{\mathrm{II}} \qquad [4]$$

If, through its functioning the states, φ_i and φ_f become orthogonal then interference is destroyed:

$$\Psi_f^+\Psi_f = \varphi_i^+\varphi_i\psi_{\mathrm{I}}^+\psi_{\mathrm{I}} + \varphi_f^+\varphi_f\psi_{\mathrm{II}}^+\psi_{\mathrm{II}} \qquad [5]$$

and the system (neutron, photon, electron) acts as a particle that goes either on path I or path II. Observation of the measuring instrument merely tells us which alternative took place and thus we replace Ψ_f by $\varphi_i\psi_{\mathrm{I}}$ or $\varphi_f\psi_{\mathrm{II}}$. This is a collapse of the wave function which simply represents a change of our knowledge and does not correspond to any real physical changes in the state of the system. If φ_i and φ_f are not orthogonal then interference persists:

$$\Psi_f^+\Psi_f = \varphi_i^+\varphi_i\psi_{\mathrm{I}}^+\psi_{\mathrm{I}} + \varphi_f^+\varphi_f\psi_{\mathrm{II}}^+\psi_{\mathrm{II}} + \varphi_i\psi_{\mathrm{I}}\varphi_f^+\psi_{\mathrm{II}}^+ + \varphi_f\psi_{\mathrm{II}}\varphi_i^+\psi_{\mathrm{I}}^+ \qquad [6]$$

and the system acts as a wave in both paths.

If by observing the apparatus we could still in fact determine the path of the particle, then in CIQM the act of observation would have to cause real physical changes in the particle's state as a consequence of a wave-packet collapse, since, if neutrons, electrons and photons are conceived as *particles* that go one way or the other, equation [6] should reduce to equation [5].

Thus CIQM concludes that all measurements capable of determining the particle's path imply orthogonality of the apparatus wave functions initially and finally. In SIQM, determination of particle path need not imply orthogonality of apparatus wave functions in order to exclude the intervention of consciousness in physical processes. What appears as a 'pseudo-collapse' is the action of a macroscopic measuring device which makes the interference terms negligible, as is consistently shown by Cini in a detailed application of the time-dependent Schrödinger equation to the interaction between a system and a measuring device. Thus there is no *a priori* impossibility of path determination and persisting interference; one has only to find an appropriate measuring device that during an interaction with the micro-system does not undergo a change to an orthogonal state, i.e. preserves the interference terms, and still offers a possibility to decode this small quantum-number change. The use of SQUIDS and no-demolition measurements could be considered in this context.

The possibility to go beyond CIQM was known to Bohm, in his original paper[3], albeit in a different context. He remarks, in relation to the loss of interference properties on measurement: 'In our interpretation, however, the destruction of the interference pattern could in principle be avoided by means of other ways of making measurements.'

3 Neutron interferometry

We propose now to take up these questions in relation to another specific quantum interference situation, neutron interferometry[23]. Neutron interferometry has the advantage that it reproduces the double-slit configuration with massive particles and introduces new possibilities for interaction through the neutron spin, thus essentially altering the situation.

(i) Spatial interference

To this purpose consider the experimental arrangement of Figure 9.4 with both spin flippers turned off. A simple calculation shows[24] that if an originally spin-up polarized beam $\psi = |\uparrow_z>$ enters the interferometer, it is subdivided in two partial beams $\psi_I = e^{i\chi}|\uparrow_z>$ and $\psi_{II} = |\uparrow_z>$ that successively recombine and yield an intensity interference behind the interferometer modulated with the phase shift factor χ:

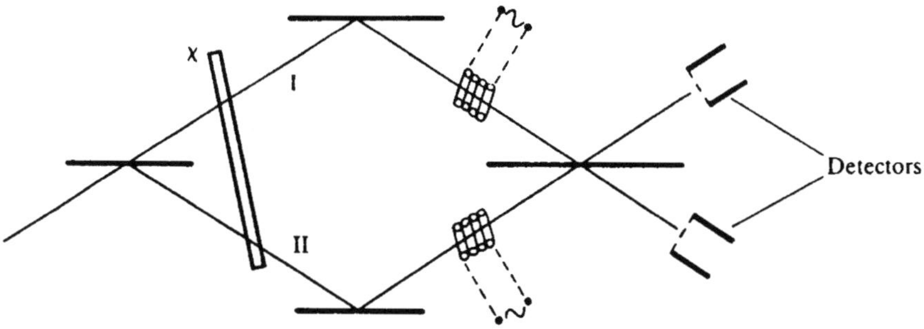

Figure 9.4 The neutron interferometer with a spin-flip coil in each arm.

$$I = (\psi_I + \psi_{II})^+(\psi_I + \psi_{II}) = 2(1 + \cos\chi) \qquad [7]$$

while the polarization remains in the z-direction, i.e.:

$$P = (0,0,1) \qquad [8]$$

Let us now turn to consider the interpretation of the results. For Bohr, the concept of the interphenomena cannot be unambiguously applied; the phenomenon is unanalysable. All we can do is calculate the interfering probability amplitudes associated with each path through the apparatus. Neutron paths and 'neutron waves' are equally ambiguous for Bohr. He states that descriptions in terms of photons and electron waves have the same ambiguity as other pictorial descriptions of the interphenomena; only the classical concepts of material particles and electromagnetic waves have an unambiguous field of application. However, if we wish to discuss the interphenomena we must set Bohr's position aside (as in fact most physicists actually do). Then a clear choice exists between two possible explanations. Either:

1 we say the neutron does not exist as a particle in the interferometer; or
2 we say the neutron actually travels along path I *or* II only, but is influenced by the physical conditions along both.

(a) The CIQM Suppose, with the usual interpretation of the quantum formalism, a particle were actually to travel along one path, then the existence of the other would be irrelevant and interference cannot occur. Interference arises not from our lack of knowledge of the path but from the fact that the neutron does not have one. Any attempt to reveal the particle between source and detector induces a wave-packet collapse, i.e. localizes the particle in one beam, and interference effects disappear. The wave and particle nature of matter are complementary aspects. Since in this view the neutron is not to be conceived of as a particle before detection localizes it, questions concerning which beam a given neutron enters at the region of superposition cannot be formulated and the question of explanation is summarily closed.

(b) The causal stochastic interpretation If, contrary to the usual interpretation outlined in (a), we believe with Einstein[25] and de Broglie[12] that neutrons are particles that really exist in space and time, then Rauch's statement, ruled out in the CIQM, can be made; namely that: 'At the place of superposition every neutron has the information that there have been two equivalent paths through the interferometer, which have a certain phase difference *causing the neutron* to join the beam in the forward or deviated direction'[26].

It is then possible to suggest physical models to explain the causation of individual events, a non-existent option in CIQM. In the SIQM neutrons can be thought of as particles accompanied by waves simultaneously; the particle travels along one path through the interferometer whilst its real wave is split and travels along both. The waves interfere in the region of superposition and give rise to a quantum potential which carries information concerning the whole apparatus and determines the particle trajectories. The changing phase relations between the waves in I and II lead to a changing quantum potential structure that determines which beam each individual neutron enters according to its initial position in the wave packet and phase shift χ. The detailed explanation provided for the two slit experiment and square potential phenomena[27] may be easily extended to this case. The details may be found in reference 28; here we simply represent the form of the effective potential (quantum potential plus classical barrier) and the associated trajectories. The results of the numerical calculation show that varying the phase shift factor χ between 0 and 2π produces the correct type of interference figure. When $\chi = 0, \pi$, the effective potential (quantum + classical), as shown in Figure 9.5, is symmetric about the barrier centre. A series of violent oscillations develops on each side of the barrier potential. These arise when the incident wave interferes with the combination of its own reflected wave and the in-phase transmitted wave from the other side. Figure 9.6 shows the associated trajectories.

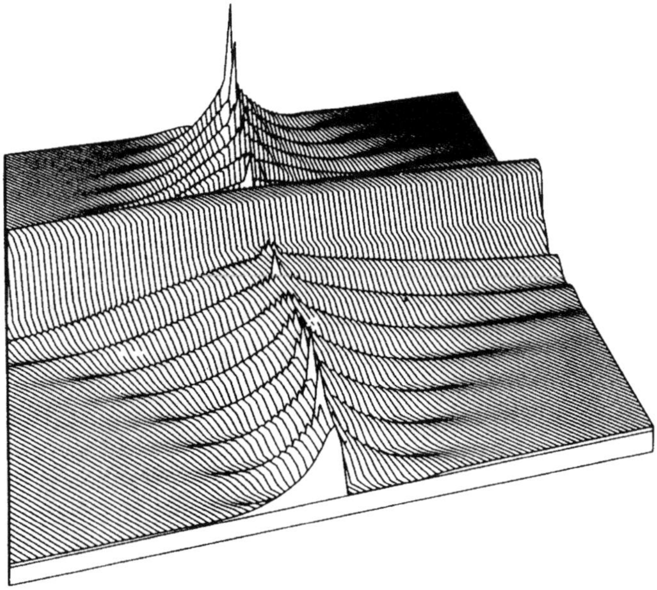

Figure 9.5 The effective potential at the last 'set of crystal planes' with phase shift π. Corresponding to region $x = 0.6$ to $x = 0.9$, $T = 4.0$ to $T = 12.0$ on trajectory plot.

With $\chi = \pi/2$ the situation is very different. In this case the quantum potential oscillations are greatly reduced on one side of the potential barrier, in the region where the density of trajectories is large, and this allows the particles to be transmitted (see upper section of Figure 9.7). In the lower section notice that the quantum potential oscillations are enhanced and occur at an earlier time, ensuring that all the trajectories constituting beam II are reflected (Figure 9.8). Those constituting beam I now enter the potential barrier and emerge after the reflection of those in beam II, both forming the single emerging beam. In this case the reflected wave from beam I is (almost completely) cancelled by the anti-phase transmitted wave from beam II.

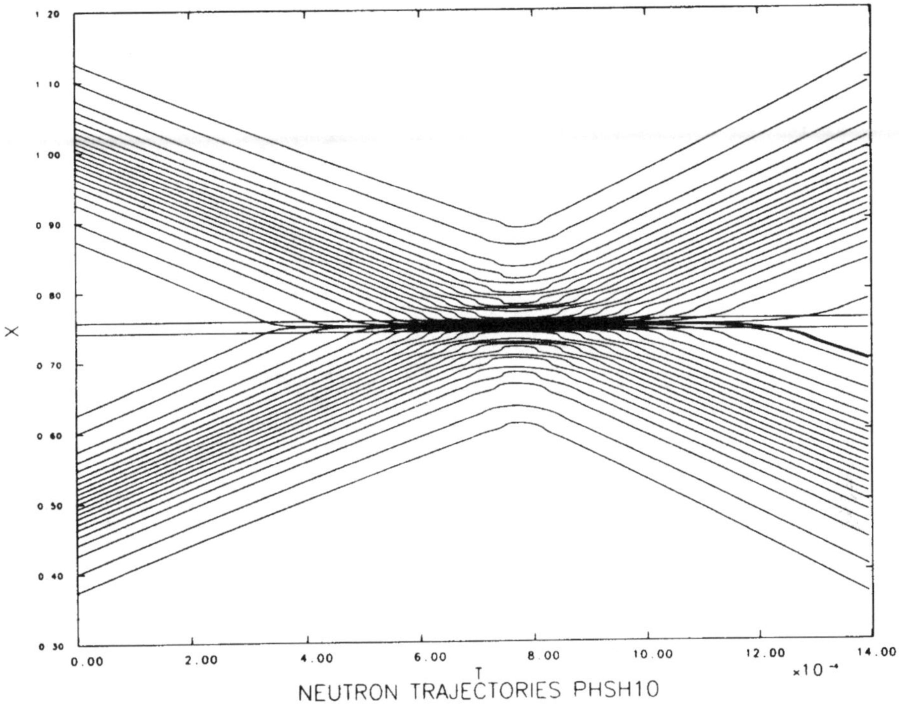

Figure 9.6 Trajectories associated with $\chi = \pi$.

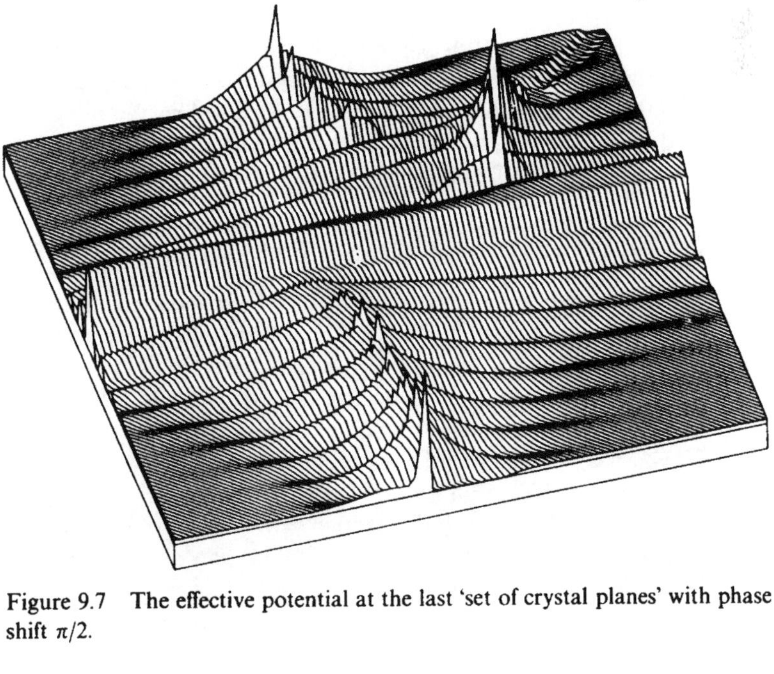

Figure 9.7 The effective potential at the last 'set of crystal planes' with phase shift $\pi/2$.

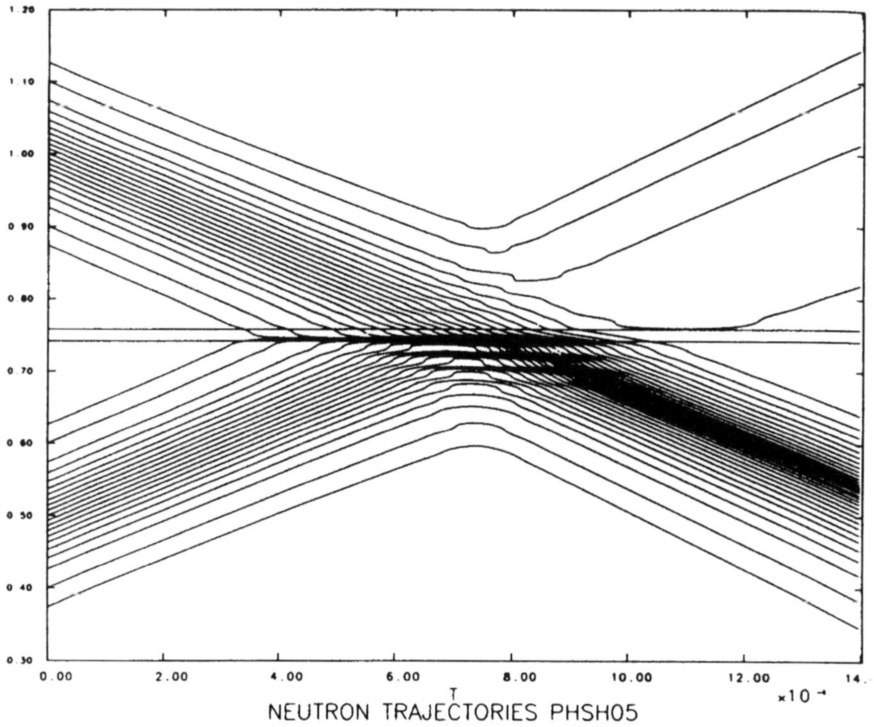

Figure 9.8 Trajectories associated with $\chi = \pi/2$.

When $\chi = 3\pi/2$ the situation is essentially reversed (see Figures 9.9, 9.10), all the trajectories and any neutron emerging in the upper section. The few trajectories which do not follow the others come from the extreme tails of the packets and so have very low probability; here they represent the effect of a finite potential width.

(ii) Time-dependent spin superposition

Now according to Badurek *et al.*[29] a completely different physical situation arises in the case of the time-dependent superposition of linear spin states using a radio-frequency spin-flip coil. Indeed, 'in that case the total energy of the neutrons is not conserved'. The detailed experimental arrangement can be schematically represented as follows. The incident neutron beam containing one neutron at a time is subsequently divided into beams I and II. On beam I acts a nuclear phase shifter represented by the action of a unitary operator $e^{i\chi}$ on ψ. Beam II is subjected to the following combination of magnetic fields:

1 a static magnetic field in the $+z$ direction $B = (0,0,B_0)$;

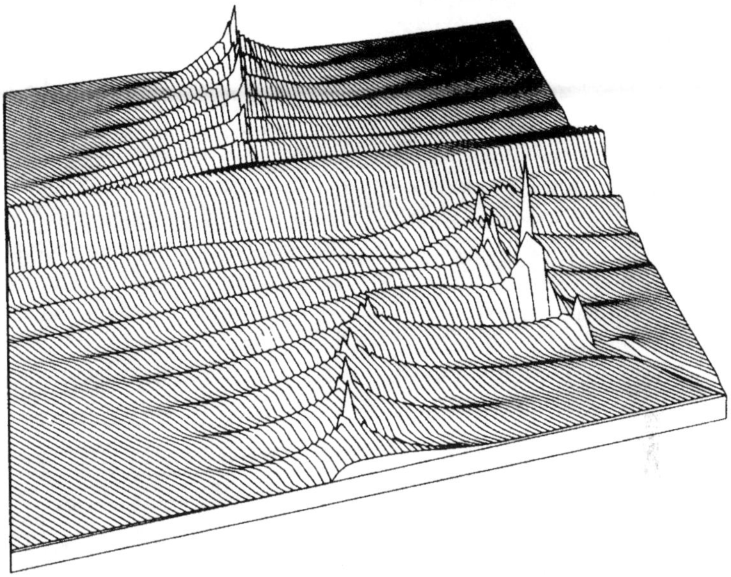

Figure 9.9 Effective potential with phase shift $\chi = 3\pi/2$.

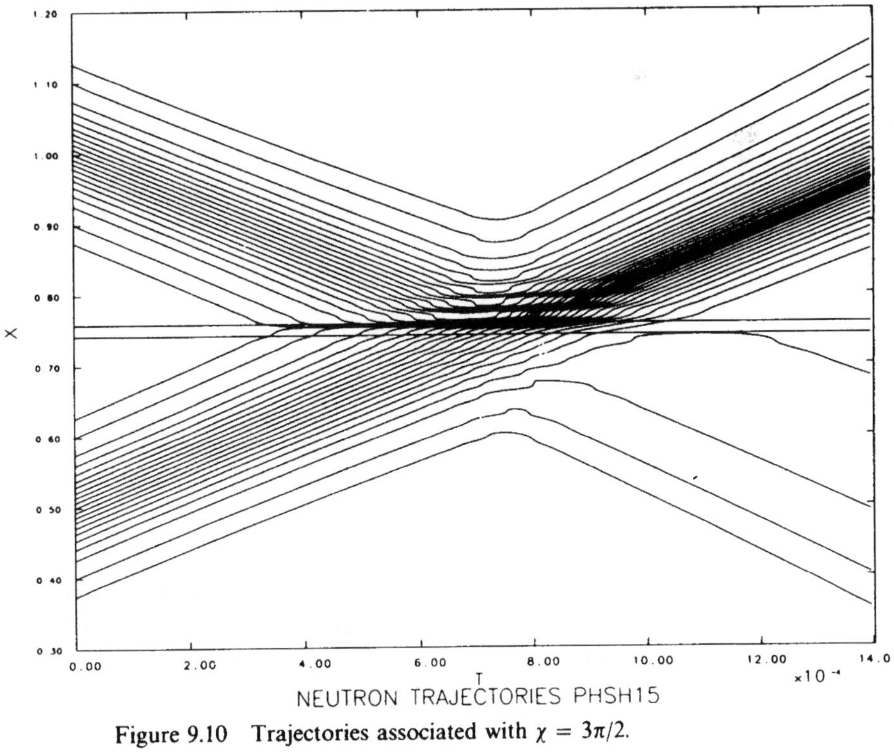

Figure 9.10 Trajectories associated with $\chi = 3\pi/2$.

2 a radio-frequency time-dependent magnetic field $B = (B_1 \cos \omega_{rf}t, B_1 \sin \omega_{rf}t, 0)$ rotating in the xy plane with a frequency ω_{rf} obeying the resonance condition, $\hbar\omega_{rf} = 2\mu B_0$, where μ is the magnetic moment of the neutron, i.e. it yields exactly the Zeeman energy difference between the two spin eigenstates of the neutron within the static field.

Neutrons passing through such a device (a spin flipper) reverse their initial $+z$ polarization into the $-z$ direction, by transferring an energy $\Delta E = 2\mu B_0$ to the coil whilst maintaining their initial momentum

The wave function in beam I after passing through the phase shifter is:

$$\psi_{\mathrm{I}} = e^{i\chi}|\uparrow_z\rangle = e^{i\chi}\binom{1}{0} \qquad [9]$$

The corresponding wave function in II after the coil should be written:

$$\psi_{\mathrm{II}} = e^{i\frac{\Delta E_t}{\hbar}}|\downarrow_z\rangle = e^{i\frac{\Delta E_t}{\hbar}}\binom{0}{1} \qquad [10]$$

since the rf-coil is shown to be almost 100 per cent efficient[30].

Let the wave function of the coil initially be φ_i and finally be φ_f. Then initially the wave function of the whole (neutron and coil) is:

$$\Psi_i = \varphi_i(a\psi_{\mathrm{I}} + b\psi_{\mathrm{II}}) \qquad [11]$$

and the final state is:

$$\Psi_f = \varphi_i a\psi_{\mathrm{I}} + \varphi_f b\psi_{\mathrm{II}} \qquad [12]$$

and the condition for the observation of interference is $\varphi_i \approx \varphi_f$; that is, the state of the coil is virtually unaltered and no measurement in the usual sense takes place. Then:

$$\Psi_f = \varphi_i(a\psi_{\mathrm{I}} + b\psi_{\mathrm{II}})$$

and intensity:

$$I = \Psi_f^+\psi_f = 2$$

with polarization:

$$\bar{p} = (\cos(\omega_{rf}t - \chi), \sin(\omega_{rf}t - \chi), 0) \qquad [13]$$

entirely in the xy plane. These are the well-known results of Badurek *et al.*[29] which are experimentally verified.

(a) The Copenhagen interpretation Now how are these results encompassed within the CIQM? The observation of interference implies the wave aspect; hence the particle cannot even be said to exist during the time between emission and absorption in the detector. A particle cannot exist in one beam (or pass through one slit in the double-slit experiment) and take part in interference. However in order to describe the functioning of the coil we must use the complementary localized particle aspect. The energy transfer that takes place giving rise to the change of ψ_{II} is described by Rauch in terms of photon exchange between the neutron and the field in the coil. Thus the neutron is conceived as a particle in one beam to explain energy transfer and simultaneously as a wave existing in both to explain interference. The complementarity of wave and particle descriptions is broken; both aspects must be used simultaneously in one and the same experimental arrangement. Complementary description is thus incomplete, or can energy be exchanged with a probability wave?

(b) The causal stochastic interpretation In the SIQM we use the Feynman–Gell-Mann equation for spin half-particles as a second-order stochastic equation for the collective excitations of the assumed underlying covariant random vacuum, Dirac's ether[31,32]. A spin half-particle is conceived as a localized entity surrounded by a real spinor wave due to perturbation of the vacuum. While the particle really travels one way (path I or II), the spinor wave propagates in both paths. In path II the interaction with the rf spin-flipper inverts the spinor symmetry of the wave while in path I the initial state is maintained. What happens in the interference region can be now represented by the action of a spin dependent quantum potential Q and a quantum torque τ which can be shown to produce a time-dependent spinor symmetry in the xy plane. The particle travelling, for example, in path I is constrained by the spinor symmetry in the interference region and its $+z$ spin is twisted into the xy plane by the quantum torque. If it travels along path II it suffers an additional spin inversion due to the rf coil, yielding this energy to the coil while in the intersection area its $-z$ spin is twisted again to the xy plane. Consequently, a coherent picture is established which accounts for both particle and wave aspects.

(iii) Measurement and time-dependent spin flippers

Now, Badurek *et al.*, who performed the experiment, have stated: 'This experiment has shown explicitly that the interference properties of beams can be preserved even when a real energy exchange takes place, which is intuitively a measurement'[29]. But does it constitute a measurement? The situation must be analysed carefully.

Clearly if the functioning of the coil is a measurement in the quantum mechanical sense, then φ_i would be orthogonal to φ_f and interference would disappear. The reasons for considering such an interaction as a quantum measurement process are the following. First, there is an energy exchange taking place unidirectionally from the passing neutron to the rf-circuit, since the energy of the initial state differs by ΔE from that of the final state. This energy exchange, if decoded and extracted from the resonator circuit, could reveal the passage of the neutron. Second, this energy transfer in the form of a photon transition establishes a one-to-one correspondence between the change of the neutron's spin state from spin up to spin down.

If φ_i and φ_f are not orthogonal, then interference and spin superposition persist, but in order to demonstrate the coexistence of particle path and interference some means must be found to decode the small quantum number change involved between φ_i and φ_f.

If, in spite of the non-orthogonality of φ_i and φ_f, the particle path could be observed, then in CIQM the act of observation itself (i.e. our knowledge) would have to destroy the interference terms (wave-packet collapse), whereas this is not ruled out *a priori* in the quantum potential approach, in which particle path and interference are not exclusive and the wave function of the apparatus does not provide a complete description of an individual apparatus. Within the quantum potential approach one could consider, as we have suggested, the possible adaptation of quantum non-demolition measurements to detect the passage of a neutron.

Does the possibility exist of detecting the passage of a neutron from the energy it transfers to the rf coil? For Bohr the question of an individual energy transfer to the coil when it is part of the interferometer set-up cannot arise, as this amounts to an attempt to subdivide the experiment. Actually changing the experiment to allow the detection of the energy transfer results in a complementary phenomenon in which interference would not be observed. In the quantum potential approach there is no contradiction between energy transfer and interference. Consider the following possibilities.

1 If the single energy transfer is detectable with certainty by inspecting the coil's state, then its final state must not overlap with the initial state. However the addition of a single photon to the field in the coil does not, even under the most favourable assumptions concerning the state of this coil, lead to any observable change (consider the field to be in a coherent state, for example).

2 If it is possible, by introducing a superconducting quantum interference device (SQUID)[33], to detect the exchange of a single photon, then according to the usual application of the quantum formalism the implied orthogonality of SQUID states destroys the interference. This experiment, if performed, would, if interference is not observed, confirm the non-separability of neutron and SQUID states. If on the other hand interference persists, then the experiment would contradict quantum non-separability; the SQUID state would be decoupled from the neutron state.

3 If a single energy transfer is not detectable, can some device be added that stores the individual unidirectional energy transfers eventually leading to a detectable amount? Is so then this energy can only have come from the passage of individual neutrons

through the coil, which implies that each individual neutron actually travels along one or the other of the paths through the interferometer *and* takes part in interference.

(iv) Energy conservation and the double-coil experiment

Consider now the situation with two time-dependent rf coils, one in each beam[34]. In this proposed experiment the doubts raised by Badurek *et al.*[29] concerning the phase-number uncertainty do not apply at all[34], since the resulting interference pattern is stationary. No theoretical objection arises for a possible detection of single photon transitions in the field of the rf spin flipper on this count.

Ignoring all common phase factors, the wave functions of the superposed beams are:

$$\psi_{I} = e^{i\chi}|\downarrow_z\rangle \qquad [14]$$

$$\psi_{II} = |\downarrow_z\rangle \qquad [15]$$

with polarization:

$$P = (0,0,-1) \qquad [16]$$

and:

$$\Psi_f^+\Psi_f = 2(1 + \cos\chi) \qquad [17]$$

Spatial interference is recovered. The results of this single apparatus are:

1 each emerging neutron has lost an energy ΔE; its spin is now 'down' in the guide field;
2 each neutron takes part in interference.

In order to explain the measured loss of energy, the neutron must pass as a particle through one or the other coil and exchange a photon with the field. In the quantum potential approach this is consistent with the observation of interference; in a description based on wave/particle duality it is not, as both particle (loss of energy) and wave (interference) properties must be manifest simultaneously in one and the same experimental arrangement.

A measurement of the polarization of the neutron behind the interferometer reveals that each neutron has suffered a spin flip. Each emerging neutron has lost an amount of energy ΔE where $\Delta E = 2\mu B$ represents the Zeeman splitting. If energy is to be conserved this energy must have gone to one or other of the coils. This is only possible if the neutron passes as a particle through one or other and gives an indivisible photon of energy $E = \hbar\omega_{rf} = \Delta E$ to the rf-field. The spatial interference can only be explained by assuming that the neutron does not pass through one or other of the coils.

Since here both interference and spin direction can be measured simultaneously, according to CIQM the neutron actually travels path I or II and at the same time does not exist as a particle at all.

In the Bohr–Einstein debate the application of particle momentum conservation in individual events always led to the consistency of CIQM. Here the energy conservation leads to the inconsistency of CIQM, since wave and particle aspects must be used together to explain the observed results. If it is insisted in CIQM that neutrons do not travel one way or the other in this experiment, no energy can be transferred to the coils and then there is no conservation of energy in individual events. Further, if a statistical ensemble of individual neutron passages is considered, we see that, even there, there is no conservation of energy in CIQM when the interference is observed.

We are confronted by a stark alternative. Either:

1 we renounce any possibility of describing what happens in the neutron interferometry experiments; there exists then no possibility of explaining quantum phenomena, not even in terms of a wave/particle duality which only leads to ambiguity; individual quantum phenomena are in principle and irreducibly indeterminist in character and there can be no form of physical determinism appropriate in the quantum domain; or

2 we adopt the quantum potential approach as the only known consistent manner in which the quantum world can be conceived and explained in terms of a physically determinist reality; then, even if the quantum potential approach is not taken as the finally satisfactory description of quantum mechanical reality, it at least shows in a clear way the features that such a description must entail.

Consider the question of energy conservation in a more general way in SIQM and CIQM. In CIQM, as emphasized by Bohr, we may only consider the energy of a system to be definite when the system is in a stationary state. The system may only be in a stationary state in the absence of perturbing forces, such as those necessarily introduced in a measuring process. Such interactions are necessary to localize the system in order to allow a space-time description. Thus, in a transition between stationary states, energy conservation can be applied to the initial and final states but this excludes the conditions necessary for a space-time description. Bohr would say that when we are in a position to speak of space-time location there can be no question of energy conservation and, when energy conservation can be applied, the concepts of space-time co-ordination lose their immediate sense.

In SIQM a stationary state means the particle energy given by $\partial S/\partial t$ is a constant. The quantum potential is time independent and the particle's motion is conservative in that:

$$\frac{(\nabla S)^2}{2m} + Q = -\frac{\partial S}{\partial t} = \text{constant}$$

The particle can gain or lose kinetic energy at the expense of quantum potential energy. For example in an S state of an H atom, the particle is stationary with energy E, this energy being held as quantum potential energy whilst the quantum force $-\nabla Q$ balances the Coulomb force $-\nabla V$. In a different example, when ψ is a plane monochromatic wave, the particle energy is a constant since the quantum potential vanishes. A more complex case is that of the double-slit experiment discussed above.

For stationary states and systems which undergo changes between them, conservation of energy may be established in both SIQM and CIQM, but SIQM can also provide a space-time description.

The case in which Ψ is a superposition of stationary states is rather different, as discussed by Bohm[3,35] and de Broglie[12,36]. Consider the state:

$$\Psi = \sum_K C_K \psi_K \quad [18]$$

where Ψ is a sum of stationary states with energy E_k.

In CIQM we simply say that the energy of the individual system is not well defined but that upon measurement the value E_n will be found with probability $|C_n|^2$. Thus any individual process that includes (as part of its initial or final state) a superposition of states of different energy cannot be described in terms of energy conservation. If we insist that infinite plane waves represent an excessive abstraction and the more realistic description of a free particle is a wave packet or superposition of plane waves of different momenta, then we are led to conclude that energy conservation cannot be applied even to the motion of an individual free particle. The particle, in so far as it exists in CIQM, has potentially all the energies E_k with probabilities $|C_k|^2$.

In SIQM a particle represented by a superposition of stationary states with different energies has a well-defined energy at each moment (dependent on its initial position) in the wave packet but this energy is continuously variable with time. (However, the mean particle energy averaged over the ensemble is equal to the usual quantum mechanical result for the energy operator.) The variation of energy is due to the influence of the quantum potential which fluctuates with time. Thus in a wave-packet representation of the free particle we find that energy is not conserved for an individual particle since $\partial S/\partial t \neq$ constant.

Let us reconsider from this point of view the neutron interferometer experiment with one rf-coil in path II. Before entering the interferometer the neutron has a well-defined energy E corresponding to $|\uparrow_z\rangle$. On leaving it is a superposition of states of different energy $E(\uparrow_z)$ on path I and $(E - \Delta E)(\downarrow_z)$ on path II. Thus the spin can be found in the xy plane, by the measurement procedure described above, clearly demonstrating the principle of spin superposition (interference), i.e. the state is not a mixture of $\uparrow_z$ and $\downarrow_z$. However this interference observation requires us to relinquish the possibility of a description in terms of energy conservation. The attempt to apply energy conservation would require the use of a definite spin state in the guide field; that is, the state would have to be either $\uparrow_z$ or $\downarrow_z$ (a mixture). In that case energy conservation could apply in individual processes since superposition and interference is lost. A neutron in beam I with $\uparrow_z$ retains its original spin energy whilst one beam II exchanges ΔE with the coil. Thus we may choose to measure the z component of the spin and apply conservation of energy or observe the superposition and deny energy conservation. The two are complementary.

A similar situation exists in the two-slit experiments. If we wish to consider conservation of momentum of an individual electron, then it must be described as passing through one slit or the other in order to exchange momentum. If it passes through one slit, or the other, then interference is not possible.

In SIQM the particles have definite positions, momenta, energy and spin at all times, their associated (spinor) waves producing interference properties through the action of the (spin-dependent) quantum potential. In the neutron interferometer experiment described above there are then two possibilities depending on the path taken at the first crystal plane when interference is observed, i.e.:

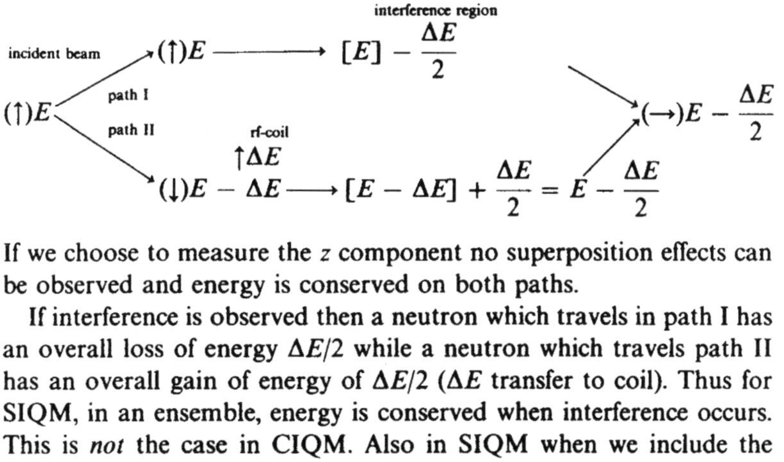

If we choose to measure the z component no superposition effects can be observed and energy is conserved on both paths.

If interference is observed then a neutron which travels in path I has an overall loss of energy $\Delta E/2$ while a neutron which travels path II has an overall gain of energy of $\Delta E/2$ (ΔE transfer to coil). Thus for SIQM, in an ensemble, energy is conserved when interference occurs. This is *not* the case in CIQM. Also in SIQM when we include the possibility of energy exchange with the ether, through the action of the quantum torque which rotates $\uparrow_z$ and $\downarrow_z$ to $\rightarrow_{xy}$ it is seen that energy may be conserved even in the individual case.

In general we see that in CIQM there is no possibility of recovering energy conservation in non-stationary situations; indeed, individual processes have no real independent existence.

Such a possibility does exist in SIQM if we assume that the particle exchanges energy with the sub-quantum Dirac ether. Indeed the recovery of conservation of energy in real individual processes is a strong reason for accepting the existence of such an ether.

We should note that the prediction of variable or non-constant energy made by SIQM does not contradict any experimental results of quantum mechanics. Indeed all the results of quantum mechanics can be reproduced by SIQM. Thus, as Bohm points out, when describing the scattering of a particle wave-packet by an atom whilst the interaction is still taking place and the wave packets overlap the particle and atomic electron energies fluctuate violently and it is only when the packets separate that the energies obtain a constant value. The corresponding feature in CIQM is given by the uncertainty relations $\Delta E \Delta t > \hbar$, and the energy of each system can only become definite after a sufficient time has elapsed to complete the scattering.

Thus the prediction of the existence of variable energies in SIQM does not contradict any result of quantum mechanics. In fact the SIQM can provide detailed information concerning the energy variation along well-determined trajectories in space-time in particular experimental situations. CIQM simply does not deny the possible existence of such energies if they are measured. The implication of this is that SIQM can make predictions which do not contradict CIQM but, in going beyond what CIQM allows to be possible, in the sense of being more precise, clearly demonstrates its incomplete character. In particular, some effects in non-linear optics experiments, i.e. the ejection of a photo-electron[37], photo-ionization of a gas[38] and fluorescence[39], occur even when the laser frequency is in fact below the necessary threshold for the process (provided the beam is put in a non-stationary state by focusing or by creation of a pulse). These effects can be interpreted in both CIQM and SIQM. However, by providing a detailed description of the individual trajectories and particle energies involved, SIQM can make testable predictions which are not possible in CIQM. In SIQM it is possible to predict at which points the particles of increased energy will be found and hence exactly where the effects should be observed[40]. If such predictions can be confirmed the CIQM would be shown to be incomplete in the original sense of Einstein.

4 Quantum statistics

In orthodox theory the wavelike density fluctuations of collections of like particles are described using Bose-Einstein or Fermi-Dirac statistics based solely on the notion of indistinguishability and the symmetry or antisymmetry of the wave function. However any interpretation of quantum mechanics which asserts the existence of individual particle trajectories is faced with a problem when the question of quantum statistical behaviour arises. Brillouin[41] had already in 1927 considered this problem. He argued that even if particles are identical *a priori*, it is easy nevertheless to distinguish them by their history. He then finds the auxiliary assumptions that enable quantum statistics to be obtained with distinguishability of elements. When the elements are assumed to be independent classical statistics result, in order to obtain quantum statistics some correlation between the distinguishable elements must be assumed. Further, as has been more recently emphasized by Feynman[42], no classical model with local interactions between the elements can ever reproduce all the results of quantum mechanics. This represents a serious problem in the statistical interpretation, and in SIQM.

In the derivation of the formulae of classical statistics with distinguishability, the assumption that the elements are free between random local collisions and that each distinct state has equal probability leads, for N elements distributed among M available discrete states, to the result that the probability of a set of occupancies $\{n_i\}$ $i = 1 \ldots M$ is proportional to the number of distinct configurations corresponding to $\{n_i\}$:

$$P\{n_i\} = M^{-N} N!/n_1! \ldots n_M! \qquad [19]$$

However, Tersoff and Bayer[43] have shown that Bose-Einstein statistics can be recovered with distinguishable particles if the assumption of equal probability distribution among available states is replaced by that of arbitrary probability weighting. It has also been shown[44] that such an arbitrary probability weighting is a natural consequence of the causal interpretation. In this interpretation the assumption of random local collisions and independent particles no longer holds. The average motions of N particles given by:

$$V_K^\mu = \frac{1}{m}\,\partial_K^\mu S \quad (K = 1 \ldots N) \qquad [20]$$

are determined from the non-local action at-a-distance quantum potential:

$$Q = \sum_K \frac{\hbar}{2m}\frac{\Box_K R}{R} = \sum_K Q_K \qquad [21]$$

This potential acts instantaneously in the centre of mass rest-frame and also implies that the interaction is causal (since the individual Hamiltonians $H_k = \frac{1}{m} P_k^\mu P_{k\mu} + Q_k$ satisfy the causality constraints $\{H_k, H_j\} = 0$) so that all colliding particles are permanently correlated and can *never* be considered free. This implies that each individual state is not identical with all others, so that we should attribute to each one a different probability weighting ω_i of course requiring $0 \leqslant \omega_i \leqslant 1$ and $\sum_{i=1}^{M} \omega_i = 1$. This weight depends on all former possible different real subquantal random motions in phase space, so that the total statistics results from an averaging over all possible ω_i in all possible configurations. Thus we should write:

$$P\{n_i\} = A_V\left[\frac{N!}{n_1!\ldots n_N!}(\omega_1)^{n_1}\ldots(\omega_M)^{n_M}\right]$$
$$= \int_0^1\ldots\int_0^1 d\omega_1\ldots d\omega_M \frac{N!}{n_1!\ldots n_M!}(\omega)^{n_1}\ldots(\omega_M)^{n_M}$$
$$\delta\left(1-\sum_{i=1}^{M}\omega_i\right) \quad [22]$$

which, as shown by Tersoff and Bayer[43], leads to the Bose-Einstein result:

$$P\{n_i\} = \frac{N!(M-1)!}{(N+M-1)!} \quad [23]$$

Fermi-Dirac statistics can also be reproduced[45] in a similar way with the constraint that $n_i = 0$ or 1.

As an illustration we now show in a particular physical situation how the individual motions of particles under the influence of the many-body quantum potential lead to different statistical results according to the type of wave function assumed[46]. The causal interpretation of quantum statistics can thus be shown to provide an intuitive understanding of quantum statistical results (in terms of correlated particle motions), classical statistics arising as a special case when the particles are not correlated by the quantum potential. The case examined here is the following. Consider a harmonic oscillator potential:

$$V = \frac{kx^2}{2} = \frac{m\omega^2x^2}{2}$$

and construct, by solving the Schrödinger equation, a wave-packet solution[35]:

$$\psi(x,t) = \exp(-i\omega t)\exp\left[-\frac{1}{2}(x - x_0\cos\omega t)^2\right]\exp$$
$$\left[\frac{i}{2}\left(x_0^2\frac{\sin 2\omega t}{2} - 2xx_0\sin\omega t\right)\right] \quad [24]$$

This wave-packet solution is non-dispersive and, depending on the time parameter t, defines in the causal interpretation a set of possible trajectories for a particle located at the position x, where x_0 is the centre of a wave packet.

Now consider the case of two particles, one in each of the wave packets $\psi_A(x_1,t)$ and $\psi_B(x_2,t)$ in the harmonic oscillator potential. The packet $\psi_A(x_1,t)$ is assumed to be centred at x_0 and, in order to simplify the calculations, the packet $\psi_B(x_2,t)$ centred at $-x_0$.

It is clear that, depending on the assumed statistics (MB, BE or FD), three wave functions can be written. These are:

$$\varphi_{MB} = \alpha_{MB}\psi_A(x_1,t)\psi_B(x_2,t) \quad [25]$$

$$\varphi_{BE} = \alpha_{BE}[\psi_A(x_1,t)\psi_B(x_2,t) + \psi_B(x_1,t)\psi_A(x_2,t)] \quad [26]$$

$$\varphi_{FD} = \alpha_{FD}[\psi_A(x_1,t)\psi_B(x_2,t) - \psi_B(x_1,t)\psi_A(x_2,t)] \quad [27]$$

where the αs are renormalization constants to be determined by the condition $\iint \varphi \mathrm{d}x_1 \mathrm{d}x_2 = 1$. A standard quantum mechanical calculation yields the mean squared separation of the particles and it can be shown that the mean squared separation of the particles is in the BE case decreased and in the FD case increased with respect to the MB case.

In SIQM individual pairs of trajectories can be calculated from suitable initial positions in the wave packets and the results are shown in Figure 9.11. This figure provides us with the basic physical features of the process. The MB particles, being independent, possess trajectories that cross one another. They propagate undisturbed and produce no interference. This is not the case for BE or FD particles. They do not cross but form interference patterns in which the two particles are on the average closer together in the BE case than in the FD case.

The correlation effects mediated by the quantum potential between the two particles determines their physical behaviour and conditions their different statistical averages of physical variables or observables. This can be easily understood in the SIQM, where particles obeying quantum statistics are constantly submitted to the stochastic random motions of the underlying subquantal medium, the Dirac ether. The symmetric or antisymmetric character of the system's wave function is a consequence of the existence (or not) of local repulsive gauge fields and not a first quantum mechanical principle.

Thus it can be seen that MB statistics (and independence) arise as a special case of the more general quantum statistics (and correlation) when the many-body quantum potential is separable in the particle variables.

5 Negative probabilities

We have seen how accepting the physical idea of particle trajectories in the quantum domain can lead to the formulation of new physical questions which one would not be led to on the basis of the CIQM. We end this paper by discussing an interpretative problem raised by relativistic quantum mechanics, namely the mathematical existence of negative probability density and negative energy solutions to second-order wave equations, which, as in all other quantum processes, we argue can only be coherently treated by assuming the real physical existence of paths. Indeed, this is an important issue in the SIQM since the very existence of paths in space-time implies positive pro-

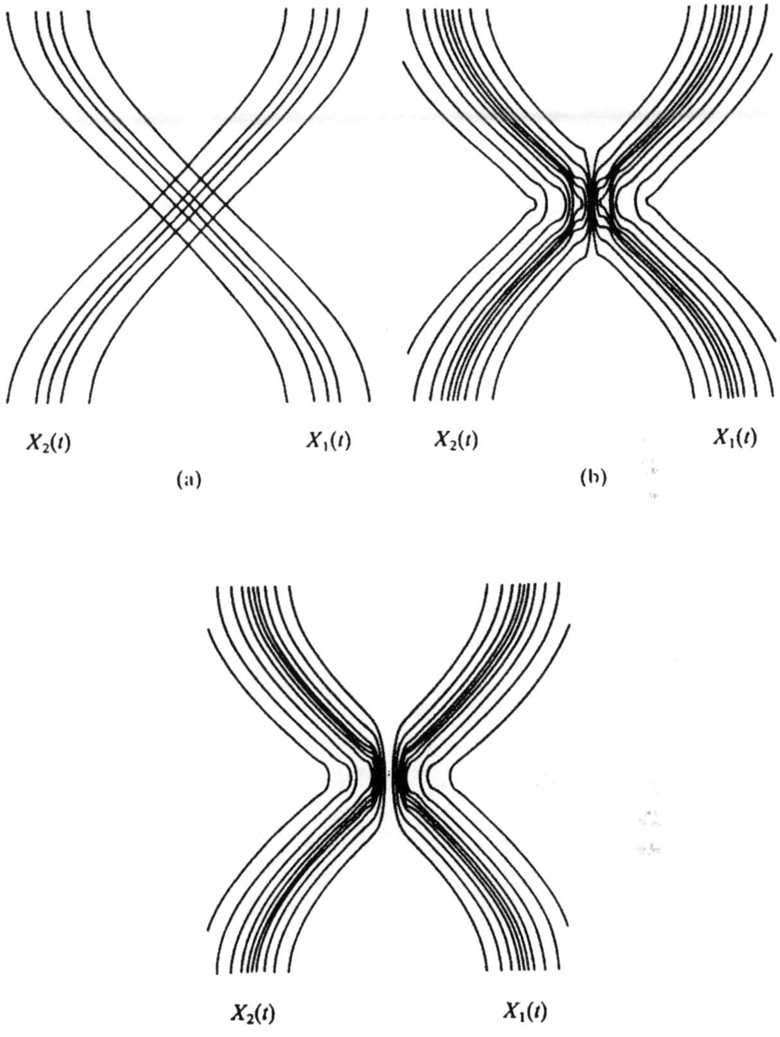

Figure 9.11 Ensemble of two particle trajectories $x_1(t)$, $x_2(t)$ with initial positions such that $x_2(0) = -x_1(0)$ and a concentration of particle trajectories around the packet maxima. (a) Maxwell-Boltzman (b) Bose-Einstein (c) Fermi-Dirac.

bability distributions and, moreover, in accordance with Einstein's basic principles, all material drift motions should be timelike and propagate positive energy forward in time.

It is sometimes erroneously stated that the only way out of the problem of negative probability solutions to the Klein-Gordon (KG) equation is to reject the first quantized formalism in favour of second quantization. In fact, this is not so and it is possible to show by a

Hamiltonian method due to Feshbach and Villars[47,48] that for certain well-behaved external potentials the KG solutions may be split into positive and negative energy parts associated respectively with positive and negative probability.

To see this, let us start from the charged scalar wave equation:

$$(D_\mu D^\mu - m^2)\psi = 0 \qquad [28]$$

where $D_\mu = \partial_\mu - ieA_\mu$, e and m are the charge and mass of the particle moving in the external field A_μ, the metric has the signature $(- + + +)$ and the units are chosen so that $\hbar = c = 1$. [28] may be expressed in the form:

$$i\dot{\Psi} = H(e)\Psi \qquad [29]$$

where Ψ is a two-component wave function and H is a 2×2 matrix Hamiltonian. One can show that for the inner product

$$<\Phi,\Psi> = \int \Phi^* \sigma_3 \Psi d^3x = \int j^0 d^3x$$

where:

$$j^\mu = \psi^*(i^{-1}\overleftrightarrow{\partial^\mu} - ieA^\mu)\psi \qquad [30]$$

is the conserved current, the mean value of H in any state is positive: $<\Psi,H\Psi> > 0$. It follows that, with $H\Psi = E\Psi$, the space of solutions of equation [29] splits into two disjoint subsets: $\{E > 0, <\Psi.\Psi> > 0\}$ and $\{E < 0, <\Psi,\Psi> < 0\}$. The latter subset of solutions may be mapped into positive-energy positive-probability anti-particle solutions by means of the charge conjugation operation:

$$\Psi \rightarrow \Psi^c(x) = \sigma_1 \Psi^*(x)$$

since from equation [29]:

$$H(-e)\Psi^c = -E\Psi^c \text{ and } j_0^c = -j_0$$

Thus, within the CIQM, one can show formally how, for stationary states, the signs of energy and integrated probability are correlated and that negative probability solutions may be physically interpreted. Note though, that the local values of probability density may become negative and that such motions remain interpreted. We shall now show how in the causal interpretation we are able to prove a stronger result than that just given, and in a way which is technically easier and physically clearer. Our approach extends some brief remarks of de Broglie[49] concerning this problem.

Substituting $\psi = e^{P+iS}$, where P, S are real scalars, in [28] yields the Hamilton-Jacobi and conservation equations:

$$(\partial^\mu S - eA^\mu)(\partial_\mu S - eA_\mu) = -M^2 \qquad [31]$$

$$\partial_\mu j^\mu = 0 \tag{32}$$

where

$$M^2 = m^2 - \Box P - \partial^\mu P \partial_\mu P$$

is de Broglie's variable rest mass and:

$$j^\mu = 2e^{2P}(\partial^\mu S - eA^\mu)$$

is the current (equation [30]).

The assumption of the SIQM is that the KG particle has a drift velocity

$$u^\mu = \mathrm{d}x^\mu/\mathrm{d}\tau$$

where τ is the proper time along paths parallel to j^μ. In terms of the momentum $P^\mu = \partial^\mu S - eA^\mu$, $u^\mu = M^{-1}P^\mu$ with $u_\mu u^\mu = -1$, from equation [31].

Defining a scalar density $\rho = Me^{2P}$ we may express equation [32] in the form[50]:

$$\frac{\mathrm{D}\rho}{\mathrm{D}\tau} \equiv \partial_\mu(\rho u^\mu) = 0$$

From this it follows that along a line of flow:

$$\omega M e^{2P} = K \tag{33}$$

where ω is a volume element of fluid and K is a real or pure imaginary constant (which, however, varies from one drift line to another). If on an initial spacelike surface the motion is timelike, then from equation [31] M is real and so is K. Now, in the rest frame, $u^0 E = M$ where the particle energy $E = \partial^0 S - eA^0$. It follows that if initially the motion is future-pointing, with $E > 0$, then $M > 0$ which implies $K > 0$ (since $e^{2P} > 0$ and $\omega > 0$ always) and we see from equation [33] that the timelike and positive energy character of the motion is preserved all along a trajectory. Moreover, the sign of the probability density $j^0 = 2e^{2P}E$ is correlated with the sign of E and will remain positive along a line of flow if the initial motion has $E > 0$.

Identical arguments lead to an association of past-pointing negative-energy motions with negative probability densities and this coupling is preserved along a line of flow if initially $E < 0$. Such solutions may be mapped on to positive energy, positive probability density anti-particle solutions by the charge conjugation given above: $\psi^c = \psi^*$.

These results, proved in the rest frame, evidently remain valid under orthochronous Lorentz transformations.

We have thus succeeded in separating the solutions to the causal

KG equation into two disjoint subsets $\{E > 0, j^0 > 0\}$ and $\{E < 0, j^0 < 0\}$ and shown that the causal laws of motion prevent the development of one type of solution into the other. This reasoning holds for all external fields A_μ which maintain the timelikeness of the momentum P^μ. Should the external potential be strong enough, pair creation may occur and the separation of the solutions breaks down. In addition, we assume that the initial motion is associated with a wave packet so that the initial total probability is unity. This is an important point, since de Broglie[12] has shown how, with a plane KG wave incident on a partially reflecting mirror, superluminal motions apparently occur in the region of the Wiener fringes. It seems that these unphysical motions are a consequence of the excessive abstraction implied by the use of plane waves.

It is emphasized that we have only been able to overcome the difficulty of negative probabilities by assuming that particles possess well-defined space-time trajectories, and that they are subject to action by the quantum potential (contained in M). With these assumptions, we can immediately associate the sign of particle energy with the sign of local probability density, an energy moreover which is well-defined and continuously variable for all possible particle motions (and not just for stationary states). The initial character of these motions is preserved for all time by the Hamilton-Jacobi and conservation equations.

If one accepts that the quantum mechanical formalism is complete then one must accept Feynman's statement[51] that there is no way of eliminating negative probabilities from the intermediate stages of, for example, an interference calculation. The problem of their physical interpretation then cannot be avoided.

However, if one accepts the introduction of trajectories in the description, then our demonstration above shows how the positive character of probability is preserved at every stage of the calculation. The association of positive probabilities with positive energy is of course in accordance with the principles of relativity theory.

We note finally that, although our discussion here has been confined to a single KG particle[52], our method may be applied to the elimination of negative probabilities from the theory of the many-body KG system, the spin-1 Proca equation, and the spin-$\frac{1}{2}$ Feynman–Gell–Mann equation[53].

Finally, we wish to stress that the causal interpretation does not reinstate the mechanistic classical world view. Particles may be described as possessing definite values of physical variables but these variables depend, through the quantum forces arising from the quantum potential and torque, on the whole quantum state which includes the influence of the environment.

The lessons of Bohm's work are clear. We can adopt Bohr's idealist epistemology and deny the very possibility of analysing what happens within quantum phenomena, such as neutron interference. However we should then be consistent and refuse to speak of the quantum world as if it actually exists. The only other known alternative, which is capable of reproducing all the results of quantum mechanics in terms of a physically determinist reality, is the non-classical causal interpretation. Far from returning to classical mechanics it shows exactly how radical a revision of our concepts quantum mechanics entails. Even if it is not taken as a fully satisfactory description of quantum mechanical reality, it at least shows in a clear way the features that such a description must entail. The interpretations of Bohr and of de Broglie-Bohm-Vigier both emphasize that the fundamentally new feature exhibited by quantum phenomena is a kind of wholeness completely foreign to the post-Aristotelean reductionist mechanism in which all of nature in the final analysis consists simply of separate and independently existing parts whose motions, determined by a few fundamental forces of interaction, are sufficient to account for all phenomena. The difference arises in the methods for dealing with the situation. One thing however is clear; the organization of nature at the fundamental level is far more complex than mere mechanistic models can encompass. The ghost cannot be exorcized from the machine.

Conclusion

Throughout this contribution we have discussed various 'interpretations' of the quantum formalism, and what has emerged is that the problem is not simply one of interpreting the same results in various ways. In fact there are good reasons for the argument that CIQM and SIQM are essentially different theories between which a choice can be made in a no arbitrary manner. Moreover:

1 They have different ontologies since the real existents are different. In SIQM individual processes are real, take place in space and time and have well-defined properties. In fact SIQM can account for all the quantum properties of matter, including all the so-called paradoxes, within this framework without conflicting with the requirements of special relativity. Further it does this in terms of a model which is immediately intuitively clear and which allows a visualization of the actual processes taking place.

2 All events occurring in space and time can be attributed to material causes which are also processes taking place in space-time, albeit non-locally. In CIQM the behaviour of matter is irreducibly indeterminate; for example, *nothing* causes the decay of an unstable nucleus.

3 In some versions of CIQM the behaviour of matter depends on the cognizance of observers. Such a possibility does not exist in

SIQM in which the material world has an existence independent of the knowledge of observers.

4 Since SIQM allows a description of the causation of individual events, it enables a deeper analysis and understanding of phenomena with the possibility of developing more penetrating theories of these events which CIQM shrouds in mystery by the dogmatic insistence in the absolute and final character of complementarity and indeterminacy.

5 The possibility exists in SIQM to make testable predictions which go beyond, by being more precise, but nevertheless do not contradict those of quantum mechanics.

6 Complementarity is inadequate in the description of time-dependent neutron interferometry and requires the renunciation of energy conservation in interference situations, whereas the description of SIQM is consistent and apparent non-conservation of energy may be explained through the possibility of energy exchange with the ether.

Acknowledgements

The authors C. Dewdney, P. R. Holland and A. Kyprianidis wish to thank the Royal Society, the SERC and the French government respectively for financial support which enabled the work reported here to be completed, and the Institut Henri Poincaré for its hospitality.

References

1 A. Garuccio, A. Kyprianidis, D. Sardelis and J.-P. Vigier, *Lettere al Nuovo Cim.*, **39**, 225 (1984).
2 A. Garuccio, V. Rapisarda and J.-P. Vigier, *Phys. Lett.*, **90A**, 17 (1982).
3 D. Bohm, *Phys. Rev.*, **85**, 166, 180 (1952); *Phys. Rev.*, **89**, 458 (1953); *Causality and Chance in Modern Physics*, Routledge & Kegan Paul, London, 1951; *Wholeness and the Implicate Order*, London, 1980; D. Bohm and J.-P. Vigier, *Phys. Rev.*, **96**, 205 (1954); D. Bohm and B. J. Hiley, *Found Phys.* **5**, 93 (1975).
4 N. Bohr, *Atomic Physics and the Description of Nature*, Cambridge University Press, 1934; *Atomic Physics and Human Knowledge*, Random House, New York, 1958; *Essays 1958–1962 on Atomic Physics and Human Knowledge*, Random House, New York, 1962.
5 N. Bohr, *op. cit.*, 1934.
6 A. Einstein, in Schilpp (ed.), *Albert Einstein: Philosopher Scientist*, Harper Torchbooks, New York, 1959.
7 D. M. Greenberger, *Rev. Mod. Phys.* **55**, 4 (1983).
8 L. E. Ballentine, *Rev. Mod. Phys.*, **42**, 358 (1970).
9 W. Duane, *Proc. Natl. Acad. Sci.*, **9**, 153 (1923).
10 A. Landé, *Am. J. Phys.*, **33**, 123 (1965); **34**, 1160 (1966); **37**, 541 (1969).
11 A. Einstein, R. Podolsky and N. Rosen, *Phys. Rev.*, **47**, 777 (1935).
12 L. de Broglie, *Non Linear Wave Mechanics*, Elsevier, 1960.
13 J.-P. Vigier, *Astr. Nachr.*, **303**, 55 (1982) and references therein.
14 D. Bohm and B. J. Hiley, *Found Phys.*, **5**, 93 (1975); **12**, 1001 (1982).

15 P. Droz-Vincent, *Phys. Rev. D.*, **19**, 702 (1979); *Ann. IHP*, **32**, 377 (1980). A. Garuccio, A. Kyprianidis and J.-P. Vigier, *Nuovo Cim.*, **B83**, 135 (1984); C. Dewdney, P. H. Holland, A. Kyprianidis, and J.-P. Vigier, *Phys. Rev. D.*, **31**, 2533, 1985.

16 D. Bohm and B. J. Hiley, *Found Phys.*, **14**, 255 (1984).

17 C. Philippidis, C. Dewdney and B. J. Hiley, *Nuovo Cim.*, **52B**, 15 (1979).

18 E. Schrödinger, *Proc. Camb. Phil. Soc.*, **31**, 555 (1935).

19 A. Aspect, P. Grangier and G. Rogier, *Phys. Rev. Lett.*, **47**, 460 (1981); **49**, 91 (1982).

20 J.-P. Vigier and N. Cufaro-Petroni, *Phys. Lett.*, **93A**, 383 (1983); *Lett. Nuovo Cim.*, **26**, 149 (1979); D. Bohm and B. J. Hiley, *Found. Phys.*, **11**, 529 (1981).

21 R. Prosser, *Int. J. Theo. Phys.*, **15**, 3 (1926).

22 D. Bohm, *Quantum Theory*, Prentice Hall (1951); M. Cini, *Nuovo Cim.*, **73B**, 27 (1983).

23 H. Rauch, *Proc. Int. Symp. Foundations of Quantum Mechanics*, Tokyo, 1983, 277; S. A. Werner and A. G. Klein in D. H. Price, K. Sköld (eds), *Neutron Scattering*, Academic Press, 1984, and references therein.

24 G. Eder and A. Zeilinger, *Nuovo Cim.*, **34B**, 26 (1976).

25 A. Einstein, *Proc. Congrès Solvay*, 1927.

26 H. Rauch, Tokyo, *Proc. Int. Symp. Foundations of Quantum Mechanics*, 1983, 277.

27 C. Dewdney and B. J. Hiley, *Found. Phys.*, **12**, 27 (1982).

28 C. Dewdney, 'Particle trajectories and interference in a time dependent model of neutron single crystal interferometry', *IHP*, preprint (1985). *Phys. Lett.*, **109A**, 377, 1985.

29 G. Badurek, H. Rauch and J. Summhammer, *Phys. Rev. Lett.*, **51**, 1015 (1983).

30 B. Alefield, G. Badurek and H. Rauch, Zeit. *Phys. B.*, **41**, 231 (1981).

31 P. A. M. Dirac, *Nature*, **168**, 906 (1951); **169**, 702 (1952).

32 N. Cufaro-Petroni, P. Gueret and J.-P. Vigier, *Phys. Rev. D.*, **30**, 495 (1984); *Nuovo Cim. B.*, **81**, 243 (1984).

33 H. J. Park, *Nuovo Cim. B.*, **55**, 15 (1980).

34 C. Dewdney, A. Garuccio, A. Kyprianidis and J.-P. Vigier, *Phys. Lett.*, **104A**, 325 (1984); L. de Broglie, *Wave Mechanics, the First Fifty Years*, Butterworths, London, 1973.

35 D. Bohm, *Quantum Theory*, Prentice Hall, (1951).

36 L. de Broglie, *Compt. Rend.*, **183**, 447 (1926); **184**, 273 (1927); **185**, 380 (1927).

37 E. M. Logothetis, and P. L. Hortman, *Phys. Rev.*, **187**, 460 (1969); G. Farkas, I. Kertesz, Z. Naray and P. Vargo, *Phys. Lett.*, **21A**, 475 (1962); E. Panarella, *Lett. Nuovo Cim.*, **3**, 417 (1972).

38 E. Panarella, *Phys. Rev. Lett.*, **33**, 950 (1974); *Found Phys.*, **4**, 227 (1974).

39 E. Panarella, *Found Phys.*, **7**, 405 (1977).

40 C. Dewdney, M. Dubois, A. Kyprianidis and J.-P. Vigier, *Lett. Nuovo Cim.*, **41**, 177 (1984); C. Dewdney, A. Garuccio, A. Kyprianidis and J.-P. Vigier, *Phys. Lett. A.*, **105**, 15 (1984).

41 L. Brillouin, *Ann. der Phys.*, **1**, 315 (1927).

42 R. P. Feynman, *Int. J. Theo. Phys.*, **21**, 612 (1982).

43 J. Tersoff, and D. Bayer, *Phys. Rev. Lett.*, **50**, 8 (1983).

44 A. Kyprianidis, D. Sardelis, and J.-P. Vigier, *Phys. Lett.*, **100A**, 228 (1984).

45 N. Cufaro-Petroni, A. Maricz, D. Sardelis and J.-P. Vigier, *Phys. Lett.*, **101A**, 4 (1984).

46 C. Dewdney, A. Kyprianidis and J.-P. Vigier, *J. Phys. A.*, **17**, L741. (1984).
47 H. Feshbach and F. Villars, *Rev. Mod. Phys.*, **30**, 24 (1958).
48 V. A. Rizov, H. Sazdjian and I. T. Todorov, 'On the relativistic quantum mechanics of two interacting spinless particles', Orsay preprint (1984).
49 L. de Broglie, *The Current Interpretation of Wave Mechanics*, Elsevier, Amsterdam, 1964.
50 F. Halbwachs, *Théories relativistes des fluides à spin*, Gauthier-Villars, Paris, 1960.
51 R. P. Feynman, 'Negative probability', this volume, p. 235.
52 N. Cufaro-Petroni *et al.*, *Phys. Lett.*, **106A**, 368 (1984).
53 N. Cufaro-Petroni *et al.*, *Lett. Nuovo Cim.*, **42**, 285 (1985); P. R. Holland *et al.*, *Phys. Lett.*, **107A**, 376 (1985); P. Gueret *et al.*, *Phys. Lett.*, **107A**, 379 (1985).

Reprinted from *Physica*, Vol. B 151, pp. 386-392, Copyright (1988)
with permission from Elsevier Science.

NEW THEORETICAL IMPLICATIONS OF NEUTRON INTERFEROMETRIC DOUBLE RESONANCE EXPERIMENTS

J.P. VIGIER
Institut Henri Poincaré, Laboratoire de Physique Théorique, 11, rue P. et M. Curie, 75231 Paris Cedex 05, France

It is shown that if one accepts Einstein's postulate that energy–momentum is conserved in all individual microprocesses, the Grenoble experiments imply that individual neutrons are waves *and* particles simultaneously. If one rejects this postulate (and thus accepts Heisenberg's statement that they are only conserved statistically) new experiments are needed to settle the Bohr–Einstein controversy.

The aim of the present contribution is to discuss some theoretical implications of the recent double resonance interferometric neutron experiments performed by Rauch's group (i.e. Badurek, Rauch and Tuppinger (BRT)) in Grenoble [1] (where two spin-flippers are present) suggested by our group in Paris [2]. The set-up is represented in fig. 1.

These implications all rest on the observation that the difference between the principle of complementarity of Bohr and Heisenberg (B–H); where micro-objects correspond to probabilistic $\Psi(x, t)$ waves *or* observed particles, never the two simultaneously and the Einstein–de Broglie (E–de B) description where they are represented by waves *and* a real particle aspect simultaneously . . . . The later beating in phase (since $E = h\nu$) with the surrounding wave.

Not everybody realizes that Bohr's assumption that quantum mechanics is a complete, final, unsurpassable description of physical phenomena implies the crucial statement that in the BRT type neutron interferometry experiments individual particle-like neutrons *do not travel in space and time* between the atomic pile and the detectors . . . . Indeed such motions would imply the existence of hidden variables (characterizing the neutron's paths) which would exist independently beyond the usual purely probabilistic description of the quantum formalism . . . so that the quantum mechanical description would be incomplete. In the B–H interpretation, quantum statistics do not result from some chaotic subquantal behaviour of hidden variables.

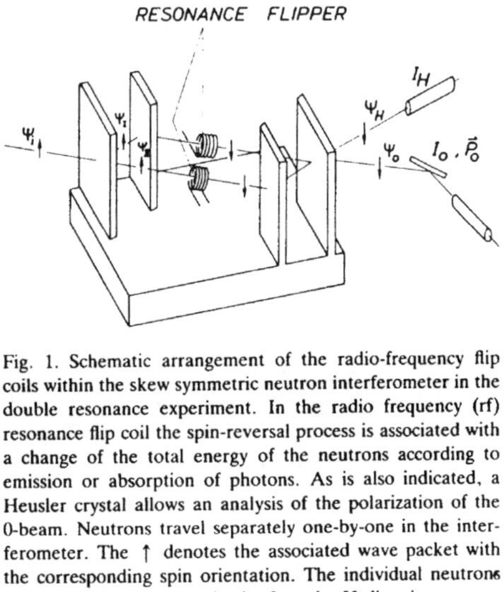

Fig. 1. Schematic arrangement of the radio-frequency flip coils within the skew symmetric neutron interferometer in the double resonance experiment. In the radio frequency (rf) resonance flip coil the spin-reversal process is associated with a change of the total energy of the neutrons according to emission or absorption of photons. As is also indicated, a Heusler crystal allows an analysis of the polarization of the 0-beam. Neutrons travel separately one-by-one in the interferometer. The ↑ denotes the associated wave packet with the corresponding spin orientation. The individual neutrons are detected one by one in the 0 or the H direction.

From this observation, one deduces the first consequence that in all types of Young's double slit interference experiments (of which the double resonance spin-flipper experiment is a particular case) the difference between the Bohr–Heisenberg (B–H) and Einstein–de Broglie (E–

de B) points of view is that for B–H, the particle does not pass through any coil while it passes through one coil only in the E–de B point of view. In other terms, if one could prove in a direct or indirect way that in Rauch's experiments the neutrons pass through one *or* the other slit, this would disprove the B–H point of view.

The second consequence is that for B–H the direct detection of the passage of the neutron through one spin-flipper only or along one of the paths, characterized by the waves Ψ_{I} and Ψ_{II}, implies a wave packet collapse of the other wave . . . so that the interference pattern would disappear on the detectors.

Before we discuss the implications of these consequences on the interpretation of the BRT results, let us briefly recall Einstein's and Bohr's opposite views on the question of direct and indirect physical knowledge of the properties of microprocesses.

For B–H, there is no such thing as indirect knowledge of a physical property of a given physical process (due for example to some necessarily conserved law) but only direct knowledge acquired through actual measurement processes.

For E–de B there are both direct and indirect knowledge. Direct knowledge results from the observed (valid) predictions of the quantum formalism. Indirect knowledge implies that some real but unobserved properties result from the fact that specific physical laws (such as the conservation of four-momentum and angular momentum) are assumed to be always satisfied, even in the absence of any actual observer or observation. This assumption (rejected by B–H) is the logical foundation of the EPR paradox [5].

We also remark here that the concept of wave packet collapse (reduction) implied by the standard [6] (B–H) quantum theory of measurement is not devoid of difficulties. It does not clearly account for the information provided by the so-called negative measurement results (NRM) which imply (as discussed by Einstein and de Broglie [8]) wave packet collapse (i.e. acquisition of information/knowledge) due to the fact that a measurement device has not reacted (i.e. interacted) with an observed particle. This is the case for example of an interference filter on the path of a photon wave packet. No direct measurement is performed at the level of the filter (constituted by a crystal slab) but the absence of reflected photon detection implies that the transmitted photon's frequency band is confined (whether it is measured or not) within the filter frequency band $2\Delta\nu$, i.e. $\nu_0 + \Delta\nu \leqslant \nu \leqslant \nu_0 + \Delta\nu$.

We can now come back to the theoretical implications of the double resonance neutron interferometry measurements and results [1].

(1) In a first set of measurements the two spin-flippers are simultaneously working and connected with the same rf oscillator (i.e. driven at equal frequencies); then, the set-up of fig. 1 never contains more than neutron at a time. In fact each time a neutron is detected, the next one has not even left the atomic pile of Grenoble.

(2) Each neutron enters the interferometer with its spin (magnetic moment) oriented upwards, i.e. parallel to the constant magnetic field which bathes the whole set-up.

(3) Each neutron leaves the set-up (and is detected by one of the detectors) with its spin oriented downward i.e. antiparallel to the constant magnetic field. As a consequence, according to the well-known theory of spin flipping, it has lost by resonance a quantum of energy $\Delta E = \hbar\omega_{rf}$: ω_{rf} being confined to the harmonic oscillator's rf frequency band.

Of course, at this stage of the BRT experiment this individual neutron's loss of energy ΔE is still indirectly known/derived as a consequence of their spin-flip. However, as we will show later, it can be established experimentally (with interference filters) in a NRM type of measurement.

If one thus assumes

(a) with Einstein and de Broglie that the neutron's energy and impulsion (i.e. mass multiplied by velocity) are always conserved in all microprocesses where thee is an exchange of energy . . . i.e. in that case that the quantum ΔE lost by the neutron must be absorbed by the spin-flippers in a resonance process;

(b) with Einstein and Bohr that all observable

(ex) change of quanta of energy is tied with the "particle" aspect of matter*.

Then one must accept (if one assumes absolute energy–momentum conservations) that this exchange of energy, i.e. the presence of the neutron's particle aspect, has necessarily happened in one *or* the other spin-flippers (i.e. in one spin-flipper only) but not in both spin-flippers simultaneously . . . where no half quantum possible resonance frequencies exist anyway in the corresponding rf harmonic oscillators.

(4) Since the BRT experiment has shown that each individual neutron localizes itself on an interference pattern (i.e., according to point 1, it interfers with itself in Dirac's sense) one can conclude, following BRT's own terms [1]: "That in the region where the waves interfere each neutron disposes of the information that there exist two associated paths Ψ_{I} and Ψ_{II} in the interferometer i.e. two paths whose phase difference obliges the neutron to appear in one of the detectors".

(5) From the combination of points 3 and 4 one deduces that between the source and the detectors each neutron manifests itself as a wave (because of the interference) and as a particle (because of the loss of energy) simultaneously. In other terms each neutron is a wave *and* a particle.

From point 5 one deduces that between the source and detector the description of the quantum probabilistic distribution in terms of waves Ψ_i then Ψ_{I} and Ψ_{II} then Ψ_0 and Ψ_{H} is correct but not complete since it does not state that each neutron has manifested itself in one of the spin-flippers. However to quote Rauch again [1]: "This experiment shows explicitly that the interference properties of Ψ_{I} and Ψ_{II} can be preserved even when real exchange of energy has occurred which is intuitively a measuring process".

The preceding argumentation is even strengthened by the quantum beat effects which result from a slight modification of the double coil experiment. In this case, the frequencies ν_{r1} and ν_{r2} of the two resonators (which drive the coils) are chosen to be slightly different, i.e. when both flip coils are operated at different frequencies:

$$\nu_{\mathrm{r1}} = 71.91 \pm 0.02\ \mathrm{kHz}\,, \qquad \nu_{\mathrm{r2}} = 72.32 \pm 0.02\ \mathrm{kHz}\,, \tag{1}$$

respectively.

As remarked by BRT [1] (and confirmed by their experiment) "in spite of the rather large frequency difference of 420 Hz the efficiencies of both flip coils remained as high as about 98% because of the broad resonance curve. By electronic means the period $T = (\nu_{\mathrm{r1}} - \nu_{\mathrm{r2}})^{-1}$ was subdivided synchronously into 8 consecutive time intervals of equal width which were used to gate inputs of 8 separate scalars. This stroboscopic registration technique transforms the interference eating into a stationary intensity distribution. As a function of the time t each interval is shifted w.r.t. an arbitrary chosen reference point". The reason for the observed occurrence of the beat effect can be described as a time-dependent phase relation in the quasistatic approximation*. We thus have $\Delta(t) = (\omega_{\mathrm{r1}} - \omega_{\mathrm{r2}})t$ between the two interfering beams and the difference of the spin-flip energy-transfer ($\hbar\omega_{\mathrm{r1}} \neq \hbar\omega_{\mathrm{r2}}$) due to the neutron wave's resonance with the two separated oscillatory fields. This implies immediately, if one assumes

(i) absolute energy–momentum conservation in all individual microprocesses,

(ii) the fact that a neutron loses one quanta of

* In the standard quantum mechanical formalism particularly in the emission/absorption of quantas every such exchange behaves as if energy momentum was carried by particles with directional motions. Compton scattering implies [9] that each electron effective in the scattering scatters a complete quantum and that these quantas of radiation are received from definite directions and are received from definite directions and are scattered in definite directions. The same is true of the double Compton scattering [10]. To quote Compton "we can find no interpretation of the scattering except in terms of the deflection of corpuscules or photons of radiation". The same is true for particle emission or absorption.

* As we shall see later, the causal interpretation provides a more physical interpretation based on different energy modifications in both beams where the time axis is related to the time-dependent Hamiltonian of the oscillating field.

energy (photon) to a coil when it flips its spin through a resonance process with the energy level difference of the rf oscillator which drives the flipper,

(iii) the possibility of indirect and NRM knowledge,

that one must accept the fact that the all neutrons which have flipped their spin, (i.e. appear in the beat process) have gone through one coil only . . . since we know from (1) that ν_{r1} is significantly distant from ν_{r2} . . . so that a neutron is a wave *and* a particle simultaneously.

This conclusion however must be qualified with the following reservations and/or restrictions.

A) As stated above at the present experimental stage we only know *indirectly* that all neutrons in the interference pattern have lost a quanta $\Delta E = \hbar\omega_{rf}$ because they have flipped their spin. One can perfectly utilize the argument that until one has measured (in the B–H sense) the passage of each neutron through one coil only one does not create any wave packet collapse on the other path. Indeed such direct individual energy loss measurements $\Delta E = \hbar\omega_{rf}$ seem practically impossible at present not only because of the Heisenberg uncertainty relation $\Delta N \cdot \Delta\phi \geqslant \frac{1}{2}$ but also because any imaginable direct measurements device on individual neutrons would apparently destroy the coherence of their associated Ψ field. However [1], "the energy transfer that is induced on passage through the rf coil can (at least in principle) be made larger than the energetic width of the beam. Nevertheless, a distinct phase relation between both rf fields is required which can be considered as an equivalent constraint against a simultaneous phase and particle number determination. Within the rf coils only quanta $\hbar\omega_{rf}$ within a narrow frequency band $\Delta\omega_k$ are excited which suggests that the energy exchange with an individual neutron takes place in one coil but in its last consequence even this statement could be in contradiction to the complementary principle of quantum mechanics".

At first sight since one cannot detect directly (individually) this exchange in one coil it would seem one cannot detect it at all. This is not true however. As we shall now show, it is possible to bypass this difficulty through the use of a specific type of NMR process. Quantum theory shows that such an experimental device is indeed possible (as will be discussed in detail in a subsequent publication). Its principle is simple and is represented in fig. 2. If we consider an interference filter built with a pure silicium plate which allows the passage of incoming neutrons, we know its reflectivity (i.e. transmitted frequency interval) depends on the distance d of its Bragg planes, i.e. on the wavelength $\lambda_0 = 2d_0 \sin\theta_{\text{Bragg}}$ which depends on its temperature T_0. This means (because we have $d = d_0 + \alpha\,\Delta T$) that one can practically shift its passing frequency band by heating the slab by a raise ΔT of its temperature T_0, i.e. create a transition $\lambda_0 \rightarrow \lambda_0 + \Delta\lambda = 2d \sin\theta_{\text{Bragg}}$ *in such a way that the two passing bands* $\lambda \pm \varepsilon$ *and* $\lambda + \Delta\lambda \pm \varepsilon$ *no longer overlap*. Considering the neutrons coherence length $\delta\lambda/\lambda \sim 10^{-5}$ one sees that one can indeed obtain such a result (with $\Delta\lambda > \delta\lambda$) if one introduces into the coils a field $B_0 \simeq 20$ kG and $\Delta T \sim 15$°C and use superconductive temperatures for the coils. This implies that the heated interference filter no longer transmits the observed neutrons of the incoming beam unless they have lost a quanta $\Delta E = 2\mu B_0 = \hbar\omega_{rf}$ in one *or* the other coil. In this way (through a NRM) we can establish the existence of individual neutron-energy losses in individual spin-flip

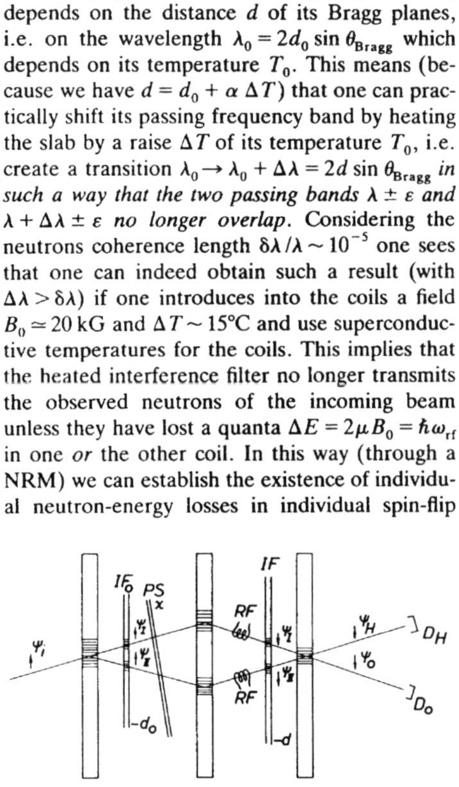

Fig. 2. Schematic arrangement of the introduction of interference filters to show that each neutron has lost a quanta $\Delta E = \hbar\omega_{rf}$ through the spin-flipping processes. Horizontal lines in the slabs and filters represent Bragg planes and I_0 and I_H the neutron detectors. IF_0 is the interference filter at temperature T_0 with Bragg distance d_0 and IF is the heated I-filter at temperature $T > T_0$ with Bragg distance d. RF are the spin-flip coils at superconducting temperature and PS is the phase shifter.

processes. Of course this argument will be strengthened if the time-beat survives once one has limited the dimensions of the incoming individual neutron wave packets ψ_i by a fast chopper so that one can be sure that the ψ_I and ψ_{II} wave packets *do not* overlap as will be done in forthcoming Grenoble experiments.

Of course if one thinks (like B–H) that the laws of conservation have only a statistical reality, then, at its present stage, the double resonance Grenoble experiments *do not* distinguish between B–H and E–de B since we cannot tell directly (i.e. measure) simultaneously through which slit the individual neutron has gone *and* observe simultaneously interference properties on the detectors.

As a consequence if one accepts energy–momentum violation in individual microprocess *then one should perform new experiments, to distinguish between the B–H and E–de B points of view.*

We conclude this discussion with three remarks.

First remark: As stated above, the BRT results strongly confirm the validity of the statistical predictions deduced from the usual formalism of quantum mechanics... a conviction shared by B–H and E–de B supporters. They raise however two crucial questions for both the B–H and E–de B supporters.

To quote BRT's own comments [1] on the one coil and the time beat (with two coils) experimental results:" How can each neutron in the spin-superposition experiment be transferred from an initial pure state in the z-direction into a pure state in the x-direction behind the interferometer, if no spin turn occurs in one beam and a complete spin reversal occurs in the other beam path? How can every neutron have information about which beam to join behind the interferometer, when a slightly different energy exchange occurs in both beams inside the interferometer and the time of flight through the system?"

Evidently these are now new serious questions for all possible interpretations of the quantum formalism. As shown before it now looks as if the B–H followers must reduce energy–momentum (and angular momentum) conservation to the status of a purely statistical property... which is not satisfied in individual microprocesses. On the other hand as we shall now show, the E–de B followers can only maintain the absolute character of these conservation laws *if they accept that the real (pilot) field is associated with a real distribution of energy–momentum density which surrounds and accompanies the particle's motions in space–time.* This implies of course

(i) that the particles exchange in general energy–momentum with the vacuum but controlled by their own Ψ fields.

(ii) that the terms "empty-wave" or "ghost waves" introduced by Einstein are misleading since they suggest that such waves not only are empty of the particle aspect of micro-objects but of energy momentum as well. As we shall now show, this is evidently not possible if one wants to preserve energy–momentum conservation in the de Broglie–Bohm trajectories (and random quantum jumps) where one knows that energy is not a constant of the motion, even in the absence of external fields of forces.

To clarify this point, let us first recall Einstein's distinction between the observability of the energy–momentum tied to fields and to particles. For Einstein the fields' energy–momentum densities are not directly observable but only detectable through modifications of the behaviour of test particles. In other terms what we observe (detect) directly is only the disturbance of the behaviour of "free" test particles imbedded in the field which is an indirect measurement in this sense. In other terms all observations are finally made on (and reducible, to observations of) the particle aspect of matter. As one knows, within the frame of the causal stochastic interpretation of quantum mechanics, the behaviour of nonrelativistic spin $\frac{1}{2}$ particles is described by the relation

$$i\hbar \frac{\partial \Psi}{\partial t} = H\Psi \,, \tag{2}$$

where the two-component Pauli spinor can be written in terms of the usual Euler angles $\theta(\boldsymbol{x}, t)$, $\phi(\boldsymbol{x}, t)$ and $\psi(\boldsymbol{x}, t)$ in the form

$$\Psi(\boldsymbol{x}, t) = R(\boldsymbol{x}, t) \begin{pmatrix} \cos \theta/2 \cdot \exp[\mathrm{i}(\psi + \phi)/2 \\ \mathrm{i} \sin \theta/2 \cdot \exp[\mathrm{i}(\psi - \phi)/2 \end{pmatrix},$$

where R denotes a real amplitude (connected to a density $\rho = R^2$) with the Hamiltonian

$$H = \frac{-\hbar^2}{2m} \left(\boldsymbol{\nabla} - \frac{\mathrm{i}e}{c} A\right)^2 + V + \frac{e\hbar}{2mc} \boldsymbol{\sigma} \cdot \boldsymbol{H}, \quad (3)$$

where the symbol $\boldsymbol{A}$ denotes the components of a vector A. We then define

– a local drift velocity field (4)

$$\boldsymbol{v} = \frac{\hbar}{2mi} \frac{(\Psi^* \boldsymbol{\nabla} \Psi - \Psi \boldsymbol{\nabla} \Psi}{\rho} - \frac{e}{mc} A$$

$$= \frac{\hbar}{2m} (\boldsymbol{\nabla}\psi + \cos \theta \boldsymbol{\nabla}\phi) - \frac{e}{mc} A, \quad (5)$$

– a local spin vector

$$s = \frac{\hbar}{2} \left(\frac{\Psi_i^\dagger \boldsymbol{\sigma} \Psi}{\rho}\right), \quad (6)$$

where $\boldsymbol{\sigma}$ denote the three usual Pauli matrices which transform like vector components.

From the preceding equations one immediately deduces

– the (local) fluid energy

$$W = \frac{mv^2}{2} + eV - \frac{\hbar^2}{2m} \frac{\Delta\rho^{1/2}}{\rho^{1/2}}$$

$$+ [(\boldsymbol{\nabla}\theta)^2 + \sin^2\theta(\boldsymbol{\nabla}\phi)^2] + \frac{e\hbar}{2mc} s \cdot \boldsymbol{H}, \quad (7)$$

– the energy–impulse tensor density

$$T_{ij} = m\rho v_i v_j + \frac{\hbar^2}{4m} \partial_i \rho \partial_j \rho$$

$$+ \frac{\hbar^2}{4m} [\partial_i \theta \partial_j \theta + \sin^2 \theta \partial_i \phi \partial_j \phi]. \quad (8)$$

As stated above the fluid and particle exchange energy in such a way that the geometric sum of all forces is zero since we get

$$\frac{\hbar}{2} \boldsymbol{\nabla} \left(\frac{\partial \psi}{\partial t} + \cos \theta \frac{\partial \phi}{\partial t}\right) - \boldsymbol{\nabla} W = 0 \quad (9)$$

where $\boldsymbol{\nabla} W$ represents the set of fluid forces which imply an expense of energy and the centrifugal force of the particle.

Moreover the existence of T_{ij} and ρW imply that a Pauli wave packet $\Psi = (\Psi_1, \Psi_2)$ carries energy–momentum and spin which are modified by the boundary conditions and external fields. The corresponding Madelung fluid can thus be represented by a fluid of spinning tops where neighbouring spins tend to become and remain parallel [12].

The (localized) spinning particle associated with this wave packet beats in phase with the surrounding wave, has its spin aligned with the spins of the surrounding wave and follows (on the average) the drift lines defined by relation (4).

Detailed calculations (utilizing computers) have recently given concrete space–time causal representations of all the quantum mechanical solutions utilized in the Grenoble experiments [12]. They show that an alternative consistent space–time description of the neutron's behaviour is a distinct possibility which does not disprove but only completes the usual predictions of quantum mechanics.

To conclude, what we now propose, within the E–de B model, is

(1) to assume that the Ψ-pilot field is associated with a real physical energy–momentum field distribution (tensor density) ρW and T_{ij} which accompanies the particle's motion,

(2) to assume that (except in special physical situations plane waves etc.) there is a real continuous energy exchange between the vacuum and the particles controlled by their Ψ-fields.

(3) to assume that the particle (neutron) moving along the drift trajectories (defined) by (4) not only beats in phase with the surrounding wave but is also endowed with the same local spin vector orientation.

Assumption (2) evidently yields [2] a simple interpretation of Rauch's questions which preserves energy–momentum (and angular momentum) conservation. Indeed, this implies that in the case of the double resonance experiments (with $\omega_{r1} \neq \omega_{r2}$)

– the wave packets Ψ_{I} and Ψ_{II} lose a different

energy density when passing though the coils since $\Delta E_1 = \hbar\omega_{r1} \neq \Delta E = \hbar\omega_{r2}$ and both waves flip their spin,
– the neutron itself loses a different energy ΔE_1 *or* ΔE_2 according to its path,
– in the interference region (i.e. the third slab) the superposition of the two states with $|z\downarrow\rangle$ but with different energies $E - \Delta E_1$ and $E - \Delta E_2$ yields an energy dependence $(\hbar\omega_{r1} - \hbar\omega_{r2})$ which generates the observed time-dependent oscillation. In other terms because both waves have really lost different energies in both coils and the neutron (which beats in phase with their superposition) "knows" that there have been two different paths and "adapts" itself to this superposition: so that the conservation laws remain valid at all times.

Second remark: The BRT experiments [1] combined with the new experiment described in fig. 2 represent a step forward in the experimental analysis of the interferences of individual particles with themselves. One knew of course that one can combine cold neutron energy losses with interferences by placing detectors higher than their sources. However one had never analyzed this loss quantitatively or associated it with one path only. The experiments analysed show indeed (i) not only that each neutron has flipped its spin but also (ii) that (if the experiment of fig. 2 succeeds . . . as predicted by quantum mechanics) each spin-flip is thus associated with the loss of a single quanta $\Delta E = 2\mu_0 B_0 = \hbar\omega_{rf}$ corresponding to a resonance with one spin-flipper only – provided of course that energy–momentum is always conserved in all individual microprocesses.

Third remark: These experiments on particles interfering individually with themselves also represent an experimental improvement on the known tests of the essential new property of quantum mechanics (postulated long ago and theorized by de Broglie and Dirac) that isolated elementary particles interfere with themselves. One knew already that one can obtain coherent interferences from sets of bosons coming from different coherent and incoherent sources (such as the Handbury–Brown–Twiss effect) but one had never done it for fermions appearing one by one. This implies that the associated Ψ wave describes not only a statistical distribution but also real properties associated with each individual neutron. Indeed one sees that with a probability 1 they never enter certain regions (i.e. the dark fringes of the interference pattern) which can only be calculated with the help of Ψ_I *and* Ψ_{II} of fig. 1. Following Einstein's definition [5] this implies that the Ψ-field describes an objective reality associated with individual micro-objects.

Acknowledgements

The author wants to thank Profs. H. Rauch and D. Greenberger and also Drs. Kyprianidis and Zeilinger for enlightening discussions and suggestions. He also wants to thank the Atominstitut des Österreichishen Universitäten for an invitation which made this research possible.

References

[1] G. Baburek, H. Rauch and D. Tuppinger, Phys. Rev. A 34 (1986) 2600.
[2] For references, see J.P. Vigier, Pramana-J. Phys. 25 (1985) 397.
[3] N. Bohr, Discussions with Einstein . . . in Einstein Philosopher Scientist, A. Schilpp and Evanston (1949).
[4] W. Heisenberg, Physics and Philosophy (Benjamin, New York, 1960).
[5] A. Einstein, B. Podolsky and N. Rosen, Phys. Rev. 47 (1935) 777.
[6] E. Wigner, Z. Phys. 133 (1932) 101.
[7] N. Cufaro-Petroni, A. Garuccio, F. Selleri and J.P. Vigier, C.R. Acad. Sci. Paris 299 (1980) 111.
[8] See the corresponding discussion in L. de Broglie, Non-Linear Wave Mechanics (Elsevier, Amsterdam, 1960).
[9] A.H. Compton, Phys. Rev. 21 (1923) 483.
[10] A.H. Compton and A. Simon, Phys. Rev. 25 (1925) 306.
[11] N. Bohr, H.A. Kramers and J.C. Slater, Phil. Mag. 47 (1924) 785.
[12] C. Dewdney, P.R. Holland and A. Kyprianidis, Phys. Lett. A 119 (1986) 259. For a review of preceding works on the causal model of the Puali equation see T. Takabayasi, Il Nuovo Cimento III (1956). D. Bohm et al. Il Nuovo Cimento I (1955) and P. Hillion, thesis no. 4014, Paris University (1957).

This page is deliberately left blank.

Reprinted from *Il Nuovo Cimento*, Vol. 88B, No. 1, pp. 20-28, Copyright (1981)
with permission from the Società Italiana di Fisica.

Positive Probabilities and the Principle of Equivalence for Spin-Zero Particles in the Causal Stochastic Interpretation of Quantum Mechanics.

P. R. HOLLAND and J. P. VIGIER

Laboratoire de Physique Théorique, Institut Henri Poincaré
11, rue Pierre et Marie Curie, 75231 Paris Cedex 05, France

(ricevuto l'11 Dicembre 1984)

Summary. — The effect of gravity on spin-0 particles is discussed in the context of the causal stochastic interpretation of quantum mechanics (SIQM). It is shown through the use of the quantum potential that, for positive probabilities which result from SIQM, the weak principle of equivalence breaks down. The theory nevertheless remains compatible with the geometrical character of quantum gravity.

PACS. 03.65. – Quantum theory; quantum mechanics.

1. – Introduction.

According to the de Broglie-Bohm [1,2] causal interpretation of quantum mechanics the wave function describes an objectively real field (the quantum potential) which guides a particle along a trajectory having simultaneously well-defined « true values » of position and velocity, the latter having reality independent of any measurement. The statistical aspects of quantum theory according to the SIQM are assumed to come about through the uncontrollable character of both subquantal fluctuations and initial conditions. They do not primarily relate to the results of measurements [3]. As a result, by recasting

(1) L. DE BROGLIE: *Non-Linear Wave Mechanics* (Elsevier, Amsterdam, 1960).
(2) D. BOHM: *Phys. Rev.*, **85**, 166 (1952).
(3) M. CINI: *Nuovo Cimento B*, **73**, 27 (1983).

the quantum-mechanical equations so that they take on the form of classical equations (*e.g.* Hamilton-Jacobi) one is able to interpret quantum phenomena in terms of readily visualizable real space-time processes. The novel feature of this theory in relation to classical mechanics and field theory presented in the same language is the appearance of the quantum potential, a highly nonlocal function of the parameters of extended particles and of their environments (including boundary conditions). It is not the use of classical concepts as such which is the virtue of this model but rather that it is able to provide an intuitive picture of phenomena in the microdomain, something which is missing from the usual approach. As a consequence the new nonlocal version of the causal interpretation is able to make more detailed predictions (for example in relation to the distribution of trajectories in the two-slit ([4]) and neutron interferometry ([5]) experiments) than the Copenhagen interpretation. In fact, it is through the use of the quantum potential that we can delineate the limits of usefulness of the classical way of thinking in a way that does not forbid further enquiry into actual quantum processes (as for example Bohr's notion of wholeness does). Thus, having a real space-time representation of the motion of particles acted on by the quantum potential before us, it is possible to introduce further assumptions as to underlying subquantum processes, for example Dirac's covariant random ether model ([6]).

In the following we shall discuss the causal theory of a spin-0 particle moving in an external gravitational field, particularly in relation to the weak equivalence principle (WEP). To do this consistently, however, we must first discuss the problem of negative probability solutions which result mathematically from the Klein-Gordon equation. Indeed the introduction of real particle paths necessarily implies a physical limitation to positive probabilities ([7,8]).

2. – The positive probability character of Klein-Gordon solutions.

Let us start from an arbitrary Riemannian space-time M_4 with line element $ds^2 = \eta_{\mu\nu} dx^\mu dx^\nu$. By writing $\psi = \exp[P + iS/\hbar]$, the Klein-Gordon equation

$$\left(\Box + \frac{m^2c^2}{\hbar^2}\right)\psi = 0\,, \quad \Box = \frac{1}{\sqrt{-\eta}}\,\partial_\mu\sqrt{-\eta}\,\eta^{\mu\nu}\,\partial_\nu\,, \qquad \eta = \det \eta_{\mu\nu}\,, \tag{1}$$

([4]) C. Philippidis, C. Dewdney and B. Hiley: *Nuovo Cimento B*, **52**, 15 (1979).
([5]) C. Dewdney, Ph. Gueret, A. Kyprianidis and J. P. Vigier: *Phys. Lett. A*, **102**, 291 (1984).
([6]) J. P. Vigier: *Astron. Nachr.*, **303**, 55 (1982).
([7]) N. Cufaro-Petroni, A. Kyprianidis, Z. Marić, D. Sardelis and J. P. Vigier: *Phys. Lett. A*, **101**, 4 (1984).
([8]) A. Kyprianidis, D. Sardelis and J. P. Vigier: *Phys. Lett. A*, **100**, 228 (1984).

may be written equivalently as

(2) $$\eta^{\mu\nu}\partial_\mu S\partial_\nu S = m^2c^2(1+Q) = M^2c^2 \text{ with } Q = \frac{\hbar^2}{m^2c^2}(\Box P + \eta^{\mu\nu}\partial_\mu P\partial_\nu P)\,,$$

(3) $$\partial_\mu(\sqrt{-\eta}\,j^\mu) = 0 \qquad \text{with } j_\mu = \exp[2P]\partial_\mu S\,,$$

where the function Q denotes the quantum potential, M de Broglie's « variable rest mass » and j^μ is the probability density current. The assumption introduced by the causal approach is that, together with a location $x^\mu = (ct, \boldsymbol{x})$, a particle has a well-defined 4-velocity

(4) $$u^\mu = \frac{dx^\mu}{ds} = (Mc)^{-1}\eta^{\mu\nu}\partial_\nu S\,, \qquad \eta_{\mu\nu}u^\mu u^\nu = 1\,,$$

where s represents the proper time along paths parallel to j^μ. From the modified Hamilton-Jacobi equation (2) we obtain the equation of motion of a quantum particle in Newtonian form:

(5) $$\frac{du^\mu}{ds} + \begin{bmatrix}\mu \\ \nu\;\sigma\end{bmatrix} u^\nu u^\sigma = \tfrac{1}{2}(\eta^{\mu\nu} - u^\mu u^\nu)\partial_\nu \log(1+Q)\,,$$

where $\left[{}^{\;\mu}_{\nu\;\sigma}\right]$ are the Christoffel symbols of $\eta_{\mu\nu}$.

Working within the spirit of general relativity we may reformulate eq. (5) in M_4 as a geodesic in a space-time of altered geometry, which we denote as Q_4, with line element $ds'^2 = g_{\mu\nu}dx^\mu dx^\nu$ [1,9].

To do this we note that the Hamilton-Jacobi equation (2) may be written as an equation for a « free » particle of mass m in Q_4:

(6) $$g^{\mu\nu}\partial_\mu S\partial_\nu S = m^2c^2\,,$$

where

(7) $$g_{\mu\nu} = (1+Q)\eta_{\mu\nu}\,.$$

Then in the geometry Q_4 we have a new causal velocity

$$V^\mu = \frac{dx^\mu}{ds'} = (1+Q)^{-\frac{1}{2}}u^\mu\,, \qquad V_\mu = g_{\mu\nu}V^\nu = (mc)^{-1}\partial_\mu S$$

with $g_{\mu\nu}V^\mu V^\nu = 1$ and $ds' = (1+Q)^{\frac{1}{2}}ds$. From (6) we find the equation of motion equivalent to (5):

(8) $$\frac{dV^\mu}{ds'} + \begin{Bmatrix}\mu \\ \nu\;\sigma\end{Bmatrix} V^\nu V^\sigma = 0\,,$$

---

(9) C. Fenech and J. P. Vigier: *C. R. Acad. Sci. Paris*, **293**, 249 (1981).

where $\left\{ {\mu \atop \nu\ \sigma} \right\}$ are the Christoffel symbols deduced from $g_{\mu\nu}$ and particles follow the geodesics of the new $g_{\mu\nu}$ metric.

Now an objection may be raised against the above theory based on the guidance formula (4). After all, during some physical process the factor $1+Q$ may become negative so that it appears that a particle motion which was initially timelike may become spacelike (for the remainder of this section we assume for ease of discussion that $\eta_{\mu\nu}$ is globally Minkowskian). Indeed DE BROGLIE (1) has given a physical example in which real superluminal motions apparently occur, in the region of the Wiener fringes of a plane Klein-Gordon wave reflected from a partially reflecting mirror. Clearly, if the momentum $(\partial_\mu S)$ in (2) resulting from a specific solution is spacelike, so is the current j^μ (3) which implies a negative probability density $(\exp[2P]\,\partial_0 S)$ accompanying negative energy $(\partial_0 S)$. However, such solutions result from the introduction of nonphysical initial conditions associated with spacelike probability currents and should, therefore, be excluded. Thus there remains the problem of timelike initial probability currents having negative energies.

This dilemma may be resolved in the following way. RIZOV *et al.* (10), following FESHBACH and VILLARS (11), have discussed recently how one may decouple the positive- and negative-energy solutions of the Klein-Gordon equation. Starting from a solution to (1), construct a two-component wave function

$$\Psi = \left(\frac{\hbar}{2mc}\right)^{\frac{1}{2}} \begin{pmatrix} i\partial_0\psi + \dfrac{mc}{\hbar}\psi \\ i\partial_0\psi - \dfrac{mc}{\hbar}\psi \end{pmatrix}.$$

(1) may then be written in Hamiltonian form as

$$i\hbar\frac{\partial\Psi}{\partial t} = H\Psi\,, \tag{9}$$

where H is a 2×2 matrix operator. With respect to the indefinite inner product

$$\langle\Phi, \Psi\rangle = \int \Phi^*\sigma_3\Psi\,\mathrm{d}^3x = \int \varphi^*\, i\hbar\overleftrightarrow{\partial}_0\psi\,\mathrm{d}^3x$$

it then follows that positive (negative) energy solutions of (9) have positive (negative) probability. We suppose now that all physical states correspond to those parts of the Klein-Gordon solutions which, following Einstein's basic

(10) V. A. RIZOV, U. SARDJIAN and I. T. TODOROV: *On the Relativistic Quantum Mechanics of Two Interacting Spinless Particles* (Orsay preprint, 1984).
(11) H. FESHBACH and F. VILLARS: *Rev. Mod. Phys.*, **30**, 24 (1958).

principles, have positive energy and probability for particles and antiparticles.

On the basis of the SIQM we shall now generalize, in a very simple manner, the above result of Rizov *et al.* By using (4), conservation equation (3) may be expressed in terms of a fluid density $\varrho = Mc \exp[2P]$ as ([12])

$$\frac{\mathrm{D}\varrho}{\mathrm{D}s} \equiv \partial_\mu(\varrho u^\mu) = 0\,.$$

From this it follows that $\omega M \exp[2P] = C$, where ω is a volume element of fluid and C is a real constant along any line of flow (C varies from one drift line to another). Thus, if initially $M > 0$ (and $\exp[2P] > 0$), then it remains so throughout a particle motion so that a positive-energy solution, associated with positive probability density, develops into the same motion —one cannot pass to a negative-energy solution. Rizov *et al.* show that the decoupling of positive- and negative-energy solutions is maintained when one inserts certain well-behaved potentials in the Klein-Gordon equation. Our demonstration, on the other hand, extends readily to the case of any vector potential introduced into eqs. (2) and (3) (provided $(P_\mu - (e/c)A_\mu)$ remains timelike) and this can also be extended to n-body systems. Moreover, this separation of positive-energy, positive-density and negative-energy, negative-density solutions is maintained in weak external gravitational fields. A further discussion of these points will be given in a separate paper.

According to the causal stochastic interpretation any Klein-Gordon fluid thus comprises particle and antiparticle motions (both being necessary in order to deduce the Klein-Gordon equation and a H-theorem ([13])). These motions are always timelike and of positive energy and probability and the guidance formula (4) refers to the average motion of an ensemble along a drift line. Any spacelike solution does not describe real space-time motions associated with real ψ waves since the latter result from Klein-Gordon propagators acting on realistic initial conditions containing no spacelike probability currents. In addition, (7) is a relation which only holds on the average, the $g_{\mu\nu}$'s being subject to real random fluctuations which do not, however, destroy the actual timelike geodesic motion of a particle in the fluid.

To summarize, we may say that we answer the objection anticipated above by showing how initial timelike motions may be separated into positive-energy, positive-density and negative-energy, negative-density subsets which are preserved by the equations of motion. All physical motions are thus associated with positive-energy, positive-density particles and antiparticles, propagating into the future and past, respectively.

([12]) F. Halbwachs: *Théorie Relativiste des Fluides a Spin* (Gauthier-Villars, Paris, 1960).

([13]) A. Kyprianidis and D. Sardelis: *Lett. Nuovo Cimento*, **39**, 337 (1984).

3. – The weak equivalence principle according to the causal interpretation.

We return now to the study of the Klein-Gordon system for a gravitational metric on the understanding that relations (4) and (7) hold only for weak fields. Our remarks will also apply to a Schrödinger particle moving in an external gravitational potential V, the weak field nonrelativistic limit of (5) being

$$\frac{d\boldsymbol{v}}{dt} = -\frac{1}{m}\nabla(V+Q)\,, \qquad Q = -\frac{\hbar^2}{2m}\frac{\nabla^2 R}{R}\,, \qquad R = e^P\,. \tag{10}$$

We see immediately from eqs. (5) and (10) that, in the absence of all external fields ($V = 0$, $\eta_{\mu\nu}$ globally Minkowskian), a spin-0 quantum particle can never, in general, be said to be « free » since it is always subject to a quantum force. Thus, if we study the motion of such a particle moving in an external gravitational field, it is in fact subject to two forces. Moreover, the quantum force, derived from the quantum potential Q, is dependent on the mass of the particle, *i.e.* the mass does not drop out of the problem as it does in classical gravitation theory. The consequence of this, of course, is that all bodies in the quantum domain do not fall with the same acceleration in a gravitational field independent of their constitution and, therefore, WEP breaks down, due to the intervention of the quantum potential.

The incompatibility of WEP with quantum mechanics (Copenhagen version) has been noted previously, but in a different sense [14].

It is argued that the quantum formalism is inherently mass dependent since it is formulated in terms of x and p and, therefore, wave phenomena are direct functions of mass. Thus any classical effect which is independent of mass cannot be extrapolated to the microdomain. Actually all that one can conclude on the basis of the Copenhagen interpretation is that « for low quantum states it is less true that the results violate equivalence than that equivalence just doesn't apply » [15]. This follows, of course, since the notion of particle trajectory is absent and it is, therefore, unclear how one would formulate a quantum analogue of WEP. In the causal stochastic formulation, on the other hand, it has meaning to pose an equivalence principle and it is then an open question which can be unambiguously solved as to whether the principle holds. For the first time we can see precisely the way in which the actual particle acceleration depends on the mass in an external gravitational field, a picture which the orthodox viewpoint cannot provide.

Inserting in (10) the Newtonian potential $V = m\varphi$ (assuming this is the correct form to use in the quantum case), we see that the mass dependence of acceleration is due just to Q, that is it is entirely of quantum-mechanical origin.

([14]) D. M. Greenberger: *Rev. Mod. Phys.*, **55**, 875 (1983).
([15]) D. Greenberger: *Ann. Phys. (N. Y.)*, **47**, 116 (1968).

solutions guide particles along trajectories which contradict the equation of motion (5) of a particle of negligible gravitational effect which follows from the Klein-Gordon equation itself, once we have introduced the trajectory assumption of the causal interpretation into the theory of the wave equation. More generally, we may say that the simultaneous use of Einstein's equations and quantum wave equations implies that the gravitational equations will hold unmodified in the small and, as we have seen, this leads to an inconsistency since two conflicting equations of motion are in operation. What is suggested here is that some modification of Einstein's equations may be called for in order that they are consistent with the causal quantum equation of motion when applied in the microdomain.

The fact that the motion of a quantum particle in a gravitational field does not depend solely on the particle's environment is considered by GREENBERGER [14] to « lend a distinctly nongeometrical cast to quantum theory » and indeed to undermine the geometrical approach to gravity. Such a conclusion is, however, not necessary if we broaden our concept of geometry. As we have seen, and in agreement with GREENBERGER, the notion of a « test particle » which « feels out » a force field without in any way imposing itself on that field, a notion that is central to classical gravitation theory, does not exist in the quantum domain. The environment cannot be separated from the characteristics of the particle. This fact, though, does not in itself imply a nongeometrical aspect to quantum gravity. We have shown in sect. **2**, for example, how we may recover geodesic motion for a quantum particle in a gravitational field. Thus, allowing that the metrical attributes of space-time are functions of all the contributing parts of a process, we can employ a geometrical theory even in the quantum domain. This is so, of course, in the pure quantum case when $\eta_{\mu\nu}$ is globally Minkowskian. Naturally, expressing the motion in the form (8) does not reintroduce a weak principle of equivalence since particles of different mass will move through regions having different $g_{\mu\nu}$'s and, therefore, their motions will differ (*).

5. – Conclusion.

This result leads us to suspect that the quantum field equations (2) and (3) (or at least their derivatives) may be reformulated as geometrical relations.

(*) Such a geometrization can be performed with other force fields, *e.g.* electromagnetic, and a notion similar to the wholeness discussed above appears in an entirely classical context [20,21].

(20) P. R. HOLLAND: *Phys. Lett. A*, **91**, 215 (1982).

(21) P. R. HOLLAND and C. PHILIPPIDIS: *Anholonomic Deformations in the Ether: A Significance for the Potentials in a New Physical Conception of Electrodynamics* (Institut Henri Poincaré, preprint, 1984).

In the relativistic case, however, it may be seen from (2) and the right-hand side of (5) that not only does the quantum potential exert a nonclassical force but also that it involves $\eta_{\mu\nu}$ and there is a coupling between $\eta_{\mu\nu}$ and Q so that we expect gravity to have a direct mass-dependent effect itself. In other words, it is a prediction of this model that in the relativistic regime the deviation of a quantum particle from a geodesic will depend not only on purely quantum parameters (including the mass) but also on the external field coupled to these parameters (whereas for weak fields and low velocities this coupling may be neglected). As emphasized in an analogous discussion (16) on the role of the electromagnetic potentials in the Aharonov-Bohm effect, as causally treated, the quantum potential plays the role of a mediator and so renders understandable the way in which a classical external potential can have nonclassical effects when there is no sign of such effects in the classical limit. In the present instance the gravitational field has a real physical effect on a Klein-Gordon particle moving through it (when classically no such effect is to be expected) through the medium of the quantum potential which provides a vehicle for such actions. As is easily seen, in the classical limit ($\hbar \to 0$) Q vanishes, the mass drops out and the weak principle of equivalence is recovered, in both (5) and (10).

4. – Gravitation and the quantum potential approach.

That the Newtonian potential is indeed the correct one to insert in the Schrödinger equation has been demonstrated by the COW neutron interferometry experiment (17). Moreover, as shown in detail by GREENBERGER and OVERHAUSER (18) and as follows from our equations (1)-(5), we expect the strong principle of equivalence to hold unaltered in quantum mechanics. Support for this contention is provided in the Schrödinger case by the COW experiment. The reasonableness of assuming a classical metric potential in the Klein-Gordon equation, however, is qualified by the following remarks. It is considered to be one of the virtues of Einstein's field equations that they imply the equation of motion of a test particle (one with no other attribute than mass which is of negligible gravitational effect), that is classical geodesic motion (19).

There appears to be some inconsistency, however, in inserting solutions to Einstein's equations into the Klein-Gordon equation, as is normally done when one studies the latter in given classical gravitational fields, since these

---

(16) C. PHILIPPIDIS, D. BOHM and R. D. KAYE: *Nuovo Cimento B*, **71**, 75 (1982).
(17) R. COLELLA, A. W. OVERHAUSER and S. A. WERNER: *Phys. Rev. Lett.*, **34**, 1472 (1975).
(18) D M. GREENBERGER and A. W. OVERHAUSER: *Rev. Mod. Phys.*, **51**, 43 (1979).
(19) C. W. MISNER, K. S. THORNE and J. A. WHEELER: *Gravitation* (W. H. Freeman, San Francisco, 1973).

Moreover, the conformally flat space-time metric (7) (neglecting gravity now) may be a special case of a general quantum metric which would be subject to a set of nonlinear field equations. In this connection it is interesting to note that de Broglie's variable rest mass (2) embodies a quantum analogue of Mach's principle. That is, the inertia of a quantum particle is determined by a scalar field which is a function of all relevant processes occurring in the environment of the particle. Q then plays the role of the field introduced by BRANS and DICKE [19] to supplement Einstein's theory with an explicit statement of Mach's principle.

We conclude that the quantum potential model, through which we may understand the irreducible coupling between particle and environment which is the outstanding feature in the microdomain, may be particularly suited to a geometrical approach (whether gravity is present or not) since

a) the quantum potential acts to determine particle motion as a field of force, albeit one with nonclassical features, and

b) the quantum potential appears directly in the conformal metric of space-time so that particle motion is a geodesic in Q_4.

* * *

One of us (PRH) thanks the Leverhulme Trust for financial support.

This page is deliberately left blank.

Reprinted from *International Journal of Theoretical Physics*, Vol. 18, No. 11, pp. 807-818, Copyright (1979) with permission from Kluwer Academic/Plenum Publishers.

Markov Process at the Velocity of Light: The Klein–Gordon Statistic

N. Cufaro Petroni

University of Bari, Italy

J. P. Vigier

Institut Henri Poincaré, Paris, France

Received June 5, 1979

The Markovian random walk of a point at the velocity of light on a two-dimensional invariant space–time lattice is shown to yield the quantum statistic associated with the Klein–Gordon equation. Quantum mechanics thus appears as a particular case of Markovian processes in velocity space: and one justifies the introduction of Dirac's invariant "ether" as a possible physical stochastic subquantum level of matter which yields a realistic mechanical basis for recent attempts to reinterpret quantum mechanics in terms of material, causal, random behavior.

1. INTRODUCTION

Recent discussions on the Einstein-Podolsky-Rosen paradox (1935) have shown that quantum mechanics implies spacelike correlations between two linear polarizers which measure the rate of coincidence between the relative orientations of pairs of photons emitted in the S state. If a forthcoming crucial experiment of Aspect (1976) confirms this then the only possible "causal" (i.e., which preserves the fundamental fact that no individual particle can leave the light cone) way out of the resulting contradiction between relativity and the quantum theory of measurement seems to lie in the direction of an extension of the stochastic interpretations of quantum mechanics in terms of subquantum random fluctuations resulting from the action of a stochastic "hidden" invariant thermostat. Indeed these models (a) deduce the form of the quantum waves from the physical assumption that the stochastic jumps occur at the velocity of light; (b) interpret the preceding superluminal interaction in terms of superluminal propagation of a "quantum potential" (Vigier, 1979) which is not carried by individual particles but results from phaselike collective motions carried by the said thermostat.

The aim of the present paper is to analyze in a more precise way the physical and mathematical implications of these stochastic interpretations in the particular case of scalar particles.

In Section 2 we shall briefly discuss the physical properties of the only possible invariant undetectable relativistic thermostat known in the literature—i.e., Dirac's "ether" model: a model that provides a realistic physical basis for the above-mentioned interpretations.

In Section 3 we shall discuss the mathematical significance of the stochastic demonstrations already given in the literature starting among others with Bohm and Vigier (1954), Nelson (1966), de Broglie (1961), and the growing number of papers dealing with stochastic electrodynamics (De la Peña and Cetto, 1975).

2. THE SUBQUANTUM THERMOSTAT

All these models imply of course a modern revival of the old "ether" idea: a concept apparently definitively destroyed by the negative result of Michelson's experiment. As one knows, however, Dirac (1951) has shown that it is not so and that one can construct at least one material covariant "ether" perfectly compatible with relativity. It rests on the idea that through any point 0 passes a flow of stochastic particles and antiparticles (described in Figure 1 as particles moving backwards in time) whose momenta have the extremities of their four-vectors P^{μ} (with $P^{\mu}P_{\mu}=m^2c^2$) distributed with a uniform surface density on the two three-dimensional surfaces of the hyperboloids H_+ and H_-. They will thus remain invariant under all Lorentz transformations.

This stochastic relativistic distribution constitutes the only possible model for a physical undetectable thermostat for spin-zero particles into which we can study the relativistic analog of the classical nonrelativistic Brownian motion. Dirac has derived this from the indeterminacy principle. However, it differs from it by two new physical properties.

(a) Since the light cone behaves like an asymptotic accumulation manifold of Dirac's stochastic distribution we can assume that the corresponding stochastic jumps of a Brownian particle, submitted to its random action, occur practically at the velocity of light. Indeed, any given exchanged energy is statistically superseded by more energetic interactions.

(b) This ultrarelativistic Brownian motion includes the possibility of pair creation and/or annihilation. This is important since the mixture of

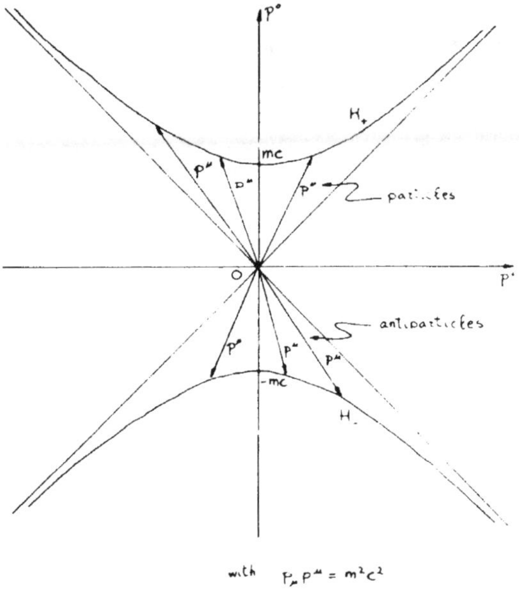

Fig. 1

particles and antiparticles has been shown to provide a realistic interpretation (Terletski and Vigier, 1961) of possible, negative, probability distributions.

The concrete analysis of this particular covariant case of stochastic motion can be carried along the two lines of demonstration utilized in nonrelativistic stochastic theory. The first line is just a relativistic generalization of the ideas introduced by Einstein and Smoluchowski into Brownian motion theory. Assuming that our particles are (1) carried along the lines of flow or a regular drift motion **v** of extended particles associated with a collective motion on the top of Dirac's thermostat, (characters in boldface type) denoting four-vectors, (2) jump stochastically at the velocity of light from one average drift line of flow to another and thus (for an ensemble of identical particles with arbitrary initial positions) reach an average mean conserved distribution $\rho(\mathbf{x})$; one can immediately demonstrate the stochastic force law first assumed by Nelson (1966), from which one deduces (Lehr and Park, 1977; Vigier, 1979; Guerra and Ruggiero, 1978) a stochastic wave $\psi(\mathbf{x})=[\rho(\mathbf{x})]^{1/2}\exp[(iS(\mathbf{x})/\hbar]$ with $\mathbf{v}=(1/m)\nabla S$, which satisfies the Klein–Gordon equation.

This demonstration, however, being based on averages taken over four-dimensional volume elements, does not connect directly the underlying particle behavior with known statistical models discussed in the mathematical literature, such as Markovian processes.

The aim of the present work is thus to extend to the preceding relativistic case the second line of approach discussed in the nonrelativistic literature, i.e., to study the random walk of a moving point on a lattice discussed by Chandrasekhar (1943) in a famous paper and later extended from elliptical to hyperbolical equations by Avez (1976). This will be done in the next section.

3. RANDOM WALK ON A COVARIANT LATTICE

To simplify our demonstration we shall limit ourselves to the study of a two-dimensional space–time case x^0, x^1. Indeed, as will be shown later, its extension to four dimensions presents no conceptual difficulty.

First one can check immediately that the points P_{nm} located at the intersection of the set of curves

$$\begin{aligned} &x^{0^2} - x^{1^2} = \lambda_n^2, \quad \lambda_0 > 0, \quad \lambda_n = \lambda_0 e^{n\delta}, \quad n, m = 0, \pm 1, \pm 2 \ldots \\ &x^1 = (\tanh\theta_m) x^0, \quad \theta_0 = 0, \quad \theta_m = m\delta, \quad \delta > 0 \end{aligned} \tag{3.1}$$

build an invariant discrete lattice (see Figure 2) in which the relativistic interval between $P_{n,m}$ and each $P_{n\pm 1, m\pm 1}$ is zero. The explicit expression for the $P_{n,m}$ coordinates is

$$x^0_{n,m} = \lambda_n \cosh\theta_m, \qquad x^1_{n,m} = \lambda_n \sinh\theta_m \tag{3.2}$$

The preceding lattice is clearly covariant since each point $P_{n,m}$ stands at the intersection of three intrisically invariant lines (i.e., a spacelike hyperbola and two isotropic light-cone-defining lines) which are transformed into themselves by any ortochronous Lorentz transformations.

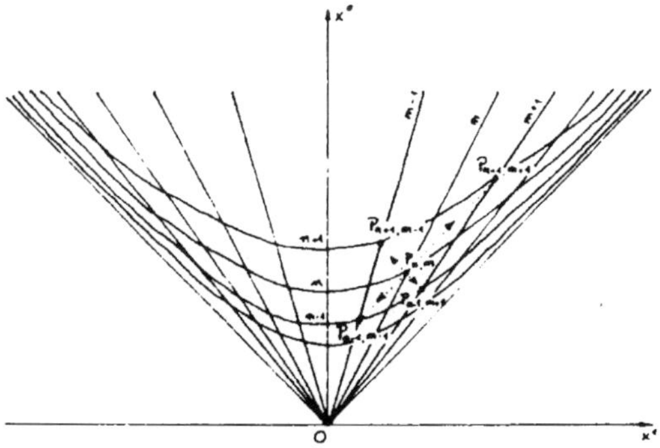

Fig. 2

The (finite) coordinate differences $\Delta_{t,s}x^0_{n,m}, \Delta_{t,s}x^1_{n,m}$ between the two points $P_{n,m}$ and $P_{n+t,m+s}$ $(t=\pm 1; s=\pm 1)$ connected by one stochastic jump will satisfy

$$\frac{\Delta_{t,s}x^1_{n,m}}{\Delta_{t,s}x^0_{n,m}} = \frac{s}{t} = \pm 1 \tag{3.3}$$

In order to describe random walks on this lattice let us now define two sets of stochastic variables $\{\varepsilon_1, \varepsilon_2, \ldots \varepsilon_j \ldots\}; \{\eta_1, \eta_2, \ldots \eta_k \ldots\}$ with $\varepsilon_j = \pm 1, \eta_k = \pm 1$ for every j, k. The sign of ε_j (η_k) determines the fact that in the corresponding jump the velocity (the time orientation) has changed its sign, $\varepsilon_j = -1$ $(\eta_k = -1)$ or has remained unchanged $\varepsilon_j = \pm 1$ $(\eta_k = \pm 1)$ with respect to the preceding jump.

One then checks immediately that the general expression for the displacement $D_N^{t,s}(n,m)$, after N jumps from the initial point $P_{n,m}$ and a first jump in the direction defined by (t,s), can be written as the development

$$D_N^{t,s}(n,m) = \frac{s}{t}\Big(\Delta_{t,s}x^0_{n,m} + \varepsilon_1 \Delta_{t\eta_1, s\eta_1\varepsilon_1} x^0_{n+t,m+s}$$

$$+ \varepsilon_1\varepsilon_2 \Delta_{t\eta_1\eta_2, s\eta_1\eta_2\varepsilon_1\varepsilon_2} x^0_{n+t(1+\eta_1), m+s(1+\eta_1\varepsilon_1)} + \cdots\Big) \tag{3.4}$$

The probabilities for the realization of the signs of ε_j, η_k (with $j=k=1$) are given by Table I.

The functions $F_N^{t,s}(n,m) = \langle f(x^1_{n,m} + D_N^{t,s}(n,m))\rangle$ are the mean values of a function f (defined on the lattice) over all random walks of N jumps; they satisfy the following system of recurrence relations [one for each value of (t,s)]:

$$F_N^{t,s}(n,m) = \big(1 - A\Delta_{-t,s}x^0_{n+t,m+s} - B\Delta_{-t,-s}x^0_{n+t,m+s} - C\Delta_{t,-s}x^0_{n+t,m+s}\big)$$

$$\times F_{N-1}^{t,s}(n+t,m+s) + A\Delta_{-t,s}x^0_{n+t,m+s} F_{N-1}^{-t,s}(n+t,m+s)$$

$$+ B\Delta_{-t,-s}x^0_{n+t,m+s} F_{N-1}^{-t,-s}(n+t,m+s)$$

$$+ C\Delta_{t,-s}x^0_{n+t,m+s} F_{N-1}^{t,-s}(n+t,m+s) \tag{3.5}$$

TABLE I

Probability	ε_1	η_1
$A\Delta_{-t,s}x^0_{n+t,m+s}$	-1	-1
$B\Delta_{-t,-s}x^0_{n+t,m+s}$	1	-1
$C\Delta_{t,-s}x^0_{n+t,m+s}$	-1	1
$1 - A\Delta_{-t,s}x^0_{n+t,m+s} - B\Delta_{-t,-s}x^0_{n+t,m+s} - C\Delta_{t,-s}x^0_{n+t,m+s}$	1	1

At the limit for $\delta \to 0$ (Heath, 1969; Kac, 1956) the lattice tends to recover all the interior of the light cone, the function $F_N^{t,s}(n,m)$ goes into the function $F^{t,s}(x^0,x^1)$ and relations (3.5) can be shown to go into a system of four differential equations ($t,s=\pm 1$), i.e.,

$$\frac{\partial F^{t,s}}{\partial x^0} = -\frac{s}{t}\frac{\partial F^{t,s}}{\partial x^1} - A(F^{-t,s}-F^{t,s}) - B(F^{-t,-s}-F^{t,s}) + C(F^{t,-s}-F^{t,s}). \tag{3.6}$$

One then sees immediately that the function

$$\Phi = (F^{1,1}+F^{1,-1}-F^{-1,1}-F^{-1,-1}) + i(F^{1,1}-F^{1,-1}+F^{-1,1}-F^{-1,1}) \tag{3.7}$$

is a solution of the free Klein–Gordon equation

$$\left[\left(\frac{\partial^2}{\partial x^{0^2}} - \frac{\partial^2}{\partial x^{1^2}}\right) - \frac{m^2c^2}{\hbar^2}\right]\Phi = 0 \tag{3.8}$$

when one writes $C=2A+4B$ and $2(A+B)^2 = m^2c^2/\hbar^2$ Q.E.D. (For details of deduction see the Appendix.)

In the preceding demonstration f is not arbitrary since it is correlated, through relation (3.7) with an average scalar density ρ and a scalar phase S (see Vigier, 1979) by the relation $\phi = \rho^{1/2}\exp(iS/\hbar)$. Indeed, one can demonstrate directly relation (3.8) with the help of a hydrodynamic picture which also yields Nelson's equation

$$m(D_c v - D_s u) = F^+ \tag{3.9}$$

This suggests three physical remarks.

(a) If one starts from a set of initial positions on a given hyperbola the function ϕ now represents an average relativistic diffusion process comparable to a sound wave (i.e., a regular collective motion) propagating within Dirac's "ether"-like vacuum and carrying a particle along v.

(b) Dirac's "ether," which creates stochastic jumps at the velocity of light, is apparently the only way to obtain such a covariant diffusion process.

(c) It also explains an essential characteristic of the said process, viz., its reversibility. As one knows, nonrelativistic stochastic processes are fundamentally irreversible, being associated with a steady loss of information about where the particle comes from. This situation is modified here by the minus sign in (3.9), which has been shown (Vigier, 1979) to result from the particle–antiparticle mixture included both in Dirac's "ether" and in our random walk. Of course our time-reversing steps just describe particle–antiparticle transitions and they both move forward in time; but they are necessary to recover an essential feature (viz., reversibility) of quantum mechanics.

4. GENERALIZATION AND CONCLUDING REMARKS

In order to achieve a generalization of our two-dimensional derivation we must remark that the preceding demonstration remains unchanged if we analyze a Markov process on simpler (not covariant) lattices. For example we can reproduce all formulas for the displacements and their consequences if we consider the following lattice:

$$\left.\begin{aligned} x^0_{n,m} &= n\Delta\tau \\ x^1_{n,m} &= m\Delta l \end{aligned}\right\} \qquad \frac{\Delta l}{\Delta\tau} = 1, \qquad n, m = 0, \pm 1, \pm 2, \ldots \tag{4.1}$$

As a consequence the covariance of the formulation at each step of a limiting process cannot be considered as a necessary requirement. However, the covariance of the result (Klein–Gordon equation) is obtained at the end of the process. This observation allows us to discuss formally the four-dimensional case as a straightforward generalization of lattices like (4.1), a procedure that evidently avoids all complications resulting from the construction of a four-dimensional covariant lattice.

To conclude, we want to make two remarks on the relation of the preceding stochastic derivation to demonstrations utilized until now in the literature.

(a) In opposition to the "classical" stochastic derivations of quantum statistics (Bohm and Vigier, 1954) it is not necessary here to consider a particle carried in a fluid wave described by the quantum mechanical equations. Indeed, our corpuscule is now just located in a sort of covariant "ether," not described a priori by a particular wave equation. A simple hypothesis on the stochastic behavior of the particle in this fluid is now sufficient to demonstrate directly the Klein–Gordon equation.

(b) The model yields a physical insight into the observed difference between classical Brownian motion and quantum Brownian motion (De la Pena and Cetto, 1975).

As one knows, in order to obtain the Schrödinger equation (nonrelativistic limit of Klein–Gordon equation) or the classical equations of Brownian motions, we must choose "a priori" a different sign in the dynamical equations obtained from Newton's law (Vigier, 1979). The present demonstration indicates that this choice has a physical basis: (1) in the existence of a Markov process at the velocity of light in velocity space, (2) in the simultaneous presence of particles and antiparticles (Feynman, 1948) in the observed statistics of quantum theory.

ACKNOWLEDGMENTS

The authors are grateful to Professors P. Malliavin and A. Avez for interesting mathematical suggestions which helped to clarify Section 3.

APPENDIX

In order to demonstrate in detail the transition from relation (3.5) to (3.8) we first remark that in the $\delta \to 0$ limit, the lattice tends to recover all the surface of the forward light cone and the number N of the jumps in our light velocity random walks becomes infinite: provided we keep fixed the space–time limits of the initial diffusion process. In this way the discrete function $F_N^{t,s}(n,m)$ becomes a continuous function depending on the initial point coordinates, namely $F^{t,s}(x^0,x^1)$.

Starting from (3.5) on the x^0 axis ($m=0$), if we subtract from both sides $F_{N-1}^{t,s}(n+t,0)$ and then we divide by the time intervals, we have

$$\left.\frac{\Delta F_N^{t,s}}{\Delta x^0}\right|_{(n,0)} = -\frac{s}{t}v\left.\frac{\Delta F_{N-1}^{t,s}}{\Delta x^1}\right|_{(n+t,0)} + \alpha A\left[F_{N-1}^{-t,s}(n+t,s) - F_{N-1}^{t,s}(n+t,s)\right]$$
$$\times \beta B\left[F_{N-1}^{-t,-s}(n+t,s) - F_{N-1}^{t,s}(n+t,s)\right]$$
$$+\gamma C\left[F_{N-1}^{t,-s}(n+t,s) - F_{N-1}^{t,s}(n+t,s)\right] \tag{A.1}$$

where

$$\Delta_t(x^0)_{n,0} = (x^0)_{n+t,0} - (x^0)_{n,0} = \lambda_0 e^{n\delta}(e^{t\delta}-1)$$
$$\Delta_s(x^1)_{n+t,0} = (x^1)_{n+t,s} - (x^1)_{n+t,0} = \lambda_0 e^{n\delta}e^{t\delta}\sinh(s\delta)$$
$$\Delta_{t,s}(x^0)_{n,m} = (x^0)_{n+t,m+s} - (x^0)_{n,m} = t\lambda_0 e^{n\delta}e^{t(ms+1)\delta}\sinh\delta$$
$$\Delta_{t,s}(x^1)_{n,m} = (x^1)_{n+t,m+s} - (x^1)_{n,m} = s\lambda_0 e^{n\delta}e^{t(ms+1)\delta}\sinh\delta$$

with
$$\frac{s}{t}v = \frac{\Delta_s(x^1)_{n+t,0}}{\Delta_t(x^0)_{n,0}} \underset{\delta\to 0}{\rightarrow} \frac{s}{t}$$

$$\left.\frac{\Delta F_N^{t,s}}{\Delta x^0}\right|_{(n,0)} = \frac{F_{N-1}^{t,s}(n+t,0) - F_N^{t,s}(n,0)}{\Delta_t(x^0)_{n,0}} \underset{\delta\to 0}{\rightarrow} \left.\frac{\partial F^{t,s}}{\partial x^0}\right|_{x^1=0} \tag{A.2}$$

$$\left.\frac{\Delta F_{N-1}^{t,s}}{\Delta x^1}\right|_{(n+t,0)} = \frac{F_{N-1}^{t,s}(n+t,s) - F_{N-1}^{t,s}(n+t,0)}{\Delta_s(x^1)_{n+t,0}} \underset{\delta\to 0}{\rightarrow} \left.\frac{\partial F^{t,s}}{\partial x^1}\right|_{x^1=0}$$

and
$$\alpha = \frac{\Delta_{-t,s}(x^0)_{n+t,s}}{\Delta_t(x^0)_{n,0}} \underset{\delta\to 0}{\rightarrow} -1$$

$$\beta = \frac{\Delta_{-t,-s}(x^0)_{n+t,s}}{\Delta_t(x^0)_{n,0}} \underset{\delta\to 0}{\rightarrow} -1$$

$$\gamma = \frac{\Delta_{t,-s}(x^0)_{n+t,s}}{\Delta_t(x^0)_{n,0}} \underset{\delta\to 0}{\rightarrow} 1;$$

so that, in the $\delta\to 0$ limit, (A.1) goes into (3.6) on the x^0 axis. From the four equations (A.1) we can now construct two complex equations:

$$\left.\frac{\Delta\varphi_N}{\Delta x^0}\right|_{(n,0)} = -v\left.\frac{\Delta\varphi_{N-1}}{\Delta x^1}\right|_{(n+tp)} - \big[\alpha A(\varphi_{N-1} - i\chi_{N-1}^*) + \beta B(\varphi_{N-1} - i\varphi_{N-1}^*)$$
$$- \gamma C(\chi_{N-1} - \varphi_{N-1})\big]\big|_{(n+t,s)}$$

$$\left.\frac{\Delta\chi_N}{\Delta x^0}\right|_{(n,0)} = V\left.\frac{\Delta\chi_{N-1}}{\Delta x^0}\right|_{(n+t,0)} - \big[\alpha A(\chi_{N-1} - i\varphi_{N-1}^*) + \beta B(\chi_{N-1} - i\chi_{N-1}^*)$$
$$- \gamma C(\varphi_{N-1} - \chi_{N-1})\big]\big|_{(n+t,s)} \tag{A.3}$$

with
$$\varphi_N = F_N^{1,1} + iF_N^{-1,-1}$$

and
$$\chi_N = F_N^{1,-1} + iF_N^{-1,1}.$$

Now, adding and subtracting (A.3) we obtain

$$\left.\frac{\Delta\xi_N}{\Delta x^0}\right|_{(n,0)} = -v\left.\frac{\Delta\xi_{N-1}}{\Delta x^1}\right|_{(n+t,0)} - \big[(\alpha A + \beta B)(\xi_{N-1} - i\xi_{N-1}^*)\big]\big|_{(n+t,s)}$$

$$\left.\frac{\Delta\xi_N}{\Delta x^0}\right|_{(n,0)} = -v\left.\frac{\Delta\xi_{N-1}}{\Delta x^1}\right|_{(n+t,0)}$$
$$-\big[(\alpha A + \beta B + \gamma C)\xi_{N-1} + (\alpha A - \beta B)i\xi_{N-1}^*\big]\big|_{(n+t,s)} \tag{A.4}$$

with
$$\xi_N = \varphi_N + \chi_N$$

and
$$\xi_N = \varphi_N - \chi_N.$$

Then we take the difference between (A.4) and obtain respectively,

$$\frac{\Delta\xi_{N-1}}{\Delta x^0}\bigg|_{(n+t',0)} = -v\frac{\Delta\xi_{N-2}}{\Delta x^1}\bigg|_{(n+t+t',0)} - \left[(\alpha A+\beta B)(\xi_{N-2}-i\xi^*_{N-2})\right]\big|_{(n+t+t',s)}$$

$$\frac{\Delta\xi_N}{\Delta x^0}\bigg|_{(n,s')} = -v\frac{\Delta\xi_{N-1}}{\Delta x^1}\bigg|_{(n+t,s')}$$

$$-\left[(\alpha A+\beta B+\gamma C)\xi_{N-1}+(\alpha A-\beta B)i\xi^*_{N-1}\right]\big|_{(n+t,s+s')} \quad \text{(A.5)}$$

Finally we divide them respectively by $\Delta_{t'}(x^0)_{n,0}$ and $\Delta_{s'}(x^1)_{n,0}$; so we have

$$\frac{\Delta^2\xi_N}{(\Delta x^0)^2}\bigg|_{(n,0)} = -av\frac{\Delta^2\xi_{N-1}}{\Delta x^0\Delta x^1}\bigg|_{(n+t,0)} - b(\alpha A+\beta B)\left[\frac{\Delta\xi_{N-1}}{\Delta x^0}\bigg|_{(n+t,s)} - i\frac{\Delta\xi^*_{N-1}}{\Delta x^0}\bigg|_{(n+t,}\right.$$

$$\frac{\Delta^2\xi_N}{\Delta x^0\Delta x^1}\bigg|_{(n,0)} = -cv\frac{\Delta^2\xi_{N-1}}{(\Delta x^1)^2}\bigg|_{(n+t,0)} - d\left[(\alpha A+\beta B+\gamma C)\frac{\Delta\xi_{N-1}}{\Delta x^1}\bigg|_{(n+t,s)}\right.$$

$$\left.+i(\alpha A-\beta B)\frac{\Delta\xi^*_{N-1}}{\Delta x^1}\bigg|_{(n+t,s)}\right] \quad \text{(A.6)}$$

where

$$\frac{\Delta^2\xi_N}{(\Delta x^0)^2}\bigg|_{(n,0)} = \left(\frac{\Delta\xi_{N-1}}{\Delta x^0}\bigg|_{(n+t',0)} - \frac{\Delta\xi_N}{\Delta x^0}\bigg|_{(n,0)}\right)\Big/\Delta_{t'}(x^0)_{n,0} \underset{\delta\to 0}{\to} \frac{\partial^2\xi}{\partial x^{1^2}}\bigg|_{x^1=0}$$

$$\frac{\Delta^2\xi_{N-1}}{(\Delta x^1)^2}\bigg|_{(n+t,0)} = \left(\frac{\Delta\xi_{N-1}}{\Delta x^1}\bigg|_{(n+t,s')} - \frac{\Delta\xi_{N-1}}{\Delta x^1}\bigg|_{(n+t,0)}\right)\Big/\Delta_{s'}(x^1)_{n+t,0} \underset{\delta\to 0}{\to} \frac{\partial^2\xi}{\partial x^{1^2}}\bigg|_{x^1=0}$$

$$\frac{\Delta^2\xi_N}{\Delta x^0\Delta x^1}\bigg|_{(n,0)} = \left(\frac{\Delta\xi_N}{\Delta x^0}\bigg|_{(n,s')} - \frac{\Delta\xi_N}{\Delta x^0}\bigg|_{n,0}\right)\Big/\Delta_{s'}(x^1)_{n,0} \underset{\delta\to 0}{\to} \frac{\partial^2\xi}{\partial x^0\partial x^1}\bigg|_{x^1=0}$$

(A.7)

(and similarly for $N\to N^{-1}, N\to n+t$).

Moreover
$$a=\frac{\Delta_{t'}(x^0)_{n+t,0}}{\Delta_{t'}(x^0)_{n,0}} \underset{\delta\to 0}{\to} 1 \qquad c=\frac{\Delta_{s'}(x^1)_{n+t,0}}{\Delta_{s'}(x^1)_{n,0}} \underset{\delta\to 0}{\to} 1$$

$$b=\frac{\Delta_{t'}(x^0)_{n+t,s}}{\Delta_{t'}(x^0)_{n,0}} \underset{\delta\to 0}{\to} 1 \qquad d=\frac{\Delta_{s'}(x^1)_{n+t,s}}{\Delta_{s'}(x^1)_{n,0}} \underset{\delta\to 0}{\to} 1$$

relation (A.6) becomes of course

$$\left.\frac{\partial^2\xi}{\partial x^{0^2}}\right|_{x^1=0} = -\left.\frac{\partial^2\xi}{\partial x^1\partial x^0}\right|_{x^1=0} + (A+B)\left.\left(\frac{\partial\xi}{\partial x^0} - i\frac{\partial\xi^*}{\partial x^0}\right)\right|_{x^1=0}$$

$$\left.\frac{\partial^2\xi}{\partial x^1\partial x^0}\right|_{x^1=0} = -\left.\frac{\partial^2\xi}{\partial x^{1^2}}\right|_{x^1=0} + \left.\left[(A+B-C)\frac{\partial\xi}{\partial x^1} + i(A-B)\frac{\partial\xi^*}{\partial x^1}\right]\right|_{x^1=0} \tag{A.8}$$

and subtracting them we obtain

$$\left.\left(\frac{\partial^2\xi}{\partial x^{0^2}} - \frac{\partial^2\xi}{\partial x^{1^2}}\right)\right|_{x^1=0} = (A+B)\left.\left(\frac{\partial\xi}{\partial x^0} - i\frac{\partial\xi^*}{\partial x^0}\right)\right|_{x^1=0} - \left.\left[(A+B-C)\frac{\partial\xi}{\partial x^1} + i(A-B)\frac{\partial\xi^*}{\partial x^1}\right]\right|_{x^1=0} \tag{A.9}$$

In order to eliminate ξ from (A.9) we observe that, in the limit $\delta\to 0$, the first equation of (A.5) is

$$\left.\frac{\partial\xi}{\partial x^0}\right|_{x^1=0} = -\left.\frac{\partial\xi}{\partial x^1}\right|_{x^1=0} + (A+B)(\xi - i\xi^*) \tag{A.10}$$

so that we have

$$\left.\left(\frac{\partial^2\xi}{\partial x^{0^2}} - \frac{\partial^2\xi}{\partial x^{1^2}}\right)\right|_{x^1=0} = (2A+2B-C)\left.\frac{\partial\xi}{\partial x^0}\right|_{x^1=0} - 2iB\left.\frac{\partial\xi^*}{\partial x^0}\right|_{x^1=0} + (A+B)(C-2B)(\xi - i\xi^*)|_{x^1=0} \tag{A.11}$$

Finally, subtracting from (A.11) the conjugate equation multiplied by i

$$i\left.\left(\frac{\partial^2\xi^*}{\partial x^{0^2}} - \frac{\partial^2\xi^*}{\partial x^{1^2}}\right)\right|_{x^1=0} = (2A+2B-C)i\left.\frac{\partial\xi^*}{\partial x^0}\right|_{x^1=0} - 2B\left.\frac{\partial\xi}{\partial x^0}\right|_{x^1=0} - (A+B)(C-2B)(\xi - i\xi^*)|_{x^1=0} \tag{A.12}$$

and requiring that

$$C = 2A+4B, \qquad 2(A+B)^2 = \frac{m^2c^2}{\hbar^2} \tag{A.13}$$

we have

$$\left[\left(\frac{\partial^2}{\partial x^{0^2}} - \frac{\partial^2}{\partial x^{1^2}}\right) - \frac{m^2c^2}{\hbar^2}\right]\Phi\bigg|_{x^1=0} = 0 \qquad \text{(A.14)}$$

which is the Klein–Gordon equation on the x^0 axis of a Lorentz frame.

This evidently implies that (A.14) is valid over all space–time. Indeed if a scalar equation such as (A.14) is valid at a point in a given frame it remains valid for the same point in all frames. Moreover, since our lattice is covariant, if (A.14) is valid along a given line $x^1=0$, it is also valid on any lattice point of another of our lines (denoted $x^{1'}=0$) which also plays the part of our x^0 axis in a different Lorentz frame, because we can evidently repeat the preceding demonstration in any Lorentz frame along any axis $x^1=0$.

REFERENCES

Aspect, A. (1976). *Physical Review Letters D*, **14**, 1944.

Avez, A. (1976). *Journées Relativistes*.

Bohm, D., and Vigier, J. P. (1954). *Physical Review*, **96**, 208.

de Broglie, L. (1961). *Compte rendus de l'Académie des Sciences de Paris*, **253**, 1078.

Chandrasekhar, S. (1943). *Reviews of Modern Physics*, **15**, 1.

De la Peña, L., and Cetto, A. M. (1975). *Foundations of Physics*, **5**, 355.

Dirac, P. M. A. (1951). *Nature*, **168**, 906.

Einstein, A., Podolski, B., and Rosen, N. (1935). *Physical Review*, **47**, 777.

Einstein, A. (1956). *Investigation on the Theory of Brownian Movement*. New York.

Feynman, R. P. (1948). *Reviews of Modern Physics*, **20**, 367.

Guerra, F., and Ruggiero, P. (1979). *Letters to Nuovo Cimento*, **23**, 529.

Heath, D. C. (1969). "Probabilistic analysis of hyperbolic systems of partial differential equations." Ph.D. thesis, University of Urbana, Illinois.

Kac, M. (1956). "Some stochastic problems in Physics and Mathematics." *Magnolia Petrolium C° Lectures in pure and applied Science*, n°2, pp. 102–122.

Lehr, W., and Park, J. (1977). *Journal of Mathematical Physics*, **18**, 1235.

Nelson, E. (1966). *Physical Review*, **150**, 1079.

Terletski, Ya. P., and Vigier, J. P. (1961). *Zhurnal Eksperimental'noi i Teureticheskoi Fiziki*, **13**, 356.

Vigier, J. P. (1979). *Letters to Nuovo Cimento*, **24**, 258, 265.

Reprinted from *Lettere al Nuovo Cimento*, Vol. 30, No. 2, pp. 57-63, Copyright (1981)
with permission from the Società Italiana di Fisica.

Description of Spin in the Causal Stochastic Interpretation of Proca-Maxwell Waves: Theory of Einstein's « Ghost Waves » (*).

A. GARUCCIO

Istituto di Fisica dell'Università - Bari, Italia
Istituto Nazionale di Fisica Nucleare - Sezione di Bari

J. P. VIGIER

Equipe Recherche Associée au C.N.R.S. n. 533
Institut Henri-Poincaré - Paris, France

(ricevuto l'1 Dicembre 1980)

Recent multiplication in the literature [1-6] of papers supporting the stochastic interpretation of quantum mechanics (*i.e.* which favour Einstein's views in the Bohr-Einstein controversy) evidently confronts Einstein's supporters with a challenge, *i.e.* to interpret in a casual way the forthcoming Aspect [7]-Rapisarda [8] experiments, provided of course (as believed by one (JPV) of the authors) that experimental results confirm the nonlocal prediction of Quantum Mechanics. Since these experiments are based on the measurement of correlated polarizations (spins) of photon pairs emitted in the singlet state this clearly implies a casual stochastic analysis of the concept of spin (and polarization) in the theory of light. The aim of the present letter is to make a first step in this direction, *i.e.* to describe the realistic foundation of the concept of photon spin (and its measurement) in Proca and Maxwell waves in the stochastic interpretation of quantum mechanics.

As indicated in the title, we start with nonzero-mass spin-1 particles. This is justified by the well-known fact (resulting from past works of de Broglie [9] Schrödinger [10])

(*) Work partially supported by collaboration between CNR (Italy) and CNRS (France).

(1) D. BOHM and J. P. WIGIER: *Phys. Rev.*, **96**, 208 (1954).
(2) E. NELSON: *Phys. Rev.*, **150**, 1079 (1966).
(3) L. DE LA PEÑA and A. M. CETTO: *Phys. Rev. D*, **3**, 795 (1971).
(4) W. LEHR and J. PARK: *J. Math. Phys.* (*N. Y.*), **18**, 1235 (1977).
(5) F. GUERRA and P. RUGGIERO: *Lett. Nuovo Cimento*, **23**, 529 (1979).
(6) J. P. VIGIER: *Lett. Nuovo Cimento*, **24**, 258, 265 (1979); N. CUFARO PETRONI and J. P. VIGIER: *Int. J. Theor. Phys.*, **18**, 807 (1979).
(7) A. ASPECT: *Phys. Rev. Lett. D*. **14**, 1944 (1976).
(8) F. FALCIGLIA, G. JACI and V. A. RAPISARDA: *Lett. Nuovo Cimento*, **26**, 327 (1979).
(9) L. DE BROGLIE: *La mécanique ondulatoire du photon* (Paris, 1940).
(10) L. BASS and E. SCHRÖDINGER: *Proc. R. Soc. London Ser. A*, **232**, 1 (1955).

Deser [11] and Vigier *et al.* [12]) that the zero-mass limit of a nonzero spin-1 Proca particle m_γ cannot be distinguished physically from a Maxwell wave. Indeed the so-called transverse waves just corresponds to $J_3 = \pm 1$, *i.e.* to opposite circular polarizations, while the longitudinal solution $J_3 = 0$ (which decouples practically from transverse waves when $m_\gamma \to 0$) describes the Coulomb field when $m_\gamma \to 0$.

As one knows the common classical counterpart of the Proca-Maxwell field is the extended classical Weyssenhoff particle [13] analysed in detail by BOHM and VIGIER [14], HALBWACHS [15], SOURIAN *et al.* [16].

This model describes its internal motions by two physical points (*i.e.* a centre of mass (c.m.) and a centre-of-matter density (c.c.)) separated by a spacelike four-vector R_μ defined by $(mc)^2 \cdot R_\mu = S_{\mu\nu} G_\nu$, where G_μ and $S_{\mu\nu}$ define the particle energy-momentum ($G_\mu G_\mu = -m^2c^2$) and internal angular momentum [14]. By introducing the (cc)'s unitary four-velocity $U_\mu (U_\mu U_\mu = -c^2)$ and its proper time τ (with $\dot{} = \mathrm{d}/\mathrm{d}\tau$) the classical set of Weyssenhoff equations is written as

$$\text{(1)} \qquad \dot{G}_\mu = 0\,, \quad \dot{S}_{\mu\nu} = G_\mu U_\nu - G_\nu U_\mu \quad \text{and} \quad S_{\mu\nu} U_\nu = 0\,.$$

They have been completely integrated by HALBWACHS [15]. They yield $R_\mu R_\mu = \text{const}$, imply the constancy of the classical spin vector

$$\text{(2)} \qquad S_\mu = \frac{i}{2c}\, \varepsilon_{\mu\nu\alpha\beta} U_\nu S_{\alpha\beta} = \tilde{S}_{\mu\nu} U_\nu$$

(so that $S_\mu S_\mu = \frac{1}{2} S_{\mu\nu} S_{\mu\nu} = \sigma_0^2 = \text{const}$) as well as that of Pauli's spin vector $S_\mu = \tilde{S}_{\mu\nu} \cdot G_\nu$ *i.e.* $S_\mu S_\mu = \text{const}$. One sees also that the extremity of R_μ moves along a circle with a constant angular velocity in the c.m. rest frame. The vector R_μ thus behaves like the needle of a de Broglie « clock » which rotates with de Broglie's frequency $\sigma_0 \nu_0 = m_0 c^2$: a classical counterpart of Schrödinger's Zitterbewegung if $\sigma_0 = h$. Moreover, the constant-length vectors U_μ, S_μ, $\dot{R}_\mu$ and R_μ are orthogonal vectors (with $\ddot{R}_\mu = -(m^2c^2/\sigma_0)^2 R_\mu$); they define a Darboux-Frenet frame which rotates once when the (c.c.) describes one circumference. In the (c.c.)'s rest frame one has $S_{\mu\nu} = R_\mu G_\nu - G_\nu R_\mu$.

The quantization of the Weyssenhoff particle's motions can be (and has been) performed in two independent ways.

A) The first is just [17] the usual quantum procedure which substitutes commutators to Poisson brackets. We briefly recall some of its aspects which will help to clarify our subsequent results. Introducing as usual $G_\mu \to i\hbar\partial_\mu$, we see [17] that R_μ can be generalized into the complex Proca-Maxwell field vector $A_\mu = R_\mu \exp[(mc^2/\hbar)\tau]$ proposed by DE BROGLIE and VIGIER [18] and that relation $G_\mu S_{\mu\nu} = (m^2c^2) R_\mu$ becomes

$$\text{(3)} \qquad \partial_\mu(\partial_\mu A_\nu - \partial_\nu A_\mu) = \frac{m^2c^2}{\hbar^2} A_\mu$$

which, completed by $G_\mu R_\mu \to \partial_\mu R_\mu = 0$, is equivalent to the Proca-Maxwell equations.

---

[11] S. DESER: *Ann. Inst. Henri Poincaré*, **16**, 79 (1972).
[12] M. MOLES and J. P. VIGIER: *C. R. Acad. Sci. Ser. B*, **276**, 697 (1973).
[13] J. V. WEYSSENHOFF and A. RAABE: *Acta Phys. Pol.*, **9**, 8 (1974).
[14] D. BOHM and J. P. VIGIER: *Phys. Rev.*, **109**, 1882 (1958).
[15] F. HALBWACHS: *Theorie relativiste des fluides à spin* (Paris, 1960).
[16] F. HALBWACHS, J. M. SOURIAN and J. P. VIGIER: *J. Phys. Radium*, **22**, 26 (1961).
[17] C. FENECH, M. MOLES and J. P. VIGIER: *Lett. Nuovo Cimento*, **24**, 56 (1979).
[18] L. DE BROGLIE and J. P. VIGIER: *Phys. Rev. Lett.*, **28**, 1001 (1972).

B) The second which develops Einstein's stochastic point of view ([19]) derives Proca's equations (3) from a fluid of classical Weyssenhoff tops endowed with random fluctuations at the velocity of light. This has been done by CUFARO-PETRONI and VIGIER ([20]). The demonstration rests on the idea that these tops are imbedded into a covariant distribution of similar tops which constitute the corresponding spin-1 Dirac aether ([21,22]).

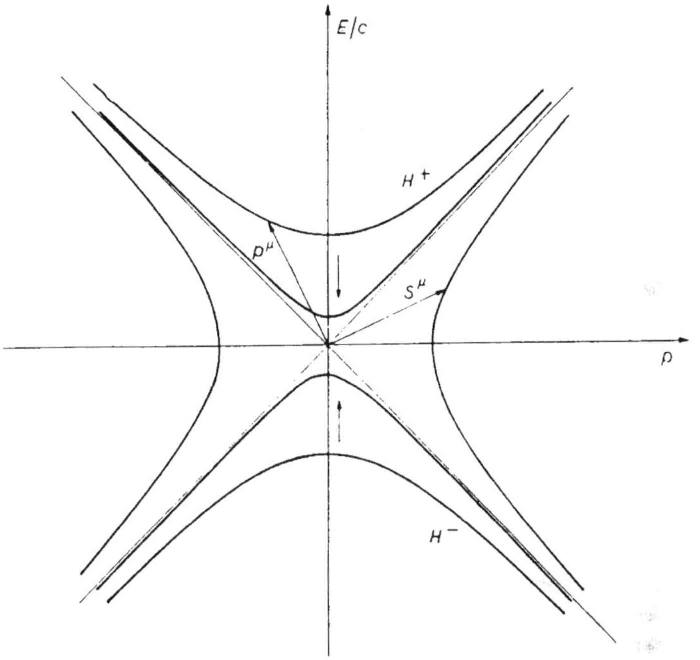

Fig. 1.

We only want to add to this result a demonstration that the diffusion coefficient D is indeed equal to $\hbar/2m$. This can be shown ([22]) by the argument that in Vigier's demonstration of Nelson's equation one finds $D = \langle \delta x_i \, \delta x_i \rangle / 2\,\Delta\tau$, where δx represents the stochastic spacelike part of the quantum jump. If one assumes that the Weyssenhoff tops must turn by one (2π) internal rotation during these jumps (so that the tops remain in phase according to de Broglie's guiding principle), we see that the distance travelled in this jump is equal to Compton wave-length *i.e.* $\lambda = \hbar/mc$, so that we find (if this distance is covered at the velocity of light) $D = c^2\,\Delta\tau^2/2\,\Delta\tau = \hbar/2m$. Of course if $m_\gamma \to 0$, H^+ and H^- transform asymptotically into the light-cone and the limiting energy density goes into the so-called zero-point electromagnetic energy density $I(\nu) = (4\pi h/c^3)\,\nu^3$, which has been used as starting point for stochastic electrodynamics ([23]).

([19]) F. VASCHLUHN: *Einstein-Centenarium*, edited by Akademie Verlag (Berlin, 1979), p. 173.

([20]) N. CUFARO PETRONI and J. P. VIGIER: *Phys. Lett. A*, **73**, 289 (1979).

([21]) P. A. M. DIRAC: *Nature (London)*, **168**, 906 (1951).

([22]) J. L. VIGIER: *De Broglie waves on Dirac's either; a testable experimental assumption*, submitted to *Lett. Nuovo Cimento*.

([23]) T. W. MARSCHALL: *Proc. Cambridge Philos. Soc.*, **61**, 537 (1965).

We now return to our initial purpose. Because of B we can assume, following EINSTEIN (19) and DE BROGLIE (19), that each individual photon is an extended Weyssenhoff top surrounded by a stochastic Proca spin wave materializing (Einstein's « Gespensterfeld » or de Broglie's « pilot wave ») and propagating on Dirac's aether. It thus behaves exactly like a plane flying at Mach 1 surrounded by its own sound wave. Its constituent (extended) de Broglie clocks remain in phase between themselves (and with the particle), so that each photon can thus be compared with a soliton (or singular region) which follows the drift lines and jumps at random (with the velocity of light) from one drift line to another. This « pilot wave » model implies of course an extension to spin of Bohm's and de Broglie's quantum potential. We shall calculate it in the hydrodynamical model of the corresponding Proca-Maxwell wave, since a general analysis has been fully developed by HALBWACHS (15).

Let us start from the Lagrangian

$$\mathscr{L} = \frac{\hbar^2}{2m}\,\partial_\mu A_\nu^*\,\partial_\mu A_\nu + \frac{mc^2}{2}\,A_\alpha^* A_\alpha + \lambda^*\,\partial_\mu A_\mu^* + \lambda\partial_\mu A_\mu - \tag{4}$$

$$+\,\lambda^*\lambda\varrho - \lambda'\,\partial_\mu S\partial_\lambda S\partial_\lambda a_\mu - \varrho'\lambda'\,,$$

where λ^*, λ, ϱ, λ', ϱ' represent Lagrange multipliers. As will be shown later the three Lagrange constraints are equivalent to the three Weyssenhoff conditions $S_{\alpha\beta}U_\beta = 0$. The field equations deduced from (4) are

$$\Box A_\mu = \frac{m^2c^2}{\hbar^2}\,A_\mu \quad \text{and} \quad \Box A_\mu^* = \frac{m^2c^2}{\hbar^2}\,A_\mu^*\,, \tag{5}$$

along with the constraints

$$\partial_\mu a_\mu = \partial_\mu S a_\mu = \partial_\mu S\partial_\lambda S\partial_\lambda a_\mu = 0\,. \tag{6}$$

As one knows (15) the photon drift current is

$$j_\mu = i\left(A_\nu\,\frac{\partial\mathscr{L}}{\partial(\partial_\mu A_\nu)} - \frac{\partial\mathscr{L}}{\partial(\partial_\mu A_\nu^*)}\,A_\nu^*\right) = \frac{\hbar}{m}\,a_\nu a_\nu\,\partial_\mu S = \frac{\hbar}{m}\,R^2\,\partial_\mu S\,, \tag{7}$$

if we write $a_\mu a_\mu = R^2$ and $a_\mu = R\cdot R_M$ with $R_\mu R_\mu = 1$.

One then sees immediately that the real and imaginary parts of (5) yield the Jacobi and conservation equations, *i.e.*

$$\partial_\mu S\partial_\mu S = \left(\frac{\hbar^2}{R^2c^2}\,a_\mu\Box a_\mu - m^2\right)c^2 = \hbar^2\,\frac{\Box R}{R} - \hbar^2\,\partial_\varrho R_\mu\,\partial_\varrho R_\mu - (m\cdot c)^2 \tag{8}$$

and

$$\partial_\mu(R^2\,\partial_\mu S) = 0\,. \tag{9}$$

Moreover, the current j_μ can be written in the classical form $j_\mu = \hbar\varrho U_\mu$ (with $U_\mu U_\mu = -c^2$), *i.e.* $\varrho = (M/m)R^2$, if we introduce the quantum potential

$$M^2 = \left(m^2 + \hbar^2\,\partial_\varrho R_\mu\,\partial_\varrho R_\mu - \hbar^2\,\frac{\Box R}{R}\right). \tag{10}$$

Moreover, the energy-momentum tensor

$$t_{\mu\nu} = \left(\frac{\partial \mathscr{L}}{\partial(\partial_\nu A_\alpha)} \partial_\mu A_\alpha + \text{c.c.}\right) - \delta_{\mu\nu}\mathscr{L} = \frac{h^2}{m}\partial_\mu a_\alpha \partial_\nu a_\alpha + \frac{R^2}{m}\partial_\nu S - \delta_{\mu\nu}\mathscr{L}$$

can be written in the classical form

$$t_{\mu\nu} = \mu_0 U_\mu U_\nu - p_\mu U_\nu + q_\nu U_\mu + \theta_{\mu\nu} \tag{11}$$

with

$$\mu_0 = \frac{1}{c^4} t_{\mu\nu} U_\mu U_\nu = \frac{1}{c^4}\left[\left(\frac{h^2}{m}\right)\mathring{a}_\alpha \mathring{a}_\alpha + c^2\mathscr{L}\right] + \frac{R^2}{m} M_0^2,$$

$$p_\mu = \mu_0 U_\mu + \frac{1}{c^2} t_{\mu\nu} U_\nu = \frac{h^2}{mc^2}\left[\frac{1}{c^2}\mathring{a}_\varrho \mathring{a}_\varrho U_\mu + \partial_\mu a_\varrho \cdot \mathring{a}_\varrho\right],$$

$$q_\nu = -\mu_0 U_\nu + \frac{1}{c^2} t_{\mu\nu} U_\mu = -\frac{h^2}{mc^2}\left[\frac{1}{c^2}\mathring{a}_\varrho \mathring{a}_\varrho U_\nu + \partial_\nu a_\varrho \mathring{a}_\varrho\right],$$

$$\theta_{\mu\nu} = t_{\mu\nu} - \mu_0 U_\mu U_\nu + p_\mu U_\nu - q_\nu U_\mu =$$

$$= \frac{h^2}{m}\partial_\mu a_\varrho \partial_\nu a_\varrho - \left(\partial_{\mu\nu} + \frac{U_\mu U_\nu}{c^2}\right)\mathscr{L} + \frac{h^2}{c^4 m}\mathring{a}_\varrho \mathring{a}_\varrho U_\mu U_\nu + \frac{h^2}{mc^2}\mathring{a}_\varrho(\partial_\mu a_\varrho U_\nu + \partial_\nu a_\varrho \cdot U_\mu),$$

where the symbol ° denotes the operator $U_\lambda \partial_\lambda$ and μ_0, p_μ, q_μ and $\theta_{\mu\nu}$ represent, respectively, Holbwach's proper mass density, transverse momentum, heat current and tension tensor. The relation $\partial_\nu t_{\mu\nu} = 0$ which results from the field equations (5) then immediately yields

$$\partial_\nu t_{\mu\nu} = \dot{g}_\mu + \partial_\nu\{(q_\nu U_\mu) + \theta_{\mu\nu}\} = 0, \tag{12}$$

if we write the particle momentum density as $g_\mu = \mu_0 U_\mu - p_\mu$ and the dot represent the operator $\partial_\lambda U_\lambda$.

This is the quantum counterpart of Weyssenhoff's first classical relation $\dot{G}_\mu = 0$: the last term representing (in Proca's case) the new (stochastic) quantum force. We now define the Proca-wave internal-momentum density $S_{\mu\nu}$ with the help of Belinfante's tensor:

$$f_{\mu\nu\lambda} = \frac{\partial \mathscr{L}}{\partial \partial_\nu A_\alpha} \mathscr{X}^{\alpha\beta}_{\mu\nu} A_\beta + \text{c.c.},$$

where

$$\mathscr{X}^{\alpha\beta}_{\mu\nu} = \tfrac{1}{2}\delta^{\alpha\beta}_{\mu\nu} = \tfrac{1}{2}(\delta_{\alpha\mu}\delta_{\beta\nu} - \delta_{\alpha\nu}\delta_{\beta\mu}).$$

We thus obtain

$$f_{\mu\nu\lambda} = \frac{h^2}{2m}(\partial_\lambda a_\mu a_\nu - \partial_\lambda a_\nu a_\mu),$$

which yelds, for the angular-momentum tensor density $S_{\mu\nu}$, the value $\frac{1}{2}S_{\mu\nu} = -(1/c^2) f_{\mu\nu\lambda} U_\lambda$, i.e.

$$S_{\mu\nu} = \frac{h^2}{c^2 m} R^2(\mathring{R}_\mu R_\nu - \mathring{R}_\nu R_\mu). \tag{13}$$

As one knows, one can deduce the spin vector density from $S_{\mu\nu}$ through the relation

$$S_\mu = \frac{i}{2c}\,\varepsilon_{\nu\alpha\beta\mu}\,U_\nu S_{\alpha\beta} = \frac{i\hbar^2}{c^3 m}\,R^2\,\varepsilon_{\nu\alpha\beta\mu}\,U_\nu R_\alpha \overset{\circ}{R}_\beta\,.$$

Since, as a consequence of (13), we also have $t_\mu = (1/c)S_{\mu\nu}U_\nu = 0$, we finally obtain $S_{\mu\nu} = (i/2c)\,\varepsilon_{\mu\nu\alpha\beta}S_\alpha U_\beta$, so that $S_{\mu\nu}S_{\mu\nu} = S_\alpha S_\alpha = 2(\hbar^2/c^2m)^2R^4\overset{\circ}{R}_\mu\overset{\circ}{R}_\mu$ and

(14) $$S_{\mu\nu}U_\nu = 0\,,$$

which corresponds to Weyssenhoff's third condition 1).

As a consequence of our nonzero rest masses we can use in each region a nonrelativistic limit and apply Bohm and Vigier's demonstration (1,15), so that the Proca particle probability and density is R^2 and we see that each individual fluid element (and particle singularity) has an individual angular momentum $\overline{S}_{\mu\nu} = (1/R^2)S_{\mu\nu}$ and a spin vector $\overline{S}_\mu = (1/R^2)S_\mu$, *i.e.*

(15) $$\overline{S}_\mu = (i\hbar^2/c^3m)\,\varepsilon_{\mu\alpha\beta\nu}\,U_\nu R_\alpha \overset{\circ}{R}_\beta\,.$$

We thus have (because $R_\alpha R_\alpha = 1$) $U_\alpha\overline{S}_\alpha = U_\beta R_\beta = U_\gamma \overset{\circ}{R}_\gamma = \overline{S}_\sigma \overset{\circ}{R}_\sigma = \overline{S}_\sigma R_\sigma = R_\nu \overset{\circ}{R}_\nu = 0$ combined with $\overset{\circ\circ}{R}_\mu = GR_\mu$ (in the rest frame $U_i = 0$) and $G = -\overset{\circ}{R}_\mu\overset{\circ}{R}_\mu = \text{const}$. This yields

(16) $$S_\mu \overset{\circ}{S}_\mu = -\left(\frac{\hbar^2}{c^3m}\right)^2 \overset{\circ}{R}_\mu \overset{\circ\circ}{R}_\mu = 0\,,$$

which shows the spin $\overline{S}_\alpha$ is a constant length vector. Clearly the vectors U_α, $\overline{S}_\alpha$, R_α and $\overset{\circ}{R}_\alpha$ now describe the individual Weyssenhoff tops which constitute our particle and the associated spin wave elements.

Of course our angular momentum (and spin) no longer remain invariant along a drift line. Since they are submitted to a quantum torque of stochastic origin. An evident calculation yields

(17) $$\overset{\circ}{\overline{S}}_\mu = (i\hbar^2/c^3m)\,\varepsilon_{\mu\alpha\beta\nu}\,\overset{\circ}{U}_\nu R_\alpha \overset{\circ}{R}_\beta$$

or

(18) $$\dot{S}_{\mu\nu} = q_\mu U_\nu - q_\nu U_\mu + \theta_{\mu\nu}$$

with $\theta_{\mu\nu} = -q_\mu U_\nu + q_\nu U_\mu + (2\overset{\circ}{R}/R)S_{\mu\nu}$, which corresponds to Weyssenhoff's second relation 1): $\theta_{\mu\nu}$ being interpreted as the torque produced by the heat flow plus a friction torque.

This completes our description of the Proca-Maxwell field.

On the basis of the real existence of the pilot wave quantum potential and quantum torque given above we can conclude this letter with a brief discussion of the photon measurement theory in the stochastic interpretation of quantum mechanics. First we can summarize the contradiction between the Copenhagen (CIQM) and stochastic (SIQM) interpretations of quantum mechanics in three antagonistic propositions which sum up the opposite starting points of Bohr and Einstein.

In CIQM

1) the quantum states are associated with individual systems and represent an ultimate statistical knowledge.; microphenomena are particles *or* waves: never them both simultaneously;

2) a measurement on a system induces an instantaneous discontinuous (space-like) collapse of the state (wave packet reduction) on one of the eigenfunctions of the measuring device;

3) the Heisenberg uncertainty principle restricts the simultaneous measurability of noncommuting observables on individual systems.

In SIQM

1') the quantum states represent real physical fields (waves) associated both with individual particles and sets of identically prepared systems; micro-objects are thus particles *and* waves simultaneously;

2') these states (particles plus waves) evolve causally in time; *there is no wave packet reduction*, but if (and when) the packet is split in an interaction with a real macroscopic physical system (*i.e.* a measuring device) into packets corresponding to given eigenvalues, the particle enters one of them (according to its initial position in the measurement) in which it is detected;

3') the Heisenberg uncertainty principle does not restrict simultaneous measurability of noncommuting observables on *individual* objects (since particles follow space-like paths in space time), but represent dispersion relations resulting from their dual (wave plus particle) character and from their subquantal stochastic motions.

These later propositions suggest interesting testable consequences. Proposition 3') can be tested as shown by FITCHARD. Proposition 1'), *i.e.* the real existence of Einstein's « gespenster Wellen » (de Broglie's pilot waves), can be checked as proposed by the authors ([27]) in a variant of Pfleegor and Mandel's experiment. Moreover, proposition 2') suggests that, since macroscopic measuring devices always contain a random part there is no pure wave packet splitting and all measurements imply a superposition of the apparatus eigenfunctions, *i.e.* a transition from Bohr's to Wigners ([24]), Araki and Yanase's ([25]) measurement theory. This is satisfactory, since it implies no violation of Einstein conservation laws ([26]) and no possibility to transmit superluminal information, despite the real superluminal propagation ([6]) of the quantum potential.

* * *

The authors want to thank Prof. F. SELLERI and Dr. B. HILEY for helpful suggestions. They gratefully acnowledge a financial support from the Italian CNR and French CNRS which made this research possible.

([24]) E. P. WIGNER: *Z. Phys.*, **133**, 101 (1952).

([25]) H. ARAKI and M. YANASC: *Phys. Rev.*, **96**, 208 (1954); M. YANASE: *Phys. Rev.*, **123**, 666 (1961).

([26]) N. CUFARO PETRONI, A. GARUCCIO, F. SELLERI and J. P. VIGIER: *C. R. Acad. Sci. Ser. B*, **290**, 111 (1980).

([27]) A. GARUCCIO and J. P. VIGIER: *Possible experimental test of the causal stochastic interpretation of quantum mechanics*, to be published on *Found. Phys.*

This page is deliberately left blank.

Reprinted from *Foundations of Physics*, Vol. 24, No. 2, pp. 61-83, Copyright (1994)
with permission from Kluwer Academic/Plenum Publishers.

Possible Test of the Reality of Superluminal Phase Waves and Particle Phase Space Motions in the Einstein–de Broglie–Bohm Causal Stochastic Interpretation of Quantum Mechanics

J. P. Vigier[1]

Received July 7, 1993; revised August 2, 1993

Recent double-slit type neutron experiments[1] *and their theoretical implications*[2] *suggest that, since one can tell through which slit the individual neutrons travel, coherent wave packets remain nonlocally coupled (with particles one by one), even in the case of wide spatial separation. Following de Broglie's initial proposal,*[3] *this property can be derived from the existence of the persisting action of real superluminal physical phase waves considered as building blocks of the real subluminal wave field packets which surround individual particle paths in the Einstein–de Broglie–Bohm interpretation of quantum mechanics.*

0. INTRODUCTION

The observed phase space coupling in neutron interference[2] now evidently strengthens the view that quantum mechanics implies nonlocal correlations/interactions even in the single-particle case. Contrary to the purely local hidden-parameters models, still defended by some followers of de Broglie (Selleri, Lochak, Santos, etc.), the only known way to save realism and causality within the frame of the Einstein–de Broglie model (which combines real particle paths with real surrounding guiding fields) is to incorporate into their model (as done by Bohm, Vigier, Bell, etc.) nonlocal quantum potential type interactions carried by some new subquantal superluminal mechanism somehow related to Dirac's aether model.[4] Accordingly, the aim of the present work is to use these experimental results of double-slit setups:

[1] Université Paris VI—CNRS/URA 769, Tour 22-12 Boîte courrier 142, 4, place Jussieu, 75005 Paris, France.

—(I) As evidence that individual neutrons have travelled through a given specific slit, so that Rauch *et al.*[1] have opened the way for an effective realizable "*Welcherweg*" (which-way) experiment in Einstein's sense, so that the particle aspect of individual microobjects follows real trajectories in phase space.

—(II) As starting point of an alternative possible interpretation of nonlocality based (as initially suggested by de Broglie himself)[3] on the real existence of superluminal phase waves related with the real particle paths associated to the causal stochastic interpretation of quantum mechanics: the variations of the nonlocal quantum potential being carried by such phase waves.

The argument can be split in the following three steps:

1. *WELCHERWEG* BEHAVIOR IN NEUTRON INTERFEROMETRY

The question of the existence (or not) of particle motions (i.e., trajectories) in the double-slit experiment has been widely discussed in the literature since the origin of quantum mechanics.

Proponents of their existence have recently underlined (a) their consequences on the probability concept[5] and (b) the possibility to test the *einweg* (one-way) part version of Einstein's reasoning[6]: a property which results from the formalism of quantum mechanics itself when/if combined with the assumption that energy-momentum conservation is always valid, even when we do not perform a measurement.

In order to show that in this type of interferometers the neutrons travel along one path only, a stroboscopic chopping of the initial wave packet containing one neutron only is introduced, polarized upward along with a macroscopic spatial separation between two coils which induce a spin-flip with an energy loss $\Delta E_{rf} = h\omega_{rf}$. Hence the separate parts of the chopped wave packet cross through coil I or coil II at different times. Moreover, flippers can be switched on and off in such a way that one coil only is working at any given time.

The frequency of the switching can be chosen such that when one packet I (II) goes through its coil, no packet goes through coil II (I). Hence, in the time interval $I_1/v \leqslant t \leqslant I_2/v$ (when packet I has passed through coil I and packet II has not yet reached coil II) the wave function will have the form

$$\mathscr{A}(r^{\mathrm{I}}, k + \Delta k)\, e^{i(\omega - \omega_{\mathrm{II}})t}\, |\downarrow_z\rangle + \mathscr{A}(r^{\mathrm{II}}, k)\, e^{i\omega t} e^{i\chi}\, |\uparrow_z\rangle \tag{1}$$

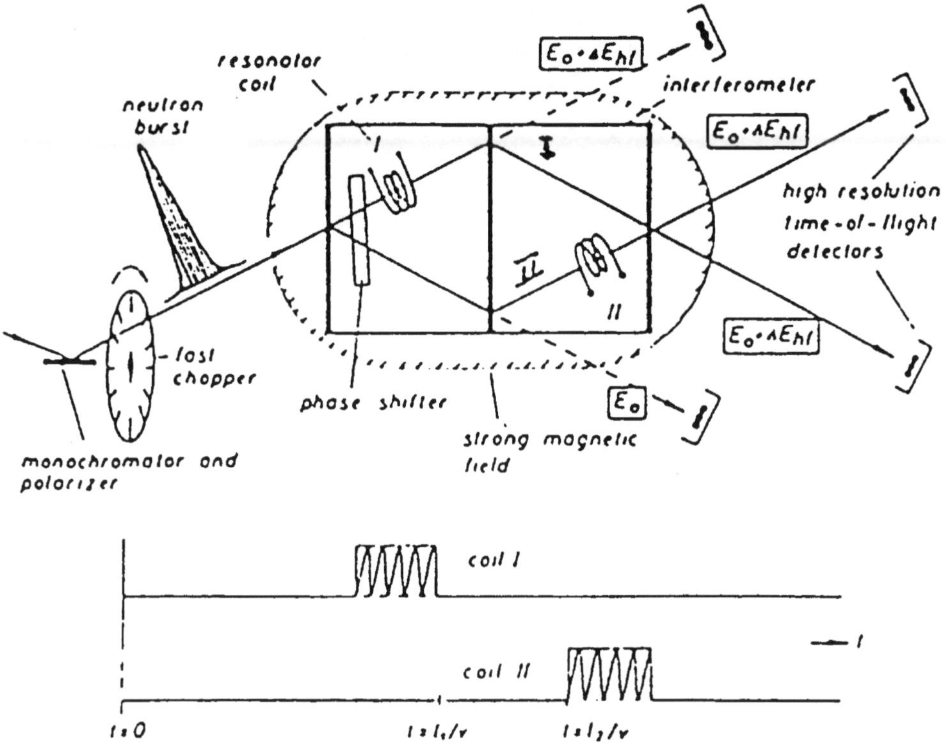

Fig. 1. Proposed experimental setup for simultaneous detection of interference and energy exchange.

Behind coil II even the second part of this wave function changes as usual. Hence, for the time interval mentioned, the energy exchange occurred in beam path I only. With a special arrangement it is also possible to measure this behavior directly.[2] We thus conclude that if experiment confirms that the neutron's energy loss is just ΔE_{rf}, i.e., confirms the prediction/validity of the quantum mechanical formalism (as believed by the author) and if one accepts Einstein's assumption that energy momentum is always locally confirmed (even in unobserved situations), then the neutron has travelled through one path only.[3] With an added measurement this behavior evidently leads to a *welcherweg* (which-way) experiment if one combines

[2] Although the authors claimed in a later comment [*Phys. Lett. A* 157, 311 (1991)] that interference disappears when the energy exchange becomes larger (and therefore measurable) than the energy width of the beam. More recent results show also, in such cases, a persistent coupling and interference properties.

[3] This experiment intended to prove that Einstein's *einweg* assumption is true if one combines the predictions of the quantum mechanical formalism with absolute local energy-momentum conservation has unfortunately been prevented by the failure of the atomic pile at Grenoble and is now being considered elsewhere.

the experiment of Fig. 1 with the probable results of an *einweg* (one-way) experiment proposed some time ago by Rauch and the author.[6]

The interest of Rauch *et al.*'s recent experiments is that, as shown by Rauch,[2] "In the course of several neutron interferometer experiments it has been established that smoothed-out interference properties at high interference order can be restored even behind the interferometer when a proper spectral filtering is applied. This postselection of states demonstrates that narrow plane-wave bands, which are components of the wave packet, remain interacting even in those cases where the wave packets do not overlap in space anymore due to a large phase shift applied to one of them. A phenomenon which appears especially for less monochromatic and less collimated beams in the same way has also been described for optical experiments.[7] The interference pattern follows from the superposition of the wave functions for both beam paths ($I = |\psi(\mathrm{I}) + \psi(\mathrm{II})|^2$) and depends on the relative phase shift ($\chi = \Delta k$) and on the momentum width δk of the beam,

$$I(\Delta) \propto 1 + \exp[-1/2(\Delta \delta k)^2] \cos(\Delta k) \tag{2}$$

Here D represents the relative spatial shift of the wave trains ($\Delta = 2\pi N b_c D/k^2$), N and D are the particle density and thickness of the phase shifting material, and b_c denotes the coherent neutron–nucleus scattering length. The damping factor $\exp[-1/2(\Delta \delta k)^2]$ can be interpreted as the real part of the mutual coherence function of the interfering beams. The corresponding experimental setup is presented in Fig. 2.

The spatially separated parts of the wave function can be interpreted as the result of the quantum superposition of two macroscopically distinguishable states, that is, a Schrödinger-cat-like state. The associated momentum distribution of the separated coherent wave packets is given by the Fourier transform which exhibit at high interference order a marked spectral modulation determined by the spatial shift of the wave trains,

$$I(k) = |a(k)|^2 \, [1 + \cos(\Delta k)] \tag{3}$$

where $a(k)$ is the amplitude function of the wave packet centered around k_0. This behavior is shown in Fig. 2 where a dimensionless quantity $m = D/D_\lambda$ ($D_\lambda = k_0/Nb_c$ is the lambda thickness) is used as a measure for the phase shift. This modulation becomes more pronounced and structured for increasing phase shifts, indicating that the disappearance of the interference pattern in ordinary space is compensated by the appearance of the spectral modulation effect in momentum space. Figure 2 also shows that an interferometric spectral narrowing can be achieved which has the characteristic features of a squeezing phenomenon where one conjugates quantity

(Δk) is below the related coherent-state value whereas the $\Delta\alpha$ value is increased accordingly. The phenomenon of complete beam modulation indicates that complete information exists at the place of beam superposition at the interferometer exit, guiding neutrons with distinct momentum values into the forward or deviated beam, respectively. This formal and experimentally proven behavior can be interpreted as the persistent action of plane-wave components ($\exp(ikr)$) outside the wave packet,

$$\psi(r) \propto \int a(k)\, e^{ikr}\, dk \tag{4}$$

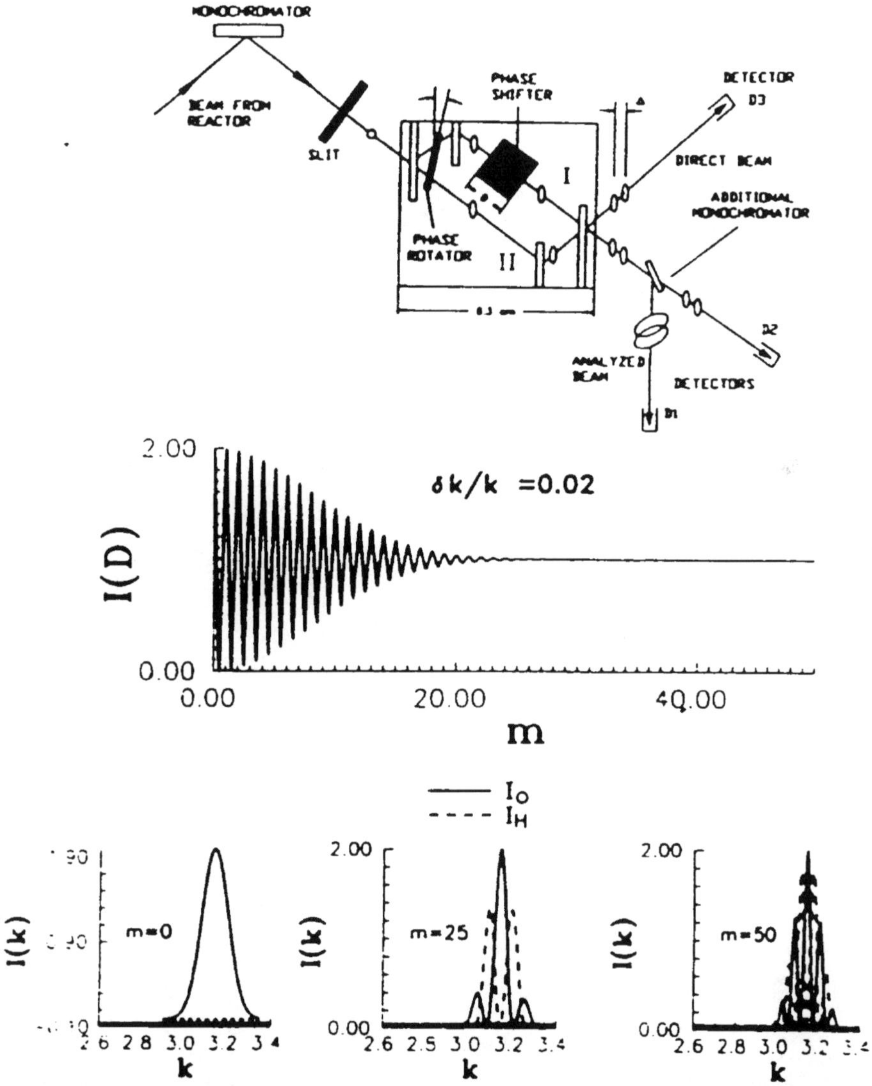

Fig. 2. Neutron interference experiments at high order. Experimental setup (above), loss of contrast at higher order (middle), and spectral modification of the beam (below) in forward (0) and deviated direction (H).

and it shows that separation in ordinary space is not sufficient to ensure separation in phase space. This far-reaching interconnection of plane wave components has interesting consequences for EPR experiments too."

As we shall now show, with an added measurement this "*Einweg*" neutron particle behavior in Fig. 1 can also be interpreted in terms of "*Welcherweg*" knowledge because in Fig. 2 the introduction of a slowing-down phase shifter implies that the two parts labelled P_I and P_{II} of the initial wave packet which result from the action of the first slab will no longer superpose in space.

This implies that, if the dimension (length) of the initial packet is small enough, the upper packet has a distinct separate entity, since it arrives in detectors with a significant time difference $\Delta l/v$; here v denotes the wave's group velocity and $\Delta l = -(D/2E)(2M\hbar^2 Nb/m)$, where E is the neutron's kinetic energy, b the scattering length, D the thickness, and m the neutron mass. In other words, if one measures the neutron's starting time with a sufficient precision (with the help of a fast chopper), one can tell whether the observed neutrons have been detected in the first P_{II} (unretarded) or the second (retarded) wave packets described in Fig. 2, i.e., whether they have travelled in the phase-shifted part or not: an evident *Welcherweg* knowledge.[4]

2. PARTICLE PHASE-SPACE MOTIONS IN THE EINSTEIN–DE BROGLIE CAUSAL STOCHASTIC INTERPRETATION OF QUANTUM MECHANICS

Along with possible *Welcherweg* measurements, the positive results of Rauch *et al.*'s experiments[1,2] thus raises new evident, epistemological problems and properties of phase space in the classical stochastic relativistic theory of quantum mechanics. For example, in the one-particle case, if the probability of the presence of a particle is zero in a given configuration space-region V_4 ($\rho = 0 \in V_4$) and no supplementary particle interaction exists, two separated wave packets should not influence each other within the Copenhagen interpretation and (apparently) in the Einstein–de Broglie–Bohm (E.d.B.) model. Indeed, if a particle has no position in space, it should have no corresponding momentum space waves in momentum space, and separated wave packets should thus be physically independent. The wave

[4] In Fig. 2 the analyzed beam is built with two nonoverlapping distinct wave packets (represented as circles in the Fig. 2); the elongated enclosing structure represents the overlap of their increased momentum length width due to the reduction of their coherence length by reducing the wavelength spread after the two packets have gone through the interferometer.

solution,[4] though formally correct, thus needs reinterpretation, as we shall later see.

To clarify this point, let us first recall the peculiar character of phase space in the causal (E.d.B.) stochastic interpretation of quantum mechanics. This character has its origin in Einstein's discovery[8] that if one considers a pointlike particle motion characterized in phase space by canonical coordinates $P_\mu(\tau)$ and $Q_\mu(\tau)$ depending on a scalar evolution parameter[9] τ with the canonical Hamilton equations

$$\dot{P}_\mu = \frac{dP_\mu}{d\tau} = -\frac{\partial H}{\partial Q^\mu}, \qquad \dot{Q}_\mu = \frac{dQ_\mu}{d\tau} = \frac{\partial H}{\partial P^\mu}$$

defined by the scalar Hamiltonian $H(P_\mu(\tau), Q_\mu(\tau), \tau)$, the only way to obtain a behavior which corresponds to the Bohr–Sommerfeld quantization laws is to utilize the special set of Jacobi canonical coordinates $p_\beta(\tau)$, $q_\mu(\tau)$, where $p_\mu = \partial S(q_\alpha(\tau), \tau)/\partial q_\mu(\tau)$ and $H = \text{const} = H(\partial_\mu S(q_\beta(\tau), q_\mu(\tau))$, which corresponds to an evident physical constraint which states (1) that quantum motions are tied to the irrotational character of the four momentum, and (2) that in an ensemble of such phase space such motions remain on a surface $\Omega(\tau)$ (denoted Einstein–Koopman[8] (surface) defined by the relation

$$P_\mu(q_\alpha(\tau), \tau) = \partial_\mu S(q_\alpha(\tau), \tau)/\partial q^\mu(\tau) = \partial_\mu S \tag{5}$$

The relation (5) implies, as we shall now see, an evident departure from the usual classical phase space (as well as of Wigner's) formalism. In the classical picture an ensemble of possible motions of identical particles (of mass m) is described in phase space by the introduction of a scalar phase-space density $f(P_\mu(\tau), Q_\mu(\tau), \tau)$ which implies that

(a) the definition of average values of any arbitrary function $A(P, Q, \tau)$ is given by the relation

$$\langle A \rangle = \int f(P_\mu, Q_\mu, \tau)\, A(P_\mu, Q_\mu, \tau)\, d^4P\, d^4Q \tag{6}$$

and one can define associated Q, P distributions in configuration and momentum space by the relations $\int (Q, \tau) = \int f(P, Q, \tau)\, dP^4$ and $g(P, \tau) = \int f(P, Q, \tau)\, d^4Q$.

(b) the phase-space density $f(p, q, \tau)$ satisfies Liouville's equation

$$\frac{df}{d\tau} + \sum_1^4 \frac{P_\mu}{m} - \sum_1^4 \frac{\partial V}{\partial Q^\mu}\frac{df}{dP_\mu} = 0 \tag{7}$$

for particles moving in an external scalar potential V so that in a cloud of representative points in phase space (whose boundary is defined by the individual phase-space orbits) the volume and total number of the ensemble elements in a volume are constant (i.e., $df/d\tau = 0$) so that f is constant along a phase-space path.

As one knows, it has not been possible (despite many attempts) to extend this formalism to the nonrelativistic formulation of quantum mechanics. The most famous (and effective) attempt is due to Wigner (1932) who introduced the function $F(Q, P, t)$ for classical phase space by the relation

$$F(Q, P, \tau) = \frac{1}{\hbar} \int_{-\infty}^{\infty} \psi^* \left(Q + \frac{1}{2} z\right) \psi \left(Q - \frac{1}{2} z\right) \cdot e^{iPz/\hbar} \, dz \tag{8}$$

where $\psi(q, t)$ satisfies the Schrödinger equation. Its marginal densities are just the usual quantum mechanical position and momentum distributions

$$\int_{-\infty}^{\infty} F(Q, P, t) \, d^3P = |\psi(Q, t)|^2 \quad \text{and} \quad \int_{-\infty}^{\infty} F(Q, P, t) \, d^3Q = |\varphi(P, t)|^2$$

and it satisfies an equation (induced by the Schrödinger equation) which looks like a generalization of the classical Liouville equation. However, F cannot be interpreted as a probability because it may take on negative values in concrete physical situations.

The situation is different in the Einstein–de Broglie model as a consequence of relation (5) which implies the irrotational character of the four momentum vector.

If one restricts oneself to the Einstein–Koopman surface Ω (defined by $p_\mu = \partial S/\partial q^\mu$) where now $q_\mu(\tau)$ denotes the configuration coordinates which map the Ω surface for a given value of the scalar evolution parameter τ, one can describe the phase-space motions on Ω in terms of real trajectories invariantly characterized by their spacelike or timelike nature. This can be conceived as a random process with conserved invariant density $\rho(q_\mu(\tau), \tau)$ on Ω, i.e., $\dot{\rho} = d\rho/d\tau = \partial\rho/\partial\tau + \partial(\rho v_\mu)/\partial q^\mu$, where v_μ denotes an average unit timelike drift velocity ($v_\mu v^\mu = -c^2$) which, when combined with Einstein's osmotic velocity $u_\mu = -D\partial(\ln \rho)/\partial q^\mu \cong (\hbar/2m)\, \partial_\mu(\ln \rho)$ (where D denotes the diffusion coefficient), yields the forward/backward derivatives $b^\mu_\pm$ for any function of a stochastic process $q_\mu(\tau)$, i.e.,

$$D_\pm = \frac{\partial}{\partial\tau} + b^\mu_\pm \, \partial_\mu \pm (h/2m) \Box \tag{9}$$

which imply that we can write for the drift and osmotic derivatives D and δD:

$$D = \frac{1}{2}(D_+ + D_-) = \frac{\partial}{\partial \tau} + v^\mu \partial_\mu \tag{10a}$$

$$\delta D = \frac{1}{2}(D_+ - D_-) = v^\mu \partial_\mu - (\hbar/2m)\Box \tag{10b}$$

On all points on Ω labelled by $q_\mu(\tau)$ (and with $p_\mu = \partial S/\partial q^\mu$) the (E.d.B.) phase-space distribution function $f(q_\mu, p_\mu, \tau)$ can now be written

$$f(p_\mu, q_\mu, \tau) = \rho(q_\mu, p_\mu, \tau) \cdot \delta(p_\mu - \partial S/\partial q^\mu) \tag{11}$$

so that the average value of an arbitrary function $A(p_\mu, q_\mu, \tau)$ on the phase space is given by

$$\langle A \rangle = \int A(p_\mu, q_\mu, \tau) f(p_\mu, q_\mu, \tau)\, d^4p\, d^4q$$

$$= \int A(\partial_\mu S, q_\mu, \tau)\, \rho(q_\mu, \tau)\, d^4q \tag{12}$$

and the momentum distribution is given by

$$g(p_\mu(\tau), \tau) = \int f(p_\mu, q_\alpha, \tau)\, d^4q$$

$$= \int \rho(q_\alpha(\tau), \tau)\, \delta(p_\mu - \partial S/\partial q^\mu)\, d^4q \tag{13}$$

Einstein's result can thus be formulated/summarized in the statement that "all quantum mechanical particle and wave motions in phase space can be represented in terms of particles or waves propagating on the Einstein–Koopman surfaces, i.e., satisfying a particle or a wave equation written in the particular configuration space variables $q_\mu(\tau)$ which map these surfaces, where $p_\mu = \partial S/\partial q^\mu$. Moreover, as noticed by John Bell, in such a model these wave equations should contain both superluminal (phase) and subluminal (particle + wave) solutions. If one assumes (neglecting spin), following Bohm and Vigier,[14] that

(1) An individual microobject is a (pilot) wave ψ and a particle simultaneously,[5] the particle behaving like an oscillator which beats in phase with the surrounding subluminal wave packet.

[5] This point of view has recently been strengthened by the discovery by Aharonov *et al.*[12] that one could measure the waves associated with the observation of single particles.

(2) These particles follow on the average the drift lines of flow tangents to the wave's momenta $\partial S/\partial q^\mu$ when one constructs a hydrodynamical representation of the wave field ψ in terms of fluid elements along the lines originally proposed by Madelung: Both particles and wave elements moving along timelike paths.

(3) That these wave elements and the "piloted" particles (which can be considered as solitons within the wave if one adds a specific nonlinear term to the wave equations[14]) follow realist random Feynman-like stochastic paths to which one can associate positive probability weights which yield the $|\psi|^2$ distribution as a consequence of an H theorem.[13]

(4) The waves themselves can be analyzed as representing the collective motion of Madelung-type elements following random paths in a subquantal stochastic Dirac-type aether.[15] Following Schrödinger, they can thus be considered as representing a diffusive Brownian type motion of such ensemble of wave elements.[16]

(5) One can justify the form given by Nelson to the stochastic process associated with quantum mechanics, i.e., deduce quantum dynamics from the Brownian recoil principle.[17]

Indeed by demanding the validity of the momentum conservation law on all conceivable scales adopted for the investigation of individual particle scattering (collisions) on the medium constituents, we are forced to incorporate the environmental recoil effects in the formalism.[18] The Brownian recoil principle thus elevates the individually negligible phenomena to the momentum conservation law on the ensemble average. The resultant dynamics of the statistical ensemble is governed by the relativistic Schrödinger equation, once the diffusion constant D is identified with $\hbar/2m$.

From these assumptions, it can be shown[10] that:

—One can deduce the relativistic wave equation on Ω in a generalized Schrödinger form

$$\frac{\partial\psi}{\partial\tau} = -\frac{1}{i\hbar} H_{\text{op}} \cdot \psi \tag{14}$$

—One can immediately interpret/justify the usual quantum operator formalism in Hilbert space in terms of real stochastic motions on the Einstein–Koopman phase space surface Ω.

3. STOCHASTIC ORIGIN OF THE QUANTUM OPERATOR ALGEBRA AND OF THE QUANTUM POTENTIAL IN THE EINSTEIN–DE BROGLIE INTERPRETATION OF QUANTUM MECHANICS

It has been known for a long time that the quantum second-order equations, such as the relativistic Klein–Gordon equation, admit a hydrodynamical analysis as originally proposed by Madelung. To this end, following de Broglie and Vigier,[14] one decomposes the Klein–Gordon equation into real and imaginary parts, thereby obtaining a Hamilton–Jacobi-type equation (with $\psi = R\, e^{iS/\hbar} = \exp(P + iS/\hbar)$,

$$\frac{1}{c^2}\partial_\mu S \partial^\mu S + m^2 - \frac{\hbar^2}{c^2}(\Box P + \partial_\mu P \partial^\mu P) = 0 \tag{15}$$

and a continuity equation

$$\Box S + 2\partial_\mu P \partial^\mu S = 0 \tag{16}$$

The latter can also be written as $\partial_\mu(\rho u^\mu) = 0$, where we have introduced a scaled scalar density $\rho = R^2(M/m)$ and the unitary 4-velocity $u_\mu = \partial_\mu S/M(v^\mu v_\mu = -c^2)$. The quantity M denotes the "variable mass" of de Broglie defined as

$$M^2 = m^2 - \frac{\hbar^2}{c^2}(\Box P + \partial_\mu P \partial^\mu P) = m^2 + Q \tag{17}$$

where Q is the relativistic generalization of the quantum potential of Bohm.

It is interesting to note that the same hydrodynamical equations of motion can be derived from a stochastic quantization procedure, along the original lines of Nelson for the Schrödinger theory and the Vigier–Guerra–Ruggiero[17] approach with an appropriate definition of a relativistic Markov process[18] which takes into account the possibility of apparent spacelike motions. The process of stochastic quantization implies a continuity equation of the type of Eq. (2) and the Hamilton–Jacobi equation[15] under the assumption that the mean drift velocity of the process is parallel to the gradient field $\partial_\mu(S)$. The analogies with standard Brownian motion are straightforward, except for the fact that quantum stochastic diffusion is frictionless, thus revealing the nondissipative character of quantum fluctuations. It should also be noted that the quantum wave equations can be equivalently derived from a stochastic variational principle in control theory where, for example, the forward velocity b_μ^+ is considered as the control field, as proposed by Guerra for the Schrödinger theory and in Ref. 19 for the Klein–Gordon theory.

It has been established[20] that in Eqs. (15) and (16) the hydrodynamical fields $\rho^{1/2}$ and S are canonical variables and constitute a phase space L of these collective coordinates. This phase space L has therefore a symplectic structure given by the 2-form

$$\omega_2(\delta\rho, \delta S; \delta'\rho, \delta'S) = \int [\delta\rho(q)\, \delta'S(q) - \delta'\rho(q)\, \delta S(q)]\, dx$$

and it can be shown that Eqs. (15) and (16) may be derived by Euler–Lagrange variations with respect to ρ and S of the relativistic Lagrangian

$$\mathscr{L} = \rho\frac{\partial S}{\partial\tau} + \frac{\rho}{2m}(\partial_\mu S\partial^\mu S + \hbar^2\partial_\mu P\partial^\mu P) \tag{18}$$

Note that (18) has an explicit τ dependence (τ is a scalar evolution parameter, called, for simplicity, the "proper time"). The Klein–Gordon limit of this relativistic "Schrödinger" theory is obtained by the *ansatz* $S(x, \tau) = S(x) - (1/2)\, mc^2\tau$. We can furthermore construct a field Hamiltonian as a phase-space function for which the equations of motion appear in a canonical form, namely[20]

$$\frac{\partial S}{\partial\tau} = \{S, \mathscr{H}\} = \frac{\delta\mathscr{H}}{\delta\rho} \qquad \frac{\partial\rho}{\partial\tau} = \{\rho, \mathscr{H}\} = -\frac{\delta\mathscr{H}}{\delta S} \tag{19}$$

More interesting, however, is a representation where the canonical variables are ψ and ψ^* (with $\psi = \sqrt{\rho}\exp(iS/\hbar)$) because the Hamiltonian becomes linear and can be put in the form

$$\mathscr{H} = \mathscr{H}(\psi, \psi^*) = \langle\psi, H_{\text{op}}\psi\rangle = \int \psi^*(x)\, H_{\text{op}}\psi(q)\, dq \tag{20}$$

which can be used to define the H_{op} of the relativistic covariant Schrödinger theory. Indeed, this reduces to $H_{\text{op}} = (2/2m)$ and casts the Hamilton equation in a linear form:

$$\frac{\partial\psi}{\partial\tau} = \{\psi, \mathscr{H}\} = \frac{-1}{ih}\frac{\partial\mathscr{H}}{\delta\psi^*} = -\frac{1}{ih}H_{\text{op}}\psi \tag{21}$$

This reveals the fact that the Hamiltonian of the system is the generator of infinitesimal proper-time translations.

What has been performed with the Hamilton operator can be transposed in a general scheme that enables us to associate any observable with the corresponding operator. In general, the observables relate to operators in the (ψ, ψ^*) representation according to the definition

$$\mathscr{A}(\rho, S) = \iint \psi^*(q)\, A(q, q')\, \psi(q')\, dq\, dq' \tag{22}$$

where $A^+(q, q') = A(q, q')$, and $A(\rho, S)$ is real. The operator A is then defined as

$$(A\psi)(q) = \int A(q, q')\,\psi(q')\,dq' \tag{23}$$

By establishing the result that Poisson Brackets are "quantum averages" of commutators

$$\{\mathscr{A}, B\} = \frac{1}{i\hbar}\iint \psi'(q)[A, B](q, q')\,\psi(q')\,dq\,dq' \tag{24}$$

one can show that the algebra of the observables is closed under Poisson pairing. Finally, in order to establish more characteristic examples of observables and the corresponding operators, we note that an observable A is the generator of an infinitesimal canonical transformation on any B according to $\delta B = \varepsilon\{BA\}$, where ε is a set of infinitesimal parameters. From this, one immediately deduces the relations between hydrodynamical phase-space variables and the quantum relativistic operator algebra, i.e.,

(a) Phase changes imply $\delta\rho(q) = 0$ and $\delta S(q) = -a_\mu q^\mu$. This is created by the generator $Q_\mu = \int \rho q_\mu\,dq$, which defines the operator $q^\mu_{\text{op}} = q^\mu$.

(b) Translations imply $x'^\mu = x^\mu - a^\mu$, which induces $\delta\rho(q) = -a_\mu \partial^\mu \rho$. The corresponding generator is $P_\mu(\rho, S) = \int \rho\partial_\mu S\,dq$, and this defines, in its turn, $p^\mu_{\text{op}} = -i\hbar\partial^\mu$.

(c) Rotations are characterized by $q'^\mu = q^\mu + \varepsilon^\mu_\nu q^\nu$, $\varepsilon_{\mu\nu} = -\varepsilon_{\nu\mu}$, which induce $\delta\rho(q) = (1/2)\,\varepsilon^{\mu\nu}(q_\nu\partial_\mu - q_\mu\partial_\nu)\rho$. Via determination of the generator $L_{\mu\nu}(\rho, S)$, we arrive at $L^{\text{op}}_{\mu\nu} = -i\hbar(q_\mu\partial_\nu - q_\nu\partial_\mu)$.

4. QUANTUM EVOLUTIONS IN PHASE SPACE REPRESENTED IN HILBERT SPACE

Using the approach sketched above, one can proceed to construct a relativistic Hilbert space in the frame of the relativistic proper time-dependent formalism. This is a consequence of the fact that one can remove the main obstacle consisting in the nonpositive-definite character of a vector norm based on the Klein–Gordon scalar product $\langle\psi, \phi\rangle = i\int \psi^* j_0 \phi\, d^3q$ in this more general formalism. In fact, the norm of a state vector is here defined on the basis of the scalar product

$$\langle\psi, \phi\rangle = \int d^4q\,\psi^*\phi \tag{25}$$

in complete analogy with the nonrelativistic Schrödinger theory. Note that the integration is now extended to the four space-time variables, a fact that poses no serious convergence problem for solutions ψ and ϕ of the relativistic Schrödinger equations, as shown in Ref. 20, and the norm of a vector ψ, i.e., $|\psi|^2$, becomes now a τ-conserved quantity.

Let us consider now the evolution in phase space and associate it with the evolution in the Hilbert space defined above. Following Koopman,[8] we remark that the canonical equations of motion resulting from a scalar relativistic Hamiltonian $H(q_i, p_i)$ which governs a dynamical system with n degrees of freedom (with q_i, p_i as canonical variables) corresponds to a motion in a $2n$-dimensional region R of the real (q, p) space represented by real analytic single-valued functions $q_k = f_k(q^0, p^0, \tau)$, $p_k = g_k(q^0, p^0, \tau)$.

The corresponding transformation in proper time S_τ: $(q^0, p^0) \to (q, p)$ has the property $S_{\tau 1} S_{\tau 2} = S_{\tau 1 + \tau 2}$, $S_0 = I$. If this motion satisfies the condition $H(q, p) = m^2$, i.e., if we pass to the usual Klein–Gordon case, where the Hamiltonian is constant on a mass shell, and Ω is a variety of points in R (i.e., a subspace in phase space, corresponding to the mass shell), then S_τ on Ω represents a one-parameter group of analytic automorphisms of Ω. S_τ creates a path curve which remains in Ω if it has a starting point in Ω and leaves the integral $\int \rho \, d\omega$ invariant when taken over an arbitrary region of Ω, provided that ρ is positive single-valued analytic function in Ω. This is a consequence of the integral invariance of $\int dq_1 \cdots dq_n \, dp_1 \cdots dp_n$.

One should note here that in the Klein–Gordon case of one spin-zero particle we can associate with the quantum system a classical free Hamiltonian with an additional nonlocal potential term, i.e., $H = (1/2)\, p^2 + U(q) = (1/2)\, m^2$, where $U(q)$ corresponds to the quantum potential, $U(x) = -(1/2)\, h^2(\partial_\mu P \partial^\mu P + \Box P)$. This equation is mapped exactly on the H–J type equation of the Madelung decomposition of the Klein–Gordon equation under the condition $p_\mu = \partial_\mu S$, which in fact reduces the dimensions of the subspace Ω to 4 (the four q variables). Now, according to Koopman, S_τ leaves $\int \rho \, d\omega$ invariant. This, in the present case, reduces to the invariance of $\int \rho \, d^4 q$, which follows from the continuity equation of the Madelung decomposition. The original Klein–Gordon spin-zero system is thus consistently reproduced by the evolution in the Ω subspace of the classical phase space R, i.e., the "drift surface" or the "Koopman surface" of the classical phase space.

Koopman's essential step is then to remark that, if $\varphi = \varphi(A)$ is a complex-valued measurable function of the point A on Ω such that the Lebesque integrals $\int \rho \, |\varphi| \, d\omega$ and $\int \rho \, |\varphi|^2 \, d\omega$ are finite, then the totality of such functions φ constitutes the aggregate points of a Hilbert space H, the metric of which is defined by the "inner product" $(\varphi, \psi) = \int_\Omega \rho \varphi \bar{\psi} \, d\omega$.

If one defines transformations in Hilbert space U_τ with the property $U_\tau \varphi(A) = j(S_\tau, A)$, these transformations are linear, continuous, and unitary (i.e., preserve the form of the scalar product) for all $\varphi \in \Omega$ and all real τ. This allows the following interpretation of the formalism: If τ represents the proper time, S_τ specifies the steady flow of a fluid of density ρ occupying the space Ω. If $\Omega(A)$ is a value attached to the point A when $\tau = 0$, these values will be carried into those of the function $U\tau\varphi(A)$, so that $U_\tau \varphi$ has at A the value which φ has at the point $S_\tau A$ in which A flows after the lapse of the proper time τ. *Canonical transformations in phase space Ω have thus been mapped on unitary transformations in Hilbert space H.*

Unitary transformations U_τ, which are automorphisms of Hilbert space (and thus preserve the scalar product as a basic constituent element in this space) are tied to the transformations S_τ in phase space, which can be understood in terms of real motions in E_4 and are a manifestation of the canonical Hamiltonian equations of motion, i.e., the symplectic structure underlying the phase space.

Along the same lines, we can develop a relativistic generalization of the results of Heslot,[21] where we can establish, in analogy with the Schrödinger case, that the unitary transformations U_τ are a canonical transformation since it preserves the Poisson brackets: $U_\tau(\{f, g\}) = \{U_\tau(f), U_\tau(g)\}$, and so the automorphisms of quantum mechanics are canonical transformations. Hilbert space must thus possess an intrinsic symplectic structure. This is a characteristic of a classical phase space.

A relativistic extension of Heslot's calculations proves the identity of the configuration space wave equation (21) and phase-space motions as follows:

Starting from the relation

$$ih d|\psi\rangle/d\tau = \hat{H}|\psi\rangle \tag{26}$$

(where ^ denotes operators), let $[|\phi_k\rangle]$ be an orthonormal basis of the Hilbert space, i.e.,

$$|\psi\rangle = \sum_k \lambda_k |\phi_k\rangle = \exp(P + iS/h) \tag{27}$$

Denoting by λ_k the complex components of $|\psi\rangle$ on this basis, we can then decompose these λ_k's in real and imaginary parts; set $\lambda_k = (q_k + ip_k)/2^{1/2}$, and write (26) as a set of equations for the canonically conjugated variables p_k and q_k in terms of the Hamiltonian function H defined as the mean value of $\hat{H}$ expressed in terms of the q_k's and p_k's.

Assuming that $|\psi\rangle$ is normalized, we have $H = \langle\psi|\, \hat{H}\, |\psi\rangle$, and we show immediately that

$$\frac{dq_k}{d\tau} = \frac{\partial H}{\partial p_k} \quad \text{and} \quad \frac{dp_k}{d\tau} = -\frac{\partial H}{\partial q_k} \tag{28}$$

so that the space of states is now a classical phase space provided with some complementary structures.

As shown by Heslot,[21] the evolution of a classical system takes place in its phase space. It is a space of even dimension, say $2n$, which is in fact the space of states of the system. In agreement with the experimental point of view, this leads to defining an observable as a real-valued regular function on that space.

The space of states is provided with a Poisson bracket, i.e., with an operation $f, g \to \{f, g\}$ on the observables, which is linear, antisymmetric, nondegenerate, and satisfies the Jacobi identity and a derivation-like product formula. One can prove that there always exists (local) systems of coordinates on the space of states, say, (q_k, p_q), $k = 1, \ldots, n$, such that $\{q_k, p_l\} = \delta_{kl}$, $\{q_k, q_l\} = \{p_k, p_l\} = 0$, $k, l = 1, \ldots, n$, where δ_{kl} is the Kronecker symbol, i.e., $\delta_{kl} = 1$, for $k = l$, and 0 otherwise. Such coordinates are called canonical; they allow the Poisson bracket to be given the familiar form

$$\{f, g\} = \sum_k \left(\frac{\partial f}{\partial x_k} \frac{\partial g}{\partial p_k} - \frac{\partial f}{\partial p_k} \frac{\partial g}{\partial x_k} \right) \tag{29}$$

The physical meaning of the Poisson bracket structure on the space of states is that the transformations of the states do not modify the nature of the system, but merely correspond to a change of point of view; e.g., a rotation, a translation, a change of inertial frame, or the time evolution, preserve the Poisson bracket, i.e., are automorphisms of that structure. More precisely, let ξ be a transformation of the states; it induces naturally a transformation $f \to \xi(f)$ of the observables. Then ξ is an automorphism of the space of states, provided with its Poisson bracket structure, if and only if, for any two observables f and g, we have

$$\mathcal{G}(\{f, g\}) = \{\mathcal{G}(f), \mathcal{G}(g)\}$$

Since the correspondence between observables and infinitesimal automorphism of the classical space of states rests on the properties of the Poisson bracket, we are led to assume the existence of some underlying Poisson bracket structure on the space of states of our generalized classical system. More precisely:

(a) The space of states is provided with a structure whose physical meaning is that the transformations of the states which correspond to a change of point of view preserve that structure, i.e., are the automorphisms of the space of states.

(b) We assume that this structure intrinsically induces a Poisson bracket structure: *The space of states is thus a classical phase space, provided in the general case with some complementary structure.*

(c) The word "intrinsically" in (b) means precisely that the automorphisms of the space of states preserve the Poisson bracket, i.e., are canonical transformations. The converse is not true in general: Not every canonical transformation is an automorphism, i.e., also preserves the complementary structure. *We define then the observables as those real-valued regular functions of the state, whose canonical transformations they generate are automorphisms of the whole structure of the space of states.*

The usual classical mechanics is characterized by the fact that there is no complementary structure beside the Poisson bracket and, therefore, any real-valued function of the state is an observable. This is not true in the general case. It can be shown, however, that the set of observables is closed under addition, product by a scalar, and the Poisson bracket. But the usual product of two observables, defined by the product of their values, no longer needs to be an observable.

The normalization condition means that part of the complementary structure of the space of states consists of a supplementary constraint. More precisely, using (25), we have

$$\langle\psi\,|\,\psi\rangle=\sum_k|\lambda_k|^2=\sum_k\frac{q_k^2+p_k^2}{2h}$$

and the normalization condition $(\psi^*\psi)=1$ is

$$\sum_k(q_k^2+p_k^2)=2h \qquad (30)$$

Let g be an observable. Since (30) is part of the structure of the space of states, it must be preserved under the infinitesimal automorphism generated by g. Now, using (29), we obtain the transformation law

$$\sum_k(q_k^2+p_k^2)\rightarrow\sum_k(q_k'^2+p_k'^2)$$
$$=\sum_k(q_k^2+p_k^2)+2\sum_k\left(\frac{\partial g}{\partial p_k}q_k-\frac{\partial g}{\partial q_k}p_k\right)\delta\alpha$$

Thus, g may be an observable only if

$$\sum_k \left(\frac{\partial g}{\partial p_k} q_k - \frac{\partial g}{\partial q_k} p_k \right) = 0 \qquad (31)$$

Notice that the Hamiltonian function H is quadratic in the qx_k's and p_k's. To reduce it to a simple form, take for $\{|\phi_k\rangle\}$ a proper basis of H. Then $H = \sum H_k |\lambda_k|^2 = \sum H_k(p_k^2 + q_k^2)/2h$, where the H_k's are the Hamiltonian levels. H appears thus as a sum of Hamiltonian functions of harmonic oscillators with pulsations $\omega_k = H_k/h$.

4. CONCLUSION

The preceding discussion evidently implies an extension of the Einstein–de Broglie model/interpretation of quantum mechanics.

What is now suggested is

(a) That "vacuum" is a real Dirac-type "aether" (along the lines suggested by Dirac( ) and Suddarshan *et al.*(22)), which carries real excited superluminal phase waves.

(b) That observed microobjects are built with subluminal wave packets ($\partial_\mu S \partial_\mu S < 0$) which result from the superposition of superluminal waves which also satisfy generalized relativistic Schrödinger equations like (14).

(c) That these subluminal wave packets pilot/control oscillator type particles (or solitons) which beat in phase with them and propagate under the influence of stochastic nonlocal potentials along real Feynman-type stochastic paths.

Such an extended model, strangely enough, represents a return to the source of de Broglie's initial discovery of wave mechanics.

To clarify this point, one can first revisit the famous paper(3) [published in the *C. R. Acad. Sci. (Paris)* **177**, 507 (1923) by Louis de Broglie, entitled "Ondes et quantas"] in which he linked relativity theory with the physical reality of superluminal phase waves for the first time ... and thus opened the way for the discovery of wave mechanics.

He starts with the idea that quantum particles of rest mass m_0 and internal rest energy m_0c^2 can be considered as real clocks (moving with a velocity $v = \beta c$ ($\beta < 1$) with a rest-mass frequency ν_0 satisfying

$$h\nu_0 = m_0c^2 \qquad (32)$$

According to relativity theory for an observer at rest the observed particle's/clock's energy corresponds to a frequency $\nu = m_0c^2/(1-\beta^2)^{1/2}$ and its observed (clock) frequency will be slowed down to a value $\nu_1 = \nu_0(1-\beta^2)^{1/2}$, i.e., vary like $\sin 2\pi\nu_1 t$, i.e., like a wave

$$\exp(i2\pi\nu_1 t)$$

De Broglie then made the now famous assumption that at $t=0$ the clock is associated (i.e., beats in phase) with a real superluminal phase wave which propagates in the same direction with the velocity $u = c/\beta$ and carries no detectable energy (since, following Einstein, all observed energy corresponds to absorption/emission of particlelike quantas) and that the phase-locking process is preserved by the particle's motion. Indeed, if at $t=0$ the particle is at a distance $x = vt$ from the origin, its internal clocklike motion is represented by a wave φ with $\varphi_{t=0} = \exp i[2\pi\nu_1(x/v)]$, while the corresponding phase wave is represented by

$$\phi = \exp\left[i2\pi\nu\left(t - \frac{x\beta}{v}\right)\right] = \exp\left[i2\pi\nu x\left(\frac{1}{v} - \frac{\beta}{c}\right)\right] \tag{33}$$

so that they are equal (i.e., phase-locked) if

$$\nu_1 = \nu(1-\beta^2)$$

an equality which results from the relativistic definitions of ν and ν_1.

In other words, if one considers the phase wave as a real space filling oscillation depending on ν_0, t_0 (t_0 being the proper time of an observer tied to the clock), the Lorentz transformation yields

$$t_0 = \frac{1}{(1-\beta^2)^{1/2}} \cdot \left(t - \frac{\beta x}{c}\right) \tag{34}$$

and if one applies to ϕ the operator $\Box = (1/c^2)\,\partial/\partial t^2 - \partial/\partial x^2$, one obtains for the phase waves the wave equation

$$\Box\phi = \frac{m_0^2 c^2}{h^2}\,\phi \tag{35}$$

which implies that they satisfy the Klein–Gordon equation which was only discovered later as defining the relativistic subluminal propagation of the scalar generalization of Schrödinger's spinless waves. In the same text[(3)] de Broglie makes the assumption that photons are just ordinary quantum particles (with $0 < m_\gamma = 10^{-65}$ gm) surrounded by superluminal phase waves—a first step in the construction of the E.d.B. experimental phase waves and a first step in the construction of the E.d.B. theory of light. [For a review of its present theoretical and experimental status, see J. P. Vigier, ISQM 92, Tokyo, and *IEEE* **18**, 64 (1990).] In this model, light particles (i.e., photons moving along the de Broglie–Bohm trajectories) are considered as clocks which remain in phase with their surrounding subluminal Maxwell-type waves as a consequence of the quantum potential, defined in relation (17).

This is an important point. As one knows, relation (35) contains both subluminal and superluminal solutions $\psi = \exp(P + iS/h)$ (where the four-vector $p_\mu = \partial_\mu S$ is timelike or spacelike, respectively), so that in the E.d.B. model to a relativistic clocklike point particle moving with a velocity v with a rest frequency given by (32) one can associate two wave fields which satisfy the same equation (35), i.e., a real superluminal phase wave solution propagating with $p_\mu p^\mu$ [which, as shown by Bell[(5)] and discussed by Kyprianidis *et al.*,[(6)] can result from the superposition of subluminal solutions of (5)], which propagates with a velocity $v = c/\beta$ and remains in phase with the particles oscillation during its motion, and a subluminal plane wave ψ which propagates with the velocity v and thus is permanently phase-locked with the particle oscillation.[6]

John Bell's argument has been discussed in detail by Kyprianidis (*Phys. Lett. A* **111**, 131 (1985)). The fact that the superposition of two subluminal plane wave solutions of relation (35) yields a superluminal impulsion current implies that one must impose initial constraints on a subluminal wave packet if it is to remain subluminal in the future.

As one knows, these plane waves effectively correspond to narrow frequency bands and, when integrated over the frequency distribution of a quantum state, yield the subluminal wave packets (where $p_\mu p^\mu < 0$ at all points) which are utilized in both the usual (Copenhagen) and in the causal stochastic (E.d.B.) interpretations.

Since both types of waves satisfy relation (5), one can develop each type of solution on a sum of solutions of the other type, i.e., the usual quantum mechanical subluminal wave packets ψ which correspond to probability distribution (or to a distribution of Madelung type particlelike elements resulting from the superposition of real superluminal phase waves which propagate on a subquantum ground-state vacuum or Dirac type "aether."[(4, 22)])

If one then recalls that an initial subluminal Madelung distribution can split into two nonoverlapping regions, while the phase waves fill all

[6] Despite the fact that some solutions of Eqs. (5) can correspond to superluminal waves, de Broglie later rejected such solutions for his "pilot" waves since he considered bosons as resulting from the fusion of two spin-1/2 Dirac particle components, a procedure which excludes solutions corresponding to spacelike currents.

space, one sees that one can construct subluminal wave packets as a sum of superluminal phase waves filling all space associated with every constitutive element of its Madelung hydrodynamical decomposition, a property which justifies the use of Rauch's relation (1) since such sums can yield two (or more) regions where the total amplitude effectively vanishes by superposition despite the fact that the corresponding constituting phase waves still exist between these regions and are thus responsible (due to the modification of their boundary conditions by external phenomena) for the observed (nonlocal) interactions between them. In other words, momentum phase waves can exist in regions where there are no subluminal configuration space particle distributions. In the E.d.B. model the quantum potential interactions are supported in empty space by real physical phase waves, and one can interpret Rauch *et al.*'s results as a consequence of the modification (tied to measurement processes) of the boundary conditions of the phase waves.[7]

As shown by de Broglie, the reality of the phase waves can be considered justified by the fact that the Bohr–Sommerfeld (orbits and quantization processes) can be derived from his model. This results from the idea that when an electron turns on a Bohr orbit, starting from a point 0, its associated phase wave (also starting from 0), travelling with a velocity c/β, catches up with him in a time τ at a point 0′ such that $\overline{00'} = \beta c\tau$, so that

$$\tau = \frac{\beta}{c}\,[\beta c(\tau + T_r)], \qquad \text{i.e.,} \quad \tau = \frac{\beta^2}{1-\beta^2}\,T_r \tag{36}$$

where T_r denotes the period of revolution of the electron on its orbit. The corresponding phase variation is

$$2\pi P_1 \tau = 2\pi \frac{m_0 c^2}{h}\, T_r \frac{\beta^2}{(1-\beta)^{1/2}} \tag{37}$$

[7] In one of his last university lectures (on his interpretation of quantum mechanics) de Broglie made two striking analogies:

(1) The relation of an oscillator-type particle with its pilot wave can be compared in a sense to that of a plane flying at Mach 1 with is surrounding sound wave.

(2) The Fourier analysis of a state of vibration of the string of a violin in terms of basic harmonics implies the reality of the total vibration *and* of the said harmonics since both can be simultaneously detected by resonance with suitable acoustic devices. In this sense one can say that the waves utilized by Rauch in relation (4) exist despite the fact that the total superposed amplitude cancels in certain space-time regions.

and the locking assumption (which evidently implies the phase wave's reality) then yields the quantum condition

$$\frac{m_0 \beta^2 c^2}{(1-\beta^2)^{1/2}} T_r = nh \tag{38}$$

This reality of phase waves is also suggested by the fact that the simultaneous existence of particle and their surrounding phase wave distribution (which constitute their usual ψ field wave packet) imply the identity of the particle paths action (Maupertuis–Hamilton principle) and of the Fermat trajectories associated with phase wave mechanics. Indeed, in its relativistic form, the least action principle can be written as

$$\delta \int J_\mu \, dx^\mu = \delta \int P^\mu \, d\dot{q}_\mu = 0 \tag{39}$$

where J_μ are the four vector components where J_4 is the particle's energy (divided by c) and J_k $(k = 1, 2, 3)$ represent its impulsion. Since, as one knows, Fermat's principle can be written in the form

$$\delta \int 0_\mu \, dx^\mu = 0 \tag{40}$$

where $0_4 = P/c$ and 0_k $(k = 1, 2, 3)$ denote a vector of length v/v tangent to the rays of the wave field, the relation $0_k = (1/h) J_k$ thus implies the equality of relations (39) and (40).

One can conclude with the remark that if one describes (following Sudarshan *et al.*[22] the subquantal "vacuum" or "aether" level as a real, covariant, chaotic, fundamental, ground-state level of matter, which can carry collective superfluid wavelike oscillations,[8] the preceding comments suggest they could be considered as a superposition of the real superluminal phase waves introduced by de Broglie when he discovered wave-mechanics. Despite his own later misgivings,[9] de Broglie was thus the first to introduce nonlocal real phenomena in quantum theory.

[8] A possible connection with Dirac's aether model is evident. If one represents the subquantal level of matter as a quasi-continuous covariant stochastic distribution of harmonic oscillators, a superluminal phase wave can travel on such a medium without covariant detectable particle/energy propagation, since it must be instantaneous (i.e., carry no energy) in one frame at least, i.e., behaves like electromagnetic phase waves in real plasmas.

[9] De Broglie was initially reluctant to accept nonlocal quantum correlations because of their possible contradiction with causality ... a problem which was later solved, in favor of causality, by Sudarshan and Droz-Vincent *et al.*; see Refs. 16 and 22.

ACKNOWLEDGMENTS

The author thanks Professor Rauch for drawing his attention to the possible epistemological consequences of Ref. 1 and 2 and M. C. Combourieu for stressing the possible importance of Ref. 3, which has practically disappeared from the literature.

REFERENCES

1. S. A. Werner, R. Clothier, H. Kaiser, H. Rauch, and H. Wölwitsch, *Phys. Rev. Lett.* **67**, 683 (1991); H. Kaiser, R. Clothier, S. A. Werner, H. Rauch, and H. Wölwitsch, *Phys. Rev. A* **45**, 31 (1992).
2. H. Rauch, "Phase space coupling in Interference and EPR experiments," *Phys. Lett. A* **173**, 240 (1993).
3. L. de Broglie, *C. R. Acad. Sci.* **177**, 507 (1923).
4. P. A. M. Dirac, *Nature (London)* **168**, 906 (1951); **169**, 702 (1952).
5. M. Bozic, Z. Maric, and J. P. Vigier, *Found. Phys.* **27**, 1325 (1992).
6. H. Rauch and J. P. Vigier, *Phys. Lett. A* **151**, 269 (1990).
7. L. Mandel, *J. Opt. Soc. Am.* **51**, 132 (1961); D. F. V. James and E. Wolf, *Opt. Comm.* **81**, 150 (1991); D. F. V. James and E. Wolf, *Phys. Lett. A* **157**, 6 (1991).
8. A. Einstein, *Sitz. Preuss Akad. Wiss.*, 606 (1917) and *Verhand. Deutsch. Phys. Ges.* **19**, 82 (1917). See also B. P. Koopman, *Quantum Theory and the Foundations of Probability* (McGraw-Hill, New York, 1955).
9. The use of a universal scalar time evolution parameter has been recently reviewed in the literature. For example, see Fanchi, *Found. Phys.* **23**, 487 (1993). For a possible physical interpretation of *t* within the E.d.B. theory of light, see M. C. Combourieu and J. P. Vigier, *Phys. Lett. A* **175**, 269 (1993).
10. For a review, see J. P. Vigier, *Proceedings, 3rd Symposium on the Foundations of Quantum Mechanics* (Physical Society of Japan, Tokyo, 1989), p. 1210.
11. D. Bohm and J. P. Vigier, *Phys. Rev.* **96**, 208 (1954).
12. Y. Aharonov and L. Vaidman, "Measurement of the Schrödinger wave of a single particle," *Phys. Lett. A*, 1993, in press.
13. A. Kyprianidis and D. Sardelis, *Lett. Nuovo Cimento* **39**, 337 (1984).
14. J. P. Vigier, *Found. Phys.* **21**, 125 (1991).
15. P. A. M. Dirac, *Nature (London)* **168**, 906 (1951).
16. J. P. Vigier, *Astr. Nachr.* **303**, 55 (1982).
17. P. Garbaczewski and J. P. Vigier, *Phys. Rev. A* **46**, 4634 (1992).
18. P. Garbaczewski and J. P. Vigier, *Phys. Lett. A* **167**, 447 (1992).
19. P. Holland, A. Kyprianidis, and J. P. Vigier, *Found. Phys.* **17**, 53 (1987).
20. N. Cufaro-petroni, C. Dewdney, P. Holland, A. Kyprianidis, and J. P. Vigier, *Phys. Rev. D* **32**, 1375 (1985).
21. A. Heslot, *Phys. Rev. D* **32**, 1341 (1985).
22. E. C. G. Sudarshan, K. P. Sinha, and J. P. Vigier, *Phys. Lett. A* **114**, 298 (1980).

This page is deliberately left blank.

Reprinted from *Lettere al Nuovo Cimento*, Vol. 39, No. 11, pp. 225-233, Copyright (1983) with permission from the Società Italiana di Fisica.

Possible Experimental Test of the Wave Packet Collapse.

A. Garuccio

Istituto di Fisica dell'Università - Bari, Italia
Istituto Nazionale di Fisica Nucleare - Sezione di Bari, Italia

A. Kyprianidis (*), D. Sardelis (*) and J. P. Vigier

Institut H. Poincaré, Laboratoire de Physique Theorique
11, rue P. et M. Curie, 75231 Paris Cedex 05, France

(ricevuto il 23 Novembre 1983)

PACS. 03.65 – Quantum theory; quantum mechanics.

Summary. – A quantum analysis is presented of combined first-order/second-order optical-interference experiments with or without the wave packet collapse concept. It is shown that the Bohr-Heisenberg model yields testable predictions which differ from those of the Einstein-de Broglie theory of light.

One of the starting points of the Bohr-Einstein controversy is evidently the wave packet collapse concept introduced by Bohr as a basis of quantum measurement theory. As one knows Einstein never accepted it, since he believed

1) that this process, assumed to be instantaneous in all frames, is evidently in contradiction with relativity theory,

2) that it contradicts his realistic conception of the wave-particle dualism; for him electromagnetic waves for example are considered to present simultaneously wave *and* particle (photons) aspects, while Bohr holds that particles are waves *or* particles, never the two simultaneously; for him the photons materialize in observation when their probability waves collapse instantaneously.

Two recent developments, however, have confirmed Einstein and de Broglie's opinions. The first is the discovery by Cini [1] that one can construct a quantum measurement theory without wave packet collapse ... so that one of the pillars of the Copenhagen interpretation is now endangered. The second is a set of recent proposals by Garuccio,

(*) On leave from the University of Crete, Physics Department, Heraklian, Crete, Greece.
(1) M. Cini: *Nuovo Cimento B*, **73**, 27 (1983).

Popper (²) Selleri (³) Rapisarda (⁴) Andrade and Vigier (⁵) to test the real independent existence of de Broglie's wave ... in conflict with the wave packet collapse concept.

These proposals have engineered a number of comments (⁶) and criticisms (⁷,⁸). The aim of the present letter is to present a complete quantum-mechanical treatment of the proposed set-ups and to show that the Bohr-Heisenberg interpretation indeed yields testable predictions which differ from those of the Einstein-de Broglie theory of light in combined first-order/second-order optical-intererence experiments ... which can thus be considered as tests for or against the introduction of the wave packet collapse in the interpretation of quantum theory.

We start with the Michelson device represented in fig. 1. The quantum theory of such a device has been given by Loudon (⁷) in some detail. We shall rediscuss it here in his formalism, since as we shall show, he incorrectly assumes the Maxwell-Boltzmann statistics for incoherent beams.

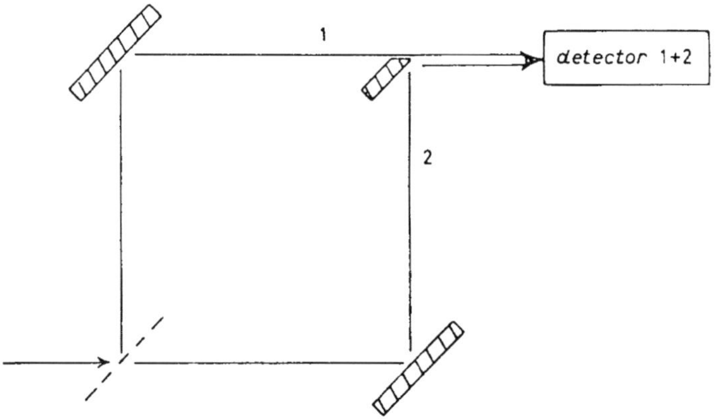

Fig. 1.

We represent by $a^\dagger, a$ the creation, annihilation operators of the input light which is split by the semi-transparent mirror into equal-amplitude components a_1, a_2 (see fig. 1). The aoutput beam at the detector d is given by the expression $\langle d^\dagger d\rangle$, where

$$a = (a_1 + a_2)\delta(n_1 + n_2 - N) ,$$

$$d = [a_1 \exp[iKz_1] + a_2 \exp[iKz_2]]\delta(n_1 + n_2 - N)$$

and

$$n_i = \langle a_i^\dagger a_i\rangle , \qquad i = 1, 2 ,$$

(²) K. R. Popper, A. Garuccio and J. P. Vigier: *Phys. Lett. A*, **86**, 326 (1982).
(³) F. Selleri: *Found. Phys.*, **12**, 1087 (1982).
(⁴) V. A. Rapisarda, A. Garuccio and J. P. Vigier: *Phys, Lett. A*, **90**, 17 (1982).
(⁵) J. Andrade e Silva, F. Selleri and J. P. Vigier: *Lett. Nvovo Cimento*, **36**, 503 (1983).
(⁶) W. M. de Muynck: *Epistem. Lett.*, **33**, 7, 13 (1982).
(⁷) R. Loudon: *Quantum theory of combined first order/second order optical interference experiments*, preprint to appear in *Opt. Commun.*
(⁸) O. Costa de Beauregard: *Nuovo Cimento B*, **42**, 41 (1977); **51**, 267 (1979).

which yields the following result:

$$\langle d^\dagger d\rangle =$$
$$= \langle [a_1^\dagger \exp[-iKz_1] + a_2^\dagger \exp[-iKz_2]][a_1 \exp[iKz_1] + a_2 \exp[iKz_2]]\delta(n_1 + n_2 - N)\rangle .$$

The restriction imposed by the δ-function, *i.e.* the definite number N of photons, can be explicitly evaluated as a factor in the composition of the state vector $|\psi\rangle$. This results in a weighing of the states, out of which $|\psi\rangle$ is composed, according to Bose-Einstein statistics appropriate for correlated particles submitted to random stochastic nonlocal actions at a distance (⁹). As we have shown (¹⁰) this leads to different predictions for the interference pattern. The Loudon formula for the intensity

$$I \sim N(1 + \cos\varphi) ,$$

which exhibits maximum fringe visibility has to be replaced by

$$I \sim N(1 + \lambda(N)\cos\varphi) , \qquad \lambda(N) \leqslant 1 .$$

Only for $N = 1$ the two formulae coincide, $\lambda(1)$ being 1. For $N = 2$ one already obtains a reduced fringe visibility corresponding to the $\frac{1}{3}$, $\frac{1}{3}$, $\frac{1}{3}$ probability for the occurrence of the states $|0, 2\rangle$, $|1, 1\rangle$ and $|2, 0\rangle$. For $N \to \infty$, $\lambda(N)$ approaches $M/4$, and the intensity pattern exhibits a reduced fringe visibility.

We now pass to the set-up of fig. 2. We introduce on arm 1, amplifier A, which is assumed to produce (when stimulated by one photon only) an outcoming photon with the same phase, generally accompanied by a spontaneous photon with a random

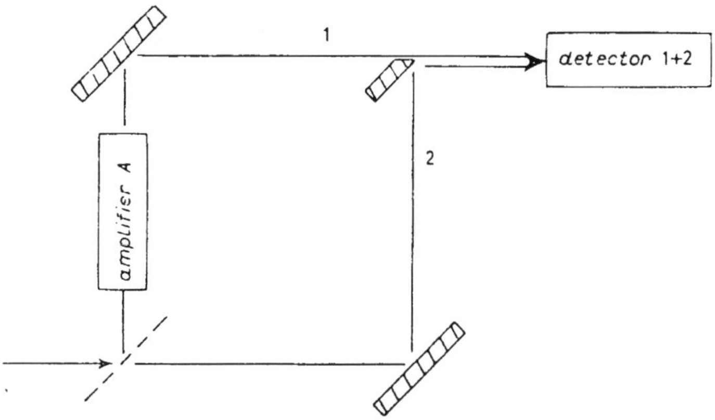

Fig. 2.

(⁹) A. Kyprianidis, D. Sardelis and J. P. Vigier: *Causal nonlocal character of quantum statistics*, submitted to *Phys. Rev. Lett.*

(¹⁰) F. De Martini, A. Kyprianidis, D. Sardelis and J. P. Vigier: *Quantum-mechanical causal actions-at-a-distance correlations in optical-beam splitting devices and interference experiments*, submitted to *Nuovo Cimento.*

phase ([11]). In the coherent state the photon operators are transformed according to

$$\exp[-A]a_1 \exp[A] = a_1 + \alpha_1\,,$$

$$\exp[-A]a_1^\dagger \exp[A] = a_1^\dagger + \alpha_1^*\,,$$

where

$$A = \alpha_1 a_2^\dagger - \alpha_2^* a_1$$

and α is determined by the geometry of the amplifier and its effective current flow.

At this stage one must introduce the physical properties of the amplifier. There are evidently two different models that can be applied to interpret the results. The first model is the usual Bohr-type Copenhagen interpretation, which implies wave-packet collapse when the photon is located somewhere, *i.e.* when it triggers the amplifier. This, however, implies the use of the amplifier explicitly as a measuring device, by means of an observer. If the amplifier is not under observation, then according to Bohr one cannot know if it has been triggered or not, and consequently there is a lack of information about the state of the system. Hence, the state of the system after the amplifier cannot be a pure state, but only a mixture of states with probabilities of occurrence. The predictability of the experimental results restricts itself to statistical predictions. On the other hand, the second model, the Einstein-de Broglie theory of light, has no wave packet collapse, which is exactly the assumption made by Cini ([1]) in his proposed description of quantum measurement theory. The Einstein-de Broglie model considers photons to be waves and particles and consequently the amplifier does change the state of the system in branch 1 of fig. 2, but does not affect our knowledge of the system, because it does not create mutually exclusive states as the Bohr-CIQM does. The system is now in a modified state with respect to the case examined previously, but still in a pure state.

Apparently, the state of the wave field depends on the physical model used. The wave packet collapse concept of Bohr introduces the following quantum mixture as the appropriate way to account for the behaviour of the system:

$$\text{state } |1\rangle = \frac{1}{\sqrt{2}}\{|1,0\rangle + |0,1\rangle\}\,, \qquad \text{with probability } p_1\,,$$

$$\text{state } |2\rangle = \frac{1}{\sqrt{1+|\alpha|^2}}\{|2,0\rangle + \alpha|\widetilde{1,0}\rangle\}\,, \qquad \text{with probability } p_2\,,$$

and with a density matrix

$$\varrho = p_1|1\rangle\langle 1| + p_2|2\rangle\langle 2|\,, \qquad p_1 + p_2 = 1\,,$$

where $|\widetilde{...}\rangle$ denotes spontaneous emission.

In the Einstein-de Broglie model, the amplifier has a nonzero probabilitF ε for any given photon to pass through without being triggered and a probability $1-\varepsilon$ to double the photon (*i.e.* stimulated emission). This process is always accompanied by spontaneous emission. The state of the system can be written now as a pure state $|\psi\rangle$

([11]) R. J. Glauber: *Phys. Rev.*, **131**, 2766 (1963).

and takes the form

$$\langle\psi\rangle = \frac{1}{\sqrt{2+|\alpha|^2}}\{\sqrt{\varepsilon}|1,0\rangle + \sqrt{1-\varepsilon}|2,0\rangle + \alpha|\widetilde{1,0}\rangle + |0,1\rangle\}$$

with a density matrix of a pure state $\varrho = |\psi\rangle\langle\psi|$.

We can proceed now to calculate the interference intensity by evaluating the following expression:

$$I \sim \mathrm{Tr}\,[\varrho\{(a_1^\dagger + \alpha_1^*)\exp[-iKz_1] + a_2^\dagger \exp[-iKz_2]\}\{(a_1+\alpha_1)\exp[iKz_1] + a_2\exp[iKz_2]\}]$$

with either of the proposed models, respectively.

A) In the Bohr model a straightforward calculation yields the following result:

$$I \sim |\alpha|^2 + \frac{2p_2}{1+|\alpha|^2} + p_1\{1+\cos\varphi\}\,.$$

This result is a statistical prediction of the interference pattern. It entails the predictions

a) that if the amplifier is triggered ($p_1 = 0$, $p_2 = 1$) no interference is observed;

b) that in the absence of the amplifier ($p_2 = 0$, $\alpha = 0$), we recover the well-known quantum-mechanical result $I \sim (1+\cos\varphi)$.

B) In the Einstein-de Broglie model we obtain

$$I \sim |\alpha|^2 + \frac{\alpha+\alpha^*}{2+|\alpha|^2}\sqrt{2}\sqrt{\varepsilon(1-\varepsilon)} + \frac{3-\varepsilon}{2+|\alpha|^2}\left\{1 + \frac{2\sqrt{\varepsilon}}{2-\varepsilon}\cos\varphi\right\},$$

which is a definite prediction depending on the efficiency of the amplifier. We can still deduce some limiting cases $\varepsilon = 0$ and $\varepsilon = 1$ and an intermediate situation $\varepsilon = \frac{1}{2}$:

$\varepsilon = 1$ (free pass-through) $\quad I \sim |\alpha|^2 + \frac{2}{2+|\alpha|^2}\{1+\cos\varphi\}$,

or with $\alpha = 0$ $\quad I \sim 1 + \cos\varphi$,

which is the quantum-mechanical result with no amplifier;

$\varepsilon = 0$ (full absorption) $\quad I \sim |\alpha|^2 + \frac{3}{2+|\alpha|^2}$,

i.e. absence of intereference, and

$$\varepsilon = 1/2\,, \quad I \sim |\alpha|^2 + \frac{\alpha+\alpha^*}{2+|\alpha|^2}\frac{\sqrt{2}}{2} + \frac{5}{2(2+|\alpha|^2)}\left\{1 + \frac{2\sqrt{2}}{5}\cos\varphi\right\}.$$

This, in our opinion, is an important result, since it shows that the Bohr and Einstein-de Broglie models give different and testable predictions in some particular simple experimental set-ups.

We conclude with a full quantum-mechanical treatment of the experiments proposed by POPPER *et al.* (2) and GARUCCIO *et al.* (4) and critisized by LOUDON (7) and COSTA DE BEAUREGARD (8). The set-up is described in fig. 3 with an input of simple isolated photons. Here also the Bohr and the Einstein-de Broglie models differ, respectively in the utilized wave field, which is now represented as a three-state ket $|a, b, c\rangle$ corresponding to the branches 1, 2 and 3 of fig. 3.

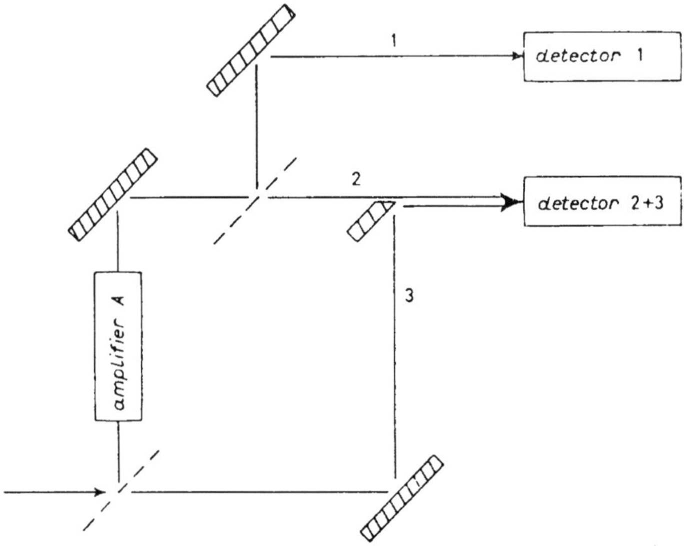

Fig. 3.

The operators acting on the three states of the ket are, respectively,

$$a_1^\dagger + \frac{\alpha^*}{\sqrt{2}}\bigg| a_1 + \frac{\alpha}{\sqrt{2}}, \quad a_2^\dagger + \frac{\alpha^*}{\sqrt{2}}\bigg| a_2 + \frac{\alpha}{\sqrt{2}} \quad \text{and} \quad a_3^\dagger | a_3 .$$

Evidently, this set-up now provides both interferences and correlations which can and will be tested experimentally (12).

A) In the Bohr model the state vector takes the form

$$\text{state } |1\rangle = \frac{1}{\sqrt{2}}\left\{|0, 0, 1\rangle + \frac{1}{\sqrt{2}}|1, 0, 0\rangle + \frac{1}{\sqrt{2}}|0, 1, 0\rangle\right\} \qquad \text{with } p_1 ,$$

$$\text{state } |2\rangle = \frac{1}{\sqrt{1+|\alpha|^2}}\left\{\frac{1}{\sqrt{3}}|2, 0, 0\rangle + \frac{1}{\sqrt{3}}|1, 1, 0\rangle + \frac{1}{\sqrt{3}}|0, 2, 0\rangle + \right.$$

$$\left. + \frac{\alpha}{\sqrt{2}}|1, \widetilde{0}, 0\rangle + \frac{\alpha}{\sqrt{2}}|0, \widetilde{1}, 0\rangle\right. \qquad \text{with } p_2 ,$$

and

$$\varrho = p_1|1\rangle\langle 1| + p_2|2\rangle\langle 2| \qquad \text{with } p_1 + p_2 = 1 .$$

---

(12) A. GOZZINI: Communication at the *Symposium on Wave-Particle Dualism, April 1982, Perugia.*

The interference intensity calculated as

$$I \sim \operatorname{Tr}\left\{\varrho\left[\left(a_2^\dagger + \frac{\alpha^*}{\sqrt{2}}\right)\left(a_2 + \frac{\alpha}{\sqrt{2}}\right) + a_3^\dagger a_3 + \left(a_2^\dagger + \frac{\alpha^*}{\sqrt{2}}\right) a_3 \exp[i\varphi] + \right.\right.$$

$$\left.\left. + a_3^\dagger\left(a_2 + \frac{\alpha}{\sqrt{2}} \exp[-i\varphi]\right]\right\}$$

yields the following result:

$$I \sim \frac{|\alpha|^2}{2} + p_1 \frac{3}{4}\left\{1 + \frac{2\sqrt{2}}{3}\cos\varphi\right\} + \frac{p_2}{1 + |\alpha|^2}.$$

This formula for the intensity is again composed of one interference term where the amplifier is not triggered ($p_2 = 0$, $p_1 = 1$) and a constant term. In the absence of amplifier ($p_2 = 0$, $\alpha = 0$), we obtain the quantum-mechanical result

$$I \sim \frac{3}{4}\left\{1 + \frac{2\sqrt{2}}{3}\cos\varphi\right\}.$$

The predictions of this model are simply of statistical nature. The correlation between two counters put on the outputs of beam 1 and beams 2 and 3 can be easily evaluated by means of the correlation function

$$C \sim \operatorname{Tr}\left\{\varrho\left[\left(a_1^\dagger + \frac{\alpha^*}{\sqrt{2}}\right)\left\{\left(a_2^\dagger + \frac{\alpha^*}{\sqrt{2}}\right)\exp[-iKz_2] + a_3^\dagger \exp[-iKz_3]\right\}\cdot\right.\right.$$

$$\left.\left.\cdot\left\{\left(a_2 + \frac{\alpha}{\sqrt{2}}\right)\exp[iKz_2] + a_3 \exp[iKz_3]\right\}\left(a_1 + \frac{\alpha}{\sqrt{2}}\right)\right]\right\}.$$

A lengthy but straightforward calculation yields

$$C \sim \frac{|\alpha|^4}{4} + p_1 \frac{3|\alpha|^2}{4} + p_2 \frac{1 + |\alpha|^2(3 + 2\sqrt{2})}{3(1 + |\alpha|^2)} + p_1 \frac{|\alpha|^2}{\sqrt{2}} \cos\varphi.$$

In the absence of an amplfier we obtain, as expected, no correlation at all. In the general case, this formula makes statistical predictions with the probability of getting an oscillatory behaviour with path length difference of beams 2 and 3.

B) The Einstein-de Broglie model has the following wave field for the arrangement of fig. 3:

$$|\psi\rangle = \frac{1}{\sqrt{2 + |\alpha|^2}}\left\{\varepsilon \frac{1}{\sqrt{2}}[|1, 0, 0\rangle + |0, 1, 0\rangle] + \right.$$

$$+ \sqrt{1 - \varepsilon}\,\frac{1}{\sqrt{3}}[|2, 0, 0\rangle + |1, 1, 0\rangle + |0, 2, 0\rangle] +$$

$$\left. + \frac{\alpha}{\sqrt{2}}[|1, \widetilde{0}, 0\rangle + |0, \widetilde{1}, 0\rangle] + |0, 0, 1\rangle\right\}.$$

The calculations can be performed a long similar lines as before and yield the following pattern for the intensity:

$$I \sim \frac{|\alpha|^2}{2} + \frac{(\alpha + \alpha^*)}{2(2 + |\alpha|^2)} \sqrt{\frac{\varepsilon(1-\varepsilon)}{6}} (1 + \sqrt{2}) + \frac{2 - \varepsilon/2}{2 + |\alpha|^2} \left\{ 1 + \frac{\sqrt{2\varepsilon}}{2 - \varepsilon/2} \cos\varphi \right\} .$$

This gives for the special cases

$$\varepsilon = 1 , \qquad I \sim \frac{|\alpha|^2}{2} + \frac{3}{2(2 + |\alpha|^2)} \left\{ 1 + \frac{2\sqrt{2}}{3} \cos\varphi \right\} ,$$

$$\varepsilon = 0 , \qquad I \sim \frac{|\alpha|^2}{2} + \frac{2}{2 + |\alpha|^2} ,$$

$$\varepsilon = \frac{1}{2} , \qquad I \sim \frac{|\alpha|^2}{2} + \frac{1 + \sqrt{2}}{2\sqrt{6}} \frac{(\alpha + \alpha^*)}{2(2 + |\alpha|^2)} + \frac{7}{4(2 + |\alpha|^2)} \left\{ 1 + \frac{4}{7} \cos\varphi \right\} .$$

The result for the interference pattern of the Einstein de Broglie model without wave packet collapse has some striking differences from the preceding one obtained with the Bohr model: Instead of predicting probabilities of occurrence, we obtain here definite predictions for the intensity depending on the amplifier characteristics ε. Furthermore, the interference term $\sim \cos\varphi$ differs essentially in the two models: while the Bohr model gives always a probabiiity of getting the pattern of the no-amplifier-case (*i.e.* $\frac{3}{4}(1 + (2\sqrt{2}/3)\cos\varphi)$ in addition to background terms, the Einstein model exhibits an interference behaviour of varying amplitude (*i.e.* $1 + (\sqrt{2\varepsilon}/(2 - \varepsilon/2))\cos\varphi$) in addition to different background terms. This is a testable prediction that could be submit:ed to expeiirental evidence.

Finally the corresponding counter correlations can be evaluated in a straightforward way. The result being rather lengthy we just quote the three interesting cases.

For $\varepsilon = 1$, we obtain as expected no correlation at all.

For $\varepsilon = 0$, the expression for G can be written as follows:

$$C \sim \frac{|\alpha|^4}{4} + \frac{3|\alpha|^2}{2(2 + |\alpha|^2)} + \frac{2\sqrt{2}}{3(2 + |\alpha|^2)} |\alpha|^2 + \frac{1}{3(2 + |\alpha|^2)} + |\alpha| \frac{\sqrt{2}}{\sqrt{3}(2 + |\alpha|^2)} \cos(\varphi + \theta_1) ,$$

where $\alpha = |\alpha| \exp[i\theta_1]$.

For the $\varepsilon = \frac{1}{2}$ case we obtain

$$C \sim f(\alpha) + \frac{|\alpha|^2}{\sqrt{2}} \frac{\cos\varphi}{2(2 + |\alpha|^2)} + \frac{|\alpha|^2}{4} \frac{\cos\varphi}{2 + |\alpha|^2} + \frac{|\alpha|}{\sqrt{3}(2 + |\alpha|^2)} \cos(\varphi + \theta_1) .$$

From the above formulae for C some striking differences compared with the Bohr model can be easily pointed out: the oscillating terms $\sim \cos(\varphi + \theta_1)$ are due to expectation values of the form

$$\langle 1, 1, 0 | a_1^\dagger a_2^\dagger a_3 \alpha | 0, 0, 1 \rangle .$$

These terms do not exist in the Bohr model. There, the oscillating term is connected with the probability P_1 ascribed to the state without wave packet collapse in the mixture. Here, the oscillating term is a natural consequence of the formalism always present in the result, while in the Bohr model it is tied with the probability of occurence of one state in the mixture. This effect manifests itself also in the general case ($\varepsilon \neq 0$ or 1) where the oscillatory pattern of the correlation has a much more complex structure.

We thus conclude this letter with the remark that the correlation pattern in the Einstein-de Broglie model without wave-packet collapse, exhibits in all cases examined (apart from the $\varepsilon = 1$ case where it vanishes) an overall oscillatory behaviour that does not vanish as we come to the full absorption case. This is of course, an important feature that yields the possibility to distinguish between the two models and should and will be tested experimentally.

Finally, one last remark: The authors do not believe that the formalism of quantum theory is erroneous, but only that the wave packet collapse concept introduced by BOHM in the Copenhagen interpretation of quantum mechanics is not correct and conflicts with experiment in certain specific situations, such as the experiments discussed or Rauch's *et al.* experiments on neutron interference ([13]). This view does not conflict with facts, since CINI has shown ([1]) that one could construct a realistic quantum measurement device without wave packet collapse, the latter being a concept which evidently conflicts with relativity theory.

The price to pay for the construction of such a reinterpretation of quantum theory is evidently the introduction of subquantal random superfluid aether along Dirac's initial suggestion ([14]). In this sense of course, quantum theory would not be complete and EINSTEIN would be right in the Bohr-Einstein controversy.

* * *

Two of the authors (AK and DS) want to thank the French Government for a grant which made this research possible.

([13]) H. RAUCH: *Proceedings of the Bari Conference on « Open Questions in Quantum Physics »* (Dordrecht, 1983) and references quoted herein.
([14]) P. A. M. DIRAC: *Nature*, **168**, 906 (1951); **169**, 702 (1952).

This page is deliberately left blank.

Reprinted from *Physics Letters*, Vol. 102A, No. 7, pp. 291-294, Copyright (1984)
with permission from Elsevier Science.

TESTING WAVE–PARTICLE DUALISM WITH TIME-DEPENDENT NEUTRON INTERFEROMETRY

C. DEWDNEY [a,1], Ph. GUERET [b], A. KYPRIANIDIS [a,2] and J.P. VIGIER [a]

[a] *Institut Henri Poincaré, Laboratoire de Physique Théorique, 11, rue P. et M. Curie, 75231 Paris Cedex 05, France*
[b] *Institut de Mathématiques Pures et Appliquées, Université P. et M. Curie, 4, place Jussieu, 75230 Paris Cedex 05, France*

Received 20 March 1984

A modified version of the time-dependent neutron spinor superposition allows a possible simultaneous detection of neutron paths and intensity self-interference. Previous theoretical doubts are removed and the Einstein–de Broglie version of the wave–particle dualism now seems to be supported by Rauch's experiment.

An important stage in single-neutron perfect-crystal interferometry has been achieved with the verification of the spinor 4π-periodicity by measuring the intensity variation behind the interferometer [1]. Moreover, recent progress in neutron interferometry has demonstrated the quantum mechanical principle of linear spin-state superposition for fermions [2]. By inverting the spin state of one of the two initially equally polarized coherent waves, propagating along different paths, and by superposing them coherently, one obtains a final spin state which lies in the plane perpendicular to the initial spin direction of the single neutron beam. Contrary to these experiments, where the spin inversion was achieved by means of a static B field and the total energy of the neutron is a constant of motion due to absence of a time-dependent interaction, a recent experiment has been performed by Rauch et al. [3], where an explicitly time-dependent interaction is introduced. Here, the spin state of one of the partial beams is inverted by means of a radio-frequency flipper. The total energy of the neutron is not conserved and quantum theory shows that a photon is exchanged between the neutron and the field which corresponds to the Zeeman energy difference of the spin eigenstates within an existing static field B_0, i.e. $\hbar\omega_{rf} = 2\mu B_0$. At the same time one obtains a time-dependent oscillating polarization

$$P(t) = (\cos(\chi - \omega_{rf}t), \sin(\chi - \omega_{rf}t), 0),$$

where $\bar{P}_{in} = (0, 0, 1)$ and χ is the nuclear phase shift introduced in one beam. As one directly sees from the above formula, $P(t)$ is in the plane perpendicular to the initial polarization, and can be detected by means of stroboscopic registration.

The interesting feature of this experiment is that one could in principle detect the path followed by the neutron (since one has a measurement associated with an energy exchange with the rf-spin flipping device [4]) and simultaneously observe the pattern of a single particle polarization interference; contrary to the customary point of view of Copenhagen quantum mechanics. This possibility has been held as doubtful by Rauch et al. [3]: For the stroboscopic detection, needed for the registration of the oscillating polarization, the phase of the rf-field has to be known with an accuracy $\Delta\phi < 2\pi$, a fact which combined with the phase-number uncertainty relation $\Delta N \Delta\phi \geqslant 2\pi$ makes the number uncertainty $\Delta N > 1$, so that single photon transitions needed for the path observations are in principle undetectable.

Since we have criticized this reasoning and the consequences of this experiment elsewhere [5] we will limit ourselves in this short letter to present a possible experimental setup, i.e. propose slight modifications of the latest Rauch experiment [3], which, in our opinion, bypass the theoretical problems and questions (based on the phase number uncertainty relation) raised by Rauch et al. [3].

[1] European Exchange.
[2] On leave from the University of Crete, Physics Department, Heraklion, Crete, Greece.

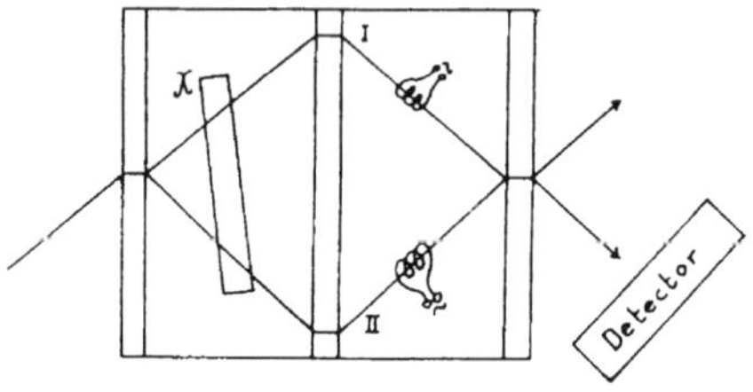

Fig. 1.

This proposal consists of the following modified experimental setup: The perfect crystal interferometer now contains two rf-flippers, one in each neutron path, the rest remaining exactly as realised by Rauch et al. [3] (see fig. 1). The two partial single neutron beams are now represented by the following wave functions [6]

beam I : $\psi_I = e^{i\chi}|\downarrow_z\rangle = e^{i\chi}\binom{0}{1}$,

beam II: $\psi_{II} = |\downarrow_z\rangle = \binom{0}{1}$.

In this representation all common phase factors have been omitted and both beams possess an energy of $E_{in} - \Delta E$ where $\Delta E = 2\mu B_0$ and the common wave vector $\boldsymbol{k}$.

One could argue on this point that, since the exchange of energy with the rf-spin-flipper actually constitutes a measuring process (because it modifies both the state of the quantum system and of the measuring device) the Bohr theory of measurement [7] would suggest the explicit assumption of a wave packet collapse in the other path, whenever an energy exchange occurs in one of the rf-flippers: Since this crucial aspect is discussed also in ref. [5] we completely omit this point in what follows and present the correct quantum mechanical calculation without wave packet collapse [8].

If the two beams are superposed coherently so that they can interfere, we can calculate in this model (without wave packet collapse) the following expressions for intensity and polarization [6] $I_{fin} = \psi^+_{fin}\psi_{fin}$ and $\boldsymbol{P}_{fin} = I^{-1}_{fin}\psi^+_{fin}\boldsymbol{\sigma}\psi_{fin}$ with $\boldsymbol{\sigma}$: Pauli spin matrices. Our Ansatz for ψ_f is the following

$$\psi_f = (\sqrt{\epsilon}|\uparrow_z\rangle + \sqrt{1-\epsilon}|\downarrow_z\rangle)\,e^{i\chi} + \sqrt{\epsilon}|\uparrow_z\rangle + \sqrt{1-\epsilon}|\downarrow_z\rangle,$$

where $1-\epsilon$ is the efficiency of the rf-spin flipper, or

$$\psi_f = (1+e^{i\chi})\sqrt{\epsilon}|\uparrow_z\rangle + (1+e^{i\chi})\sqrt{1-\epsilon}|\downarrow_z\rangle.$$

With this the intensity is calculated to be

$$I_f = 2(1+\cos\chi)(\sqrt{\epsilon})^2 + 2(1+\cos\chi)(\sqrt{1-\epsilon})^2 = 2(1+\cos\chi),$$

which, independent of the efficiency of the coil, exhibits an oscillatory behaviour dependent on the nuclear phase shift factor χ.

The polarization $\boldsymbol{P}$ can now be written as

$$\boldsymbol{P}_f = 2I_f^{-1}(1+\cos\chi)\{[\sqrt{\epsilon}(1,0) + \sqrt{1-\epsilon}\,(0,1)] \times (\sigma_x, \sigma_y, \sigma_z)\,[\sqrt{\epsilon}\tbinom{1}{0} + \sqrt{1-\epsilon}\tbinom{0}{1}]\}$$

which yields

$$\boldsymbol{P}_f = ([\sqrt{\epsilon}(1,0) + \sqrt{1-\epsilon}(0,1)]\,[\sqrt{\epsilon}\tbinom{0}{1} + \sqrt{1-\epsilon}\tbinom{1}{0}],$$
$$[\sqrt{\epsilon}(1,0) + \sqrt{1-\epsilon}(0,1)]\,[\sqrt{\epsilon}\tbinom{0}{i} + \sqrt{1-\epsilon}\tbinom{-i}{0}]$$
$$[\sqrt{\epsilon}(1,0) + \sqrt{1-\epsilon}(0,1)]\,[\sqrt{\epsilon}\tbinom{1}{0} + \sqrt{1-\epsilon}\tbinom{0}{-1}])$$
$$= (2\sqrt{\epsilon(1-\epsilon)}, 0, -1+2\epsilon),$$

which does no more exhibit a time dependent rotation in the xy-plane as in ref. [3] but has a constant value mainly along the $-z$ direction and a small part in the x direction. For a perfect rf-flipper ($\epsilon = 0$) the polarization thus reduces to a constant z-directed magnitude

$$\boldsymbol{P}_{ideal} = (0, 0, -1).$$

In this proposed experimental setup a very interesting feature arises: Neutron paths can be detected due to single photon energy transfer to the spin flipper while at the same time a time-independent constant intensity interference pattern can be observed.

One should immediately stress that in this proposed experimental setup no doubts of the form presented by Rauch et al. [3] can be raised for a simultaneous detection of interference and path followed by the neutrons. Indeed since the resulting interference pattern is stationary, the argument concerning the phase-number uncertainty (which is in fact at least ambiguous [5] and theoretically incorrect [9]) does not apply at all in the present case: so that no theoretical objection arises for a possible detection of single photon transitions in the field of the rf-spin flipper. With this modification all related objections are obviously removed, and an energy transfer is in principle detectable, possibly with the use of a SQUID superconducting device [5]. Two more points, merely of practical concern have still to be clarified. If the single photon energy transfer to the rf-flipper is not detectable due to insufficient resolution capability of the instrument then the following solution can be sought. After having a passage of sufficient number of neutrons through the interferometer the energy transferred to each coil is summed up to an amount that has been shown to be detectable [4]. One thus obtains statistical evidence for the passage of a number of neutrons through each path if the energy detected by the measuring device is an integral multiple of the Zeeman splitting energy difference, i.e.

$$E_{\text{detected}} = n \cdot 2\mu B_0, \quad n \in \mathrm{N}.$$

Since the coils can be made ≈100% efficient [10] one can obtain from this reasoning a number n_{I} corresponding to the neutrons passed through path I and a number n_{II} for path II and $n_{\mathrm{I}} + n_{\mathrm{II}}$ should equal to the total neutron number measured on the interferometer. If in such a setup the statistical information concerning neutron paths in the interferometer still coexist (as believed by the authors) with the persistence of interference pattern of the intensity of the beams containing one neutron at a time, this result is clearly incompatible with the wave packet collapse concept which would imply the use of mixtures [5] and destroy possible interferences. Of course if we finally consider the case where a single photon transition to the rf-flipper is resolvable then one would obtain direct accurate evidence for a single neutron passage obtaining at the same time an intensity interference. This would yield, in our opinion, a direct proof for the incompleteness of a quantum description using the complementarity argument (particle *or* wave) while it establishes the simultaneous existence of both properties in the interferometer. If following Rauch we really accept that neutron self-interference establishes the fact that "every neutron in the area of interference knows simultaneously what has happened in both paths" [10] then we would have to conclude that our new proposed version of Rauch's experiment unambiguously establishes the real physical existence of the de Broglie waves [11].

If and when individual rf-photons are measured we can finally add a slight "gedanken" modification of the proposed experiment by adding an anticoincidence counter between the two rf-flippers, which will give information about the passage of a neutron through the interferometer. If a 100% anticoincidence is established, then one would have another direct proof of the fact that, while the de Broglie waves propagate on both paths in the interferometer, the particle follows only one of the beams with a probability given by the reflexion transmission coefficient in the first plane of the perfect crystal interferometer (in the ideal case 1:1). In this case we would obtain an experimental confirmation for the localized particle structure of every individual neutron involved in the interference, in addition to the evidence for the accompanying de Broglie pilot wave.

To conclude, the proposed modification of the time dependent superposition of spinors has the basic merit that it decouples the possibility of simultaneous detection of the path followed by a neutron together with interference phenomena from a possible contradiction with the (contested [5,9]) so-called fifth uncertainty relation, the phase-number uncertainty relation. It thus enables in principle the separate examination of the "impossible" simultaneous path/interference detection, strictly forbidden by the Copenhagen interpretation of quantum mechanics. It can also be used (if this possibility is established by experiment) along the lines indicated in this letter to check the phase-number uncertainty relation: Indeed once a single photon transition is measured, and if the interference pattern of the intensity is not affected by this measurement, this implies that one has directly disproved by experiment the phase number uncertainty: For single photon transitions would have $\Delta N = 0$ as a consequence and an unaffected interference pattern would imply well defined phase relations within the beam and hence also in the rf-field, i.e. $\Delta\phi \approx 0$. If one also considers that these types of experiments represent a possible test of Bohr's wave packet collapse concept [5], then one can really appreciate the crucial possible contribution of neutron interferometry in testing the foundations of quantum mechanics.

The authors wish to thank Professor H. Rauch for many discussions and helpful suggestions. One of the authors (A.K.) wants to thank the French government for a grant and another (C.D.) wishes to thank the Royal Society for the European Exchange Fellowship award which enabled him to do this research.

References

[1] H. Rauch et al., Phys. Lett. 54A (1975) 425.

[2] J. Summhammer, G. Badurek, H. Rauch and U. Kishko, Phys. Lett. 90A (1982) 110;
G. Badurek, H. Rauch, J. Summhammer, U. Kischko and A. Zeilinger, J. Phys. A16 (1983) 1133;
J. Summhammer, G. Badurek, H. Rauch, U. Kischko and A. Zeilinger, Phys. Rev. A27 (1983) 2523.

[3] G. Badurek, H. Rauch and J. Summhammer, Phys. Rev. Lett. 51 (1983) 1015.

[4] B. Alefeld, G. Badurek and H. Rauch, Z. Phys. B41 (1981) 231.

[5] C. Dewdney, Ph. Guéret, A. Kyprianidis and J.P. Vigier, Time dependent neutron interferometry: Evidence against wave packet collapse, IHP preprint (1984).

[6] G. Eder and A. Zeilinger, Nuovo Cimento 34B (1976) 76.

[7] N. Bohr, Atomic physics and human knowledge (Wiley, New York, 1958).

[8] M. Cini, Nuovo Cimento 73B (1983) 27.

[9] P. Carruthers and M. Nieto, Rev. Mod. Phys. 40 (1968) 411.

[10] H. Rauch, private communication.

[11] L. de Broglie, Non-linear wave mechanics (Elsevier, Amsterdam, 1960).

Reprinted from *Physics Letters*, Vol. 104A, No. 6,7, pp. 325-328, Copyright (1984) with permission from Elsevier Science.

ENERGY CONSERVATION AND COMPLEMENTARITY IN NEUTRON SINGLE CRYSTAL INTERFEROMETRY

C. DEWDNEY [1]
Institute Henri Poincaré, Paris, France

A. GARUCCIO
Istituto di Fisica dell'Universitá, Bari, Italy and INFN Sezione di Bari, Bari, Italy

and

A. KYPRIANIDIS [2] and J.P. VIGIER
Institut Henri Poincaré, Paris, France

Received 5 July 1984

The complementarity principle is shown to conflict with the energy conservation laws in neutron single crystal interferometry. Its shortcomings are revealed in specific performed or proposed neutron interferometry experiments.

The quantum formalism provides the correct predictions for the experiments performed in neutron interferometry [1] but nevertheless fundamental questions, open since the Bohr–Einstein debate 50 years ago, are raised again concerning their possible interpretations. In a recent letter [2] we discussed the problems related with time dependent neutron interferometry [3] and the possibility to consider the radiofrequency spin flipper as a measuring device, a fact that, if established, should enable the "impossible" simultaneous path/interference detection. In this letter we are mainly concerned with aspects of the complementarity principle and its contradiction with the fundamental energy conservation law.

To this purpose we consider the experimental arrangement of fig. 1 with both spin flippers turned off. A simple calculation shows [4] that if an originally spin up polarized beam $\psi = |\uparrow_z\rangle$ enters the interferometer, it is subdivided in two partial beams $\psi_I = e^{i\chi}|\uparrow_z\rangle$ and $\psi_{II} = |\uparrow_z\rangle$ that successively recombine and yield an intensity interference behind the interferometer modulated with the phase shift factor χ.

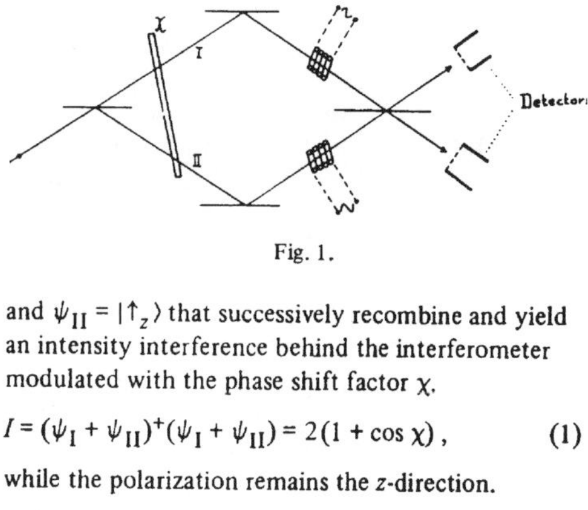

Fig. 1.

$$I = (\psi_I + \psi_{II})^+(\psi_I + \psi_{II}) = 2(1 + \cos\chi)\,, \qquad (1)$$

while the polarization remains the z-direction.

$$P = (0, 0, 1)\,. \qquad (2)$$

This "double-slit" like situation offers two possible explanations:

(a) Either we say the neutron actually travels along path I *or* II only, but is influenced by the physical conditions along both;

(b) Or we say the neutron does not exist as a particle in the interferometer.

[1] European Exchange Fellowship.
[2] On leave from the University of Crete, Physics Departement, Heraclion, Crete, Greece.

According to the Copenhagen interpretation of quantum mechanics (CIQM) if the particle actually travels along one path the existence of the other is therefore irrelevant and interference cannot occur. Interference arises not from our lack of knowledge of the path but from the fact that the neutron does not have one. Thus certain experimental apparatus make the non-localized or wave nature of the neutron manifest whereas other mutually exclusive apparatus make its particle nature manifest. This is the complementarity of wave and particle.

The first option (a) was that favoured by Einstein [5] and de Broglie [6] in the causal stochastic interpretation of quantum mechanics (SIQM). The neutron always travels along one path whilst its real guiding wave travels along both. In the region of superposition the waves combine information about both paths and guide the particle accordingly. Determinate individual particle trajectories can be calculated in this model in interference experiments (for a detailed calculation of the particle trajectories in a two slit experiment see ref. [7]).

In the original Bohr–Einstein debate Bohr was able to defend the complementarity principle by showing that attempts by Einstein, to use detailed energy or momentum conservation in individual processes to determine particle trajectories and give a fuller description, required a change of the experimental arrangement which resulted in a loss of the wave aspect [8].

The argument is often put in the following way in CIQM. The introduction of a device capable of determining the particle trajectory induces a collapse of the wave function in the rest of the apparatus and a consequent loss of interference. Such a collapse is a consequence of the purely probabilistic interpretation of the wave function and follows from the requirement that wave and particle pictures are complementary. In fact one should note that the collapse concept (projection postulate) need never be used in quantum calculations; whether or not observable interference persists depends on the actual interaction that has taken place with the apparatus, as shown by Cini [9]. For example consider the wave function of an apparatus introduced in one path to be ϕ_i initially and ϕ_f finally, then we have:

$$\Psi_i = \phi_i\psi_I + \phi_i\psi_{II} \rightarrow \Psi_f = \phi_i\psi_I + \phi_f\psi_{II} \,. \qquad (3)$$

If through its functioning the states ϕ_i and ϕ_f become orthogonal then interference is destroyed

$$\Psi_f^+\Psi_f = \phi_i^+\phi_i\psi_I^+\psi_I + \phi_f^+\phi_f\psi_{II}^+\psi_{II} \,, \qquad (4)$$

and the neutron acts as a particle that goes either on path I or path II. Observation of the measuring instrument merely tells us which alternative took place and thus we replace Ψ_f by $\phi_i\psi_I$ or $\phi_f\psi_{II}$. This is a collapse of the wave function which simply represents a change of our knowledge and does not correspond to any real physical changes in the state of the neutron. If ϕ_i and ϕ_f are not orthogonal then interference persists:

$$\Psi_f^+\Psi_f = \phi_i^+\phi_i\psi_I^+\psi_I + \phi_f^+\phi_f\psi_{II}^+\psi_{II} + \phi_i\psi_I\phi_f^+\psi_{II}^+ + \phi_f\psi_{II}\phi_i^+\psi_I^+ \,, \qquad (5)$$

and the neutron acts as a wave in both paths.

If by observing the apparatus we could still in fact determine the path of the neutron then the act of observation in CIQM would have to cause real physical changes in the neutron's state as a consequence of a wave packet collapse. Since if neutrons are conceived as *particles* that go one way or the other, eq. (5) should reduce to eq. (4).

Thus CIQM concludes that all measurements capable of determining the neutron's path imply orthogonality of the apparatus wave functions initially and finally. In SIQM determination of particle path need not imply orthogonality of apparatus wave functions in order to exclude the intervention of consciousness in physical processes. What appears as a "pseudo-collapse" is the action of a macroscopic measuring device which makes the interference terms negligible as is consistently shown by Cini [9]. Thus, there is no a priori impossibility of path determination and persisting interference; one has only to find an appropriate measuring device that during an interaction with the microsystem does not undergo a change to an orthogonal state, i.e. preserves the interference terms, and still offers a possibility to decode this small quantum number change (e.g. one could envisage the possibility of using a "quantum non-demolition measurement" process for such a purpose [10,11]).

Consider now the set up with the spin flip device in path II operating (at 100% efficiency). The intensity modulation in the emerging beams disappears, there is no spatial interference. This lack of interference need not however imply that the coil acts as measuring device localising the neutron in one beam since interference persists in the spin superposition yielding a final polarization [3]

$$\bar{P} = (\cos(\omega_{rf}t - \chi), \sin(\omega_{rf}t - \chi), 0)\,.$$

Each neutron emerging from the interferometer is polarized in the x–y plane. The spin up $|\uparrow_z\rangle$ and spin down $|\downarrow_z\rangle$ states are superposed and hence it is argued along CIQM lines that the neutron actually does not exist as a particle with spin in either beam, if it did a mixture of spin states would result. Now in order to explain the change of spin state when the polarized neutron acts with the rf coil it is argued [3] that the neutron emits a photon of energy $E = \hbar\omega_{rf}$ to the time dependent field and hence the energy of the neutron and the coil are altered. No description using the purely wave-like aspect can explain these results. If the neutron actually is only a wave (or in neither beam) during the experiment no such energy exchange could be described: How can the coil exchange a photon with a neutron that does not exist! In order to explain the change of spin states produced by the coil we require a localized particle, but a description in terms of localized neutrons in one or other beam cannot explain the superposition. It would seem that we must use here wave and particle aspects of neutrons simultaneously. Alternatively it may be argued that when the neutron is said to behave like a wave we should not imagine a physical wave, all we see in this case is an interference of the probability amplitudes for an event to happen in indistinguishable ways. The amplitude to travel path I without spin flip "interferes" with the amplitude to travel path II with spin flip since the two are said to be indistinguishable. While this pattern seems to be consistent it still maintains a fundamental ambiguity: the neutron must exist in one beam or another in order to exchange energy, while this statement must be denied in order to preserve interference.

One should note the implications of this contradiction: If this scheme is extended from a recipe for predictions to an explanatory pattern, its failure becomes evident. In order to preserve interference when having energy transfer the rf-coil wave functions ϕ_i and ϕ_f must be indistinguishable from the interference point of view and distinguishable concerning the energy transfer omitting the case of orthogonality.

Now consider the apparatus with both spin flippers operational. Since now a spin flip takes place in both beams spatial interference is recovered. A measurement of the polarization of the neutron behind the interferometer reveals that each neutron has suffered a spin flip. Each emerging neutron has lost an amount of energy ΔE where $\Delta E = 2\mu B$, the Zeeman splitting. If energy is to be conserved this energy must have gone to one or other of the coils, this is only possible if the neutron passes as a particle through one or other and gives an indivisible photon of $E = \hbar\omega_{rf} = \Delta E$ to the rf-field. The spatial interference can only be explained by assuming that the neutron does not pass through one or other of the coils. The change of energy can only be explained by the particle aspect.

Since both interference and spin direction can be measured simultaneously, according to CIQM the neutron actually travels path I or II and at the same time does not exist as a particle at all.

In the Bohr–Einstein debate the application of particle momentum conservation in individual events always led to the consistency of CIQM. Here the energy conservation leads to the inconsistency of CIQM since wave/particle aspects appear together. If it is insisted in CIQM that neutrons do not travel one way or the other, no energy can be transferred to the coils and then there is no conservation of energy in individual events. Further if a statistical ensemble of individual neutron passages is considered we see that, even there, there is no conservation of energy in CIQM.

If we wish to consider the mechanism of spin flip and the conservation of energy, then the neutrons must travel on one path and through a coil. If we wish to consider spin superposition then the neutrons must travel along neither.

Is the consequence of the above presentation to renounce complementarity for the CIQM? Probably not, because the Heisenberg uncertainty relations could provide a means to escape the conclusiveness of the presented reasoning. In fact one could argue that since the energy uncertainty δE introduced to the neutron energy E due to the Bragg scattering of the crystal is greater than the energy transfer ΔE to the rf-coil (10^{-6} and 10^{-8} respectively) no energy conservation could be established. Still a problem remains: Since the final spin polarization (i.e. $-z$) is detectable, the Zeeman energy loss of the neutron in B is known with respect to its initial polarization (i.e. $+z$). Because of this, the magnetic field energy of the neutron spin is accurately known. One can now interpret the uncertainty in E as inherent in the corresponding operator, in which case the problem of a "hypothetical" energy transfer to the rf-coil inferior to δE is not legitimate but the simultaneous sharp value of the Zeeman part remains incomprehensible. One could then still escape to a formulation of the kind: "the energy transfer cannot be measured because $\Delta E < \delta E$". This operational aspect of the uncertainty clearly promotes an existence of sharp instantaneous values which are steadily perturbed. This explains the experimental detection of a spin state but in CIQM fails to account for interference because it yields a definite neutron path. In both aspects/versions of the energy uncertainty interpretation in CIQM contradictions arise that do not perhaps affect the complementarity principle as a useful recipe in most of the cases but which do reveal its shortcomings as an explanatory pattern in specific situations as the performed or proposed experiments on neutron interferometry.

The authors want to thank Professor Rauch for the useful discussions. One of us (C.D.) wants to thank the British Royal Society for a European Exchange Fellowship, another (A.K.) the French Government for a grant and another (A.G.) the Administration Council of the University of Bari for a grant which made this collaboration possible.

References

[1] J. Summhammer, G. Badurek, H. Rauch and O. Kischko, Phys. Lett. 90A (1982) 110;
G. Badurek, H. Rauch, J. Summhammer, U. Kischko and A. Zeilinger, J. Phys. A16 (1983) 1133;
J. Summhammer, G. Badurek, H. Rauch, U. Kischko and A. Zeilinger, Phys. Rev. A27 (1983) 2523.
[2] C. Dewdney, Ph. Gueret, A. Kyprianidis and J.P. Vigier, Phys. Lett. 102A (1984) 291.
[3] G. Badurek, H. Rauch and J. Summhammer, Phys. Rev. Lett. 51 (1979) 15.
[4] G. Eder and A. Zeilinger, Nuovo Cimento 34B (1976) 76.
[5] A. Einstein, Proc. Congres Solvay (1927).
[6] L. de Broglie, Une tentative d'interprétation causale et non-linéaire de la mécanique ondulatoire (la théorie de la double solution) (Gauthier-Villars, Paris, 1956).
[7] J.C. Philippidis, C. Dewdney and B.J. Hiley, Nuovo Cimento 52B (1979) 15.
[8] N. Bohr, Atomic physics and human knowledge (Wiley, New York, 1958).
[9] M. Cini, Nuovo Cimento 73B (1983) 27.
[10] C.W. Caves, R.W. Drever, V. Sandberg, K.S. Thorne and M. Zimmermann, Phys. Rev. Lett. 40 (1978) 667; Rev. Mod. Phys. 52 (1980) 341.
[11] W.G. Unruh, Phys. Rev. D17 (1978) 1180; D18 (1978) 1769; D19 (1979) 2888.

Reprinted from *Lettere al Nuovo Cimento*, Vol. 40, No. 16, pp. 481-487, Copyright (1984) with permission from the Società Italiana di Fisica.

Time-Dependent Neutron Interferometry: Evidence in Favour of de Broglie Waves.

C. DEWDNEY (*), A. KYPRIANIDIS (**) and J. P. VIGIER

Institut Henri Poincaré - Paris

A. GARUCCIO (***)

Istituto di Fisica dell'Università - Bari
Istituto Nazionale di Fisica Nucleare - Sezione di Bari

PH. GUERET

Institut de Mathematique Pure et Appliquée, Université P. et M. Curie - Paris

(ricevuto il 2 Maggio 1984)

PACS. 03.65. – Quantum theory; quantum mechanics.

Summary. – Time-dependent spinor superposition in neutron interferometry by means of radio frequency spin flippers enables a possible simultaneous path and interference detection and provides evidence for the real physical existence of de Broglie « pilot » waves.

The research of the Vienna group (1) on neutron interferometry opens new exciting experimental possibilities to discuss the different interpretations of quantum statistics and to answer the age-old question of whether the neutrons (or any other massive particle) really travel along a path in space-time between their source and the observer (as believed by EINSTEIN and DE BROGLIE) or if such a space-time co-ordination does not exist (as believed by BOHR and HEISENBERG). For this reason we intend to discuss the most recent experiment of the Vienna group (2) on the time-dependent neutron interferometry.

(*) European Physical Society Fellow.
(**) On leave from the University of Crete, Physics Department, Heraclian, Crete, Greece.
(***) Work partially supported by M.P.I. and I.N.F.N.
(1) J. SUMMHAMMER, G. BADUREK, H. RAUCH and O. KISCHKO: *Phys. Lett. A*, **90**, 110 (1982); G. BADUREK, H. RAUCH, J. SUMMHAMMER, O. KISCHKO and A. ZEILINGER: *J. Phys. A*, **16**, 1133 (1983); J. SUMMHAMMER, G. BADUREK, H. RAUCH, O. KISCHKO and A. ZEILINGER: *Phys, Rev. A*, **27**, 2523 (1983).
(2) G. BADUREK, H. RAUCH and J. SUMMHAMMER: *Phys. Rev. Lett.*, **51**, 1015 (1983).

The experimental arrangement can be schematically represented as follows: an incident neutron beam containing one neutron at a time (fig. 1) is subsequently divided into beams I and II. On beam I there acts a nuclear phase shifter represented by the action of a unitary operator $\exp[i\chi]$ on ψ with $\chi = -N\cdot\lambda\cdot b_c\cdot D$, where b_c is the coherent scattering length, λ the neutron wave-length, D the thickness of the phase shifter and N the number of lattice elements/volume element. Beam II is subjected to the following combination of magnetic fields: a) a static magnetic field in the $+z$ direction $B = (0, 0, B_0)$; b) a radiofrequency time-dependent magnetic field $B_{\text{r.f.}} = (B_1\cdot\cos\omega_{\text{r.f.}}\cdot t,$ $B_1\cdot\sin\omega_{\text{r.f.}}\cdot t,\ 0)$ rotating in the xy-plane with a frequency $\omega_{\text{r.f.}}$, obeying the resonance condition $\hbar\omega_{\text{r.f.}} = 2\mu B_0$, where μ is the magnetic moment of the neutron, *i.e.* it yields exactly the Zeeman energy difference between the two-spin eigenstates of the neutron within the static field. Neutrons passing through such a device (a spin flipper) reverse their initial $+z$ polarization into the $-z$ direction, by transferring an energy $\Delta E = 2\mu B_0$ to the coil, whilst maintaining their initial momentum [2].

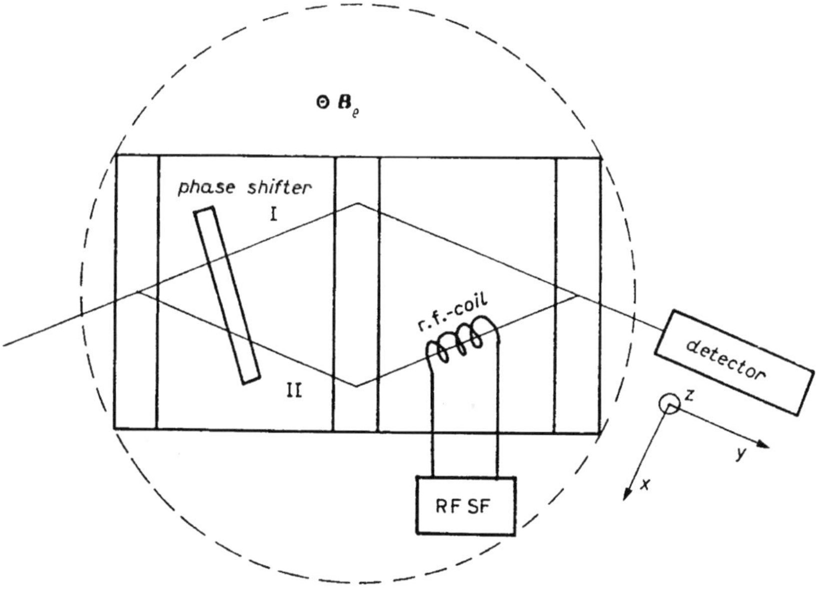

Fig. 1. – Sketch of the spin superposition experiment with a radiofrequency spin-flip device.

Thus, while a wave function of a neutron in beam I after passing through the nuclear phase shifter is represented by [3]

$$\Psi_{\text{I}} = \exp[i\chi]\,|\uparrow_z\rangle = \exp[i\chi]\begin{pmatrix}1\\0\end{pmatrix},$$

the corresponding wave function in beam II after a spin-flip should be written as

$$\Psi_{\text{II}} = \exp\left[i\frac{\Delta E}{\hbar}t\right]|\downarrow_z\rangle = \exp\left[i\frac{\Delta E}{\hbar}t\right]\begin{pmatrix}0\\1\end{pmatrix}.$$

---

[3]) G. Eder and A. Zeilinger: *Nuovo Cimento B*, **34**, 76 (1976).

If we assume a 100% efficient radio-frequency spin flipper (RFSP) (4) the polarization behind the interferometer lies entirely in the xy-plane of fig. 1 and has the following pattern (2):

$$\bar{P} = \frac{1}{I}\,\psi_f^{+}\,\sigma\psi_f = (\cos(\omega_{r.f.}\cdot t - \chi),\ \sin(\omega_{r.f.}\cdot t - \chi),\ 0)\,,$$

while the final intensity is constant. Since $\bar{P}$ is time dependent, a stroboscopic registration is needed to convert the pattern stationary, and, by performing such a detection, one obtains the pattern of fig. 2 as a function of the nuclear phase shift χ.

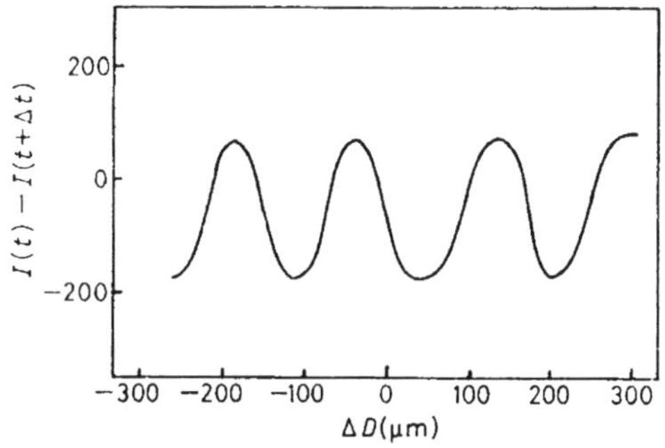

Fig. 2. – Stroboscopic picture of the interference pattern when the polarization component along x-axis is measured. Observed intensity difference between two phase-locked subintervals separated in time by half a period of r.f. field *vs.* the path difference ΔD of the interfering beams.

In the actual experiment it is not possible to measure the individual neutrons energy transfer, but only a cumulative energy transfer which correspond to a well-known (but different from one) number of « particles » travelling along the second path and passing through the RFSF.

The interaction of a neutron with the RFSF constitutes a quasi-classical microscopic measurement process because

a) it involves energy (signal) exchange,

b) it modifies in a predictable way the state of the measured and of measuring device,

c) the exchanged energy ΔE is small compared to the neutron rest energy.

The importance of these kinds of experiments is now evident, because they reproduce the well-known situation of Young's double slit set-up with additional information derived from the interaction between the neutron and the RFSF. Because of this additional fact it is of interest to test the explanatory ability of the different interpretations of quantum mechanics on the grounds of this recent experiment of the Vienna

(4) H. Rauch: *Z. Phys.*, **197**, 373 (1966); W. G. Williams and J. Penfold: *Measurement of the efficiencies of Mezei thermal neutron spin flippers*, NBRU, Jan. 1973; B. Alefeld, G. Badurek and H. Rauch: *Z. Phys. B*, **41**, 231 (1981).

group. We distinguish here three such approaches: the Copenhagen ([5]), the statistical ([6]) and the Einstein-de Broglie model ([7]).

a) The Copenhagen interpretation. Adherents of the Copenhagen interpretation argue that what happens between source and detection when intereference is observed cannot be conceived in quantum description, which only concerns the statistical prediction of results in well-defined experiments. The fundamental unit for description in these terms is the whole « phonomenon », constituted by the system and the experimental apparatus which together form an indivisible and unanalysable whole.

Thus in spinor superposition of neutrons either we design an apparatus to observe interference and forgo a description in terms of space-time co-ordination or we design an incompatible arrangement to determine the space-time motion and forgo the possibility of observing interference. Any attempt to subdivide these phenomena leads to ambiguities. The two are complementary phenomena. Complementary is to be understood in this manner, not according to the dictum that matter never reveals its particle and wave aspects together.

The quantum world has no independent real existence. The wave function Ψ is the most complete description of an individual that can be given. It is merely a probability amplitude which states the odds on various results and is subject to instantaneous changes on measurement. If some preparation device (source, shutter, collimator) is designed to produce a wave packet then all we can say is that the wave packet represents the fact that a single particle has a probability of appearing at a position $\bar{r}$ given by $|\Psi(\bar{r})|^2$ *if a measurement is made.* Until such a time it is not legitimate even to conceive of a particle, let alone its properties. For the Copenhagen interpretation it is thus impossible to use this experiment for a simultaneous detection of both a path of particle and a self-interference pattern. This view is founded also on the grounds of the Heisenberg uncertainty relations, *i.e.* in the present case the so-called fifth, phase-number uncertainty relation $\Delta N \cdot \Delta\varphi \geqslant 2\pi$. Indeed, this argumentation claims that the knowledge of the accurate phase of the radio-frequency field needed for the stroboscopic registration of the oscillating polarization pattern destroys the possibility of the detection of a single-photon transition because ΔN becomes indeterminate.

b) The Statistical interpretation. As emphasized by Ballentine ([6]) the statistical interpretation is to be distinguished from the Copenhagen interpretation. He asserts that the wave function simply represents an ensemble of similarity systems and does not provide a complete description of an individual system.

« In general, quantum theory predicts nothing which is relevant to a single measurement » ([6]).

The interpretation of a wave packet is that although each particle has always a definite position $\bar{r}$, each position is realized with relative frequency $|\Psi(\bar{r})|^2$ in an ensemble of similarily prepared experiments. It follows that each particle has a well-defined trajectory, but its specification is beyond the statistical quantum theory, probabilities arising in the predictions of the theory are to be interpretated as in classical theory.

c) The Einstein-de Broglie model. In this interpretation it is argued that the quantum-mechanical description, through the wave function, of an individual is incomplete in the sense of Einstein. The entities of the micro-world are thought of as being particles *and* waves, in the sense given by de Broglie in his model of oscillators accompained by a real physical wave (the de Broglie « pilot » wave) beating in phase. Quantum

---

(5) N. Bohr: *Atomic Physics and the Description of Nature,* CUP (1934).
(6) L. E. Ballantine: *Rev. Mod. Phys.,* **42**, 358 (1970).
(7) L. de Broglie: *Nonlinear Wave Mechanics* (Elsevier Publ. Co., Houston, Tex., 1960).

phenomena can be described in this model by means of space-time pictures and the neutron interference experiments can be conceived in a straightforward way: while particles travel along *one* of the paths in the interferometer the « pilot » wave propagates on both.

We now claim that the experimental facts established by the Vienna group indeed support this interpretation, namely along the following lines of reasoning.

i) A detection of an energy amount E_{det} by the RFSF implies that this amount of energy has been transferred to the coil by the neutrons involved in the experiment.

ii) A coil absorb energy only at its reasonance frequency $\omega_{r.f.}$, *i.e.* the energy transfer has occurred as a series of single-energy transfers $\hbar\omega_{r.f.} = \Delta E$, *i.e.* as a series of energy transfers, corresponding to the Zeeman energy splitting. This implies that E_{det} is a sum of equal individual energy transfers corresponding to a spin-flip of each individual neutron, *i.e.* $E_{det} = n\,\Delta E$.

iii) Consequently the energy E_{det} corresponds to a sum of n spin-flip, hence n neutrons have passed through the path containing the RFSF coil.

iv) Therefore, if N neutrons are successively involved in the experiment, $N - n$ neutrons have passed through the path without the RFSF coil.

v) By means of this measurement one cannot tell which neutron has gone through which path, but one establishes the following: Out of N neutrons involved in the experiment n neutrons pass through path II and $N - n$ through path I. Every neutron has a probability given by the transmission/reflection coefficient of the first incident plane in the interferometer of going in I or II, *but* it either goes through path I *or* through path II.

vi) Since now neutron self-interference persists and shows that « each neutron in the area of interference knows simultaneously what has happened in both paths » ([8]), this implies that something which has a real physical existence independent of the particle travels along both paths and contributes to the forming of the interference.

This at least prooves the incompleteness of the quantum-mechanical Copenhagen description because the persistence of an interference pattern is combined with the existence of a definite trajectory for each particle, a fact forbidden in Copehangen interpretation.

Of course the experiment only represent an indirect argument in favour of the Einstein-de Broglie point of view and one can legitimately feel that a final proof of the existence of such paths requires the individual detection of passage of each neutron in the RFSF coil, *i.e.* the detection of photons of energies ~ 1 µev; this may indeed be possible by using superconducting quantum interference detectors (SQUID).

If every individual neutron energy transfer is measured, the RFSF will behave as a yes-no device and two conflicting results are possible.

The Copenhagen interpretation of quantum mechanics in this case implies the wave packet collapse ([9]). This also implies the use of a mixture because each particle either follows path I or path II which yields, for every individual case, either $p_I = 1$, $p_{II} = 0$ or $p_I = 0$, $p_{II} = 1$, with a relative frequency determinated by the reflexion/transmission coefficient of the first incident plane. This means that, since each particle

---

([8]) H. Rauch: *Seminar given at the Institut H. Poincaré, Paris*, 21.2.1984.

([9]) N. Bolev: *Atomic Physics and Human Knowledge* (Wiley and Sons, New York, N.Y., 1958).

has a definite trajectory in the apparatus, the probability wave collapses in the other path.

This situation is calculated as follows. When energy is transferred to the coil, the neutron has passed and the wave is described by the state $\psi^{I} = \exp[i(\Delta E/\hbar)]\,|\downarrow_z\rangle$ with no wave corresponding to path I. When no energy is transferred, the wave is given by $\psi^{II} = \exp[i\chi]\,|\downarrow_z\rangle$ with no wave corresponding to path II.

The corresponding probabilities are p^{I}, p^{II} with $p^{I} + p^{II} = 1$, and if the transmission/reflection coefficient in the first plate of the interferometer is equal, one has to assume $p^{I} = p^{II} = \frac{1}{2}$. Under the assumption we can calculate this interference intensity

$$I = p^{I}\langle\psi^{I}|\psi^{I}\rangle + p^{II}\langle\psi^{II}|\psi^{II}\rangle = p^{I} + p^{II}$$

and correspondingly the polarization

$$\bar{P} = p^{I}(0, 0, 1) + p^{II}(0, 0, -1)\,.$$

One can immediately see that, within this kind of Ansatz, the oscillating pattern of the interference intensity vanishes completely and the polarization pattern consists again of two constant contributions directed along the z-axis (in the $+z$ and $-z$ directions). One should also notice that the theoretically predicted pattern exhibits no interference effects (due to the lack of overlap between the wave functions on each path) and consists of two distinct sets of points, the relative frequency of each set being specified by the assumed probabilities p^{I} and p^{II}. Finally it should be stressed that this complete destruction of any interference is a straightforward consequence of a perfectly working measuring device (100% efficient coil) in Bohr's wave packet collapse theory.

The Einstein-de Broglie model excludes the possibility of collapse of the real de Broglie « pilot » wave and, therefore, in this second case predicts the detected interference (*i.e.* polarization in the xy-plane), and implies a simultaneous detection of path and interference pattern (*).

As for the theoretical objections, based on the phase-number uncertainly, concerning such a simultaneous detection, it is by no means clear that this fifth uncertainty relation is correct. Indeed this relation, already contested by DE BROGLIE [11], is shown to be incorrect by CARRUTHERS and NIETO [12] and must be substituted by a more complex one which does not reduce to $\Delta N \cdot \Delta\varphi \geqslant 2\pi$ in the case examined here [13]. Furthermore, following a suggestion presented elsewhere [14] we can by-pass this relation and the related problems.

(*) This problem remains in the statistical interpretation, which also admits definite (but unknown) particle trajectories in each individual case, the wave referring only to ensemble probabilities, statistical frequences. In Young's double-slit experiment the same problem is resolved by reference to Duane's extension [10] of the Bohr-Sommerfeld theory. In the present case it is not clear how a time ensemble of spin-up and spin-down particles results in an ensemble in which the polarization lies in the xy-plane.

[10] W. DUANE: *Proc. Natl. Acad. Sci. USA*, **9**, 158 (1923).

[11] L. DE BROGLIE: *Wave Mechanics, the First 50 Years* (Bathemorths, London, 1973), Chapt. 5.

[12] P. CARRUTHERS and M. NIETHO: *Rev. Mod. Phys.*, **40**, 411 (1968).

[13] C. DEWDNEY, A. GARUCCIO, PH. GUERET, A. KYPRIANIDIS and J.-P. VIGIER: *Time-dependent neutron interferometry: Evidence Against Wave Packet Collapse*, to be published.

[14] C. DEWDNEY, PH. GUERET, A. KYPRIANIDIS and J.-P. VIGIER: *Testing Wave-Particle Dualism with Time Dependent Neutron Interferometry*, *Phys. Lett. A.*, **102**, 291 (1984).

We wish to conclude with the following remark. Even if the existing experimental evidence has not directly proved the real physical existence of the de Broglie waves, it has at least shown that neither the Copenhagen nor the statistical interpretation provide an adequate description of neutron spinor superposition interferometry. On the other hand, the Einstein-de Broglie model can provide a satisfactory intuitive description and it seems that further experiments will provide a positive confirmation of this model.

* * *

The authors wish to thank Prof. H. Rauch for many discussion and helpful suggestions. One of the authors (AK) wants to thank the French government for a grant and another (CD) wishes to thank the Royal Society for the European Exchange Fellowship award which enabled him to do this research.

This page is deliberately left blank.

Reprinted from *Lettere al Nuovo Cimento*, Vol. 41, No. 6, pp. 177-185, Copyright (1984) with permission from the Società Italiana di Fisica.

Causal Stochastic Prediction of the Nonlinear Photoelectric Effects in Coherent Intersecting Laser Beams.

C. DEWDNEY, A. KYPRIANIDIS (*) and J. P. VIGIER

Institut Henri Poincaré, Laboratoire de Physique Théorique
11, rue P. et M. Curie, 75231 Paris Cedex 05

M. A. DUBOIS

Association Euratom C.E.A., D.R.F.C., C.E.N., F.A.R.
B.P. No. 6, 92260 Fontenay-Aux-Roses

(ricevuto il 6 Giugno 1984)

PACS. 03.65. – Quantum theory; quantum mechanics.

Summary. – The introduction of a causal quantum potential (which vanishes for coherent light beams) in the Einstein-de Broglie theory of light implies an enhancement of some photon individual energy $E = h\nu$ in the intersection region of coherent laser beams. This property not only interprets nonlinear effects already observed in highly focused laser beams, but also yields new predictions which can be tested in simple intersecting coherent-laser-beam experiments.

As shown in a set of experiments [1-3] nonlinear effects have already been observed in photoelectric emission and gas photoionization of highly focused laser beams. Indeed everything goes as if some photons undergo an increase of energy due to the focusing so that photoionization occurs even though the original energy of the photons before focusing is below the photoionization potential of the target. Three interpretations of these effects have been attempted

i) in terms of simultaneous multiphoton absorption [4,5],

ii) in terms of Panarella's «effective» photon assumption [6-8],

---

(*) On leave from the University of Crete, Physics Department, Greece.
(1) E. M. LOGOTHETIS and P. L. HARTMAN: *Phys. Rev.*, **187**, 460 (1969).
(2) GY. FARKAS, I. KERTESZ, ZS. NARAY and P. VARGO: *Phys. Lett. A*, **21**, 475 (1967).
(3) E. PANARELLA: *Lett. Nuovo Cimento*, **3**, 417 (1972).
(4) H. BARRY BEBB and A. GOLD: *Phys. Rev.*, **143**, 1 (1966).
(5) R. L. SMITH: *Phys. Rev.*, **128**, 2225 (1962).
(6) E. PANARELLA: *Phys. Rev. Lett.*, **33**, 950 (1974).
(7) E. PANARELLA: *Found. Phys.*, **4**, 227 (1974).
(8) E. PANARELLA: *Found. Phys.*, **7**, 405 (1977).

iii) in terms of the Heisenberg uncertainty relations by ALLEN (9).

The aim of the present letter is

a) to propose a fourth interpretation in terms of the quantum potential's action which appears in the stochastic version of the Einstein-de Broglie theory of light (10) (where photons are waves *and* particles simultaneously), since the energy $E = h\nu$ of the individual photons which follow various lines of flow is no longer constant along these lines;

b) to propose and discuss simple experimental tests based on observations performed in the overlap region of intersecting coherent laser beams which could discriminate between these various interpretations and eventually satisfy new specific predictions of the Einstein-de Broglie model.

Since all known data seem to preclude strong photon-photon interactions, the first possible quantum explanation of how and why focussing should increment photon energy is to utilize the quantum electrodynamical theory of multiphoton absorption processes.

This approach yields the following relations for the electron current i and the maximum kinetic energy of the emitted electrons E_{max}:

$$i \sim I^n , \qquad E_{max} = n \cdot h\nu - W ,$$

where I is the light power density, n is the number of photons of energy $h\nu$ involved in the process and W is the work function of the target material.

These predictions of the multi-photon QED are contested by PANARELLA on the basis of an experiment produced with a highly focused laser beam (3) where the electron current after subtracting the minor influence of a thermionic emission effect turns out to be strictly proportional to the laser power density, *i.e.* $i \sim I$. Furthermore, it was verified that, by increasing the laser power density by means of focusing, the maximum kinetic energy of the electrons E_{max} is augmented in the same sense. These results seem to establish unambiguously that the photoelectric effect is due to single-photon processes and that the photon (and electron) energy depend on the focusing procedure. Of course a final experimental test of the nonvalidity of this multiphoton process would be to see if some above threshold processes are observed (when the laser's $E = h\nu < W$) even when the laser intensity is so reduced that simultaneous arrival of more than one photon at a time becomes completely improbable.

Based on these results, PANARELLA proposed the so-called « effective photon » hypothesis as a theoretical account for the observed energy increase of the photons in the focused laser beam (7). He introduced an *ad hoc* modification of the photon energy relation $E = h\nu$ in the form

$$E = h\nu \exp [B_\nu f(I)] \approx \frac{h\nu}{1 - \beta_\nu f(I)} ,$$

which implies a frequency modulation with increasing light intensity. The origin of this modulation lies in a not further specified *ad hoc* photon-photon interaction, significant in high-density regions only (5). This scheme can be fitted well to describe the

(9) A. D. ALLEN: *Found. Phys.*, **7**, 609 (1977).

(10) L. DE BROGLIE: *La Mécanique ondulatoire du photon* (Hermann, Paris, 1940); A. EINSTEIN: *Ann. Phys.*, **17**, 132 (1905); **18**, 639 (1905); *Z. Phys.*, **18**, 121 (1917).

experimental results of both the photoelectric effect and the gas ionization by a focused laser beam.

Despite the reproduction of the experimental results, a severe theoretical objection against this approach still exists. In fact, as ALLEN correctly points out: « If Panarella's equation is to replace Planck's, then we must also change Einstein's equation for photon energy in order to have a de Broglie wave-length for photons, or, alternatively, we must change de Broglie's equation in order to continue to write $E = mc^2$ » ([9]). Furthermore, ALLEN shows that one can derive the energy increase of the photons from first quantum-mechanical principles, namely as a consequence of the position-momentum uncertainty relation which manifests itself in the focusing process, avoiding thus any introduction of intensity modulation of the photon frequency. ALLEN concludes by proposing an experiment where two lasers of the same photon energy, but different power are focused in a way that both obtain the same intensity, but different beam cross-sections at the focal point. Then the uncertainty principle predicts that the laser with the smaller cross-section would produce photoionization, while the other one remains inactive.

In addition to Allen's critique, we wish to argue that the new postulated photon-photon interaction of Panarella is either a totally unknown effect or, if it is of the well-known nature (*i.e.* pair creation/anihilation), it is completely irrelevant because it has a vanishingly small cross-section for the given experimental conditions ([11]). Here also (as in the conventional multiphoton interpretation) the « effective photon » theory predicts that all photoelectric phenomena (with an initial photon energy $h\nu < W$) should disappear in the focal region when the initial laser intensity is sufficiently reduced ... a prediction which can (and should) evidently be tested with focused laser beams or similar devices.

As initially stated, we wish now to show

a) that all such nonlinear effects can be simply interpreted in the frame of the causal stochastic interpretation of the quantum theory of the Einstein-de Broglie theory of light;

b) that we can deduce from this interpretation new experimental predictions which

i) are consistent but go beyond the usual statistical predictions of the Copenhaguen interpretation now held to be incomplete in our point of view,

ii) can be tested in a simple realizable experimental set-up.

As one knows ([12]), in the Einstein-de Broglie theory of light the photon is considered as an oscillating localized particle with a nonzero mass $m_\gamma \neq 0$ ($m_\gamma \ll 10^{-48}$ g), which moves (on the average) along the lines of flow of a continuous wave field described by the complex four-vector wave field ([13]) $A_\mu = \exp[P + i(S/\hbar)]a_\mu$, where P and S are real functions of the co-ordinates X_μ and a_μ is a real four-vector with $a_\mu a^\mu = 1$. The motions can be derived from the spin-1 Lagrangian

$$L = -\tfrac{1}{4}F^*_{\mu\nu}F_{\mu\nu} - \tfrac{1}{2}\mu^2_\gamma A^*_\mu A^\mu \,, \tag{1}$$

where one has added to the usual Maxwell term $-\tfrac{1}{4}F^*_{\mu\nu}F_{\mu\nu}$ (with $F_{\mu\nu} = \partial_\mu A_\nu - \partial_\nu A_\mu$) a mass term ($\mu_\gamma = mc^4/\hbar$) so that the field equations

$$\partial_\mu F^{\mu\nu} = \mu^2_\gamma A^\nu \tag{2}$$

([11]) EULER: *Ann. Phys.*, **26**, 398 (1936); EULER and W. HEISENBERG: *Z. Phys.*, **98**, 714 (1936).

([12]) M. MOLES and J. P. VIGIER: *Compt. Rend.*, **276**, 697 (1973).

([13]) A. GARUCCIO and J. P. VIGIER: *Lett. Nuovo Cimento*, **30**, 57 (1981).

imply the transverse gauge $\partial_\mu A^\mu = 0$. It has recently been shown that, in this stochastic interpretation of quantum mechanics (SIQM),

1) the individual photons (except when involved in random stochastic fluctuations) move along paths tangent to the conserved four-current ($\partial_\mu j^\mu = 0$) deduced from (2) given by the relation

$$j_\mu = i\left(A_\nu \frac{\partial L}{\partial(\partial_\mu A_\nu)} - \frac{\partial L}{\partial(\partial_\mu A^*_\mu)} A^*_\nu\right) = \frac{\hbar}{m} \exp[2P]\,\partial_\mu S\,. \tag{3}$$

they have a velocity V very close to c (if $m_\gamma \ll 10^{-48}$ g), which satisfies the Einstein-de Broglie relation

$$E = h\nu = m_\gamma c^2/\sqrt{1 - v^2/c^2} \tag{4}$$

and a four-momentum $P_\mu = \partial_\mu S$;

2) their stochastic motions induced by the real chaotic subquantum « vacuum » level (*i.e.* Dirac's aether ([14,15])) occur at the velocity of light c so that photons are statistically distributed (according to a relativistic H theorem ([16])) with a probability $p = A^*_\mu A^\mu = \exp[2P]$) and the uncertainty relations are now interpreted as real physical dispersion relations;

3) the real part of the wave equation (2) contracted by A^*_ν immediately yields the relativistic Hamilton-Jacobi equation

$$\partial_\mu S \partial^\mu S + m^2_\gamma c^2 + \hbar^2 \partial_\mu a_\nu \cdot \partial_\mu a_\nu - \hbar^2(\Box P + \partial_\mu P \partial^\mu P) = 0\,, \tag{5}$$

i.e.

$$P_\mu P^\mu + m_\gamma c^2 + Q + \tau = 0\,,$$

where $Q = -(\partial_\mu P \partial_\mu P + \Box P)$ represents the usual de Broglie-Bohm quantum potential and $\tau = \partial_\mu a_\nu \cdot \partial_\mu a_\nu$ a spin-quantum torque which vanishes in any linearly polarized beam, where $a_\nu = \text{const}$.

This relation (5) is important, since it shows that the energy $E = h\nu = \partial S/\partial\tau$ is not, in general, a constant of the motion along a photon path in this theory except when Q and τ vanish like in the case of parts of linearly polarized beams, where neighbouring current lines are both straight and parallel so that V and thus ν remains constant.

As an example we see that an unperturbed laser beam is characterized by the phase coherence and can be represented by a plane-wave solution for which the quantum potential Q vanishes. Hence, the photons in the plane-wave solution should yield a constancy of their energy along the lines of flow represented by $\partial_\mu S$ and, if the laser frequency ν is below the threshold W, no photoelectric effect should occur no matter how high the intensity of the beam is.

The situation is completely different if we disturb the phase relations in the laser beam, a fact that is produced by the focusing procedure. In fact, if we analyse the focusing procedure, then we realize that it can be thought of as an interference of infini-

---

([14]) P. A. M. Dirac: *Nature (London)*, **168**, 906 (1951).
([15]) P. A. M. Dirac: *Nature (London)*, **169**, 702 (1952).
([16]) A. Kyprianidis and D. Sardelis: *An H-theorem in the causal stochastic interpretation of quantum mechanics*, in *Lett. Nuovo Cimento*, in press.

tesimal parts of the beam symmetric with respect to the cental line of the lens. In that case the physical boundary conditions on the phase relations are changed and the plane-wave solution does no longer reproduce the experiment. The resulting interference pattern is calculated by DEWDNEY ([17]) and is shown in fig. 1*a*.

Fig. 1*a*. – Trajectories for interfering beams.

In this computer simulation we clearly recognize that the originally parallel lines in each beam are distorted and bent so as to form the fringe pattern and subsequently rejoin in a scheme of parallel lines after a certain distance from the interference region. In fact, it should be pointed out that this fringe pattern is independent of the number of interfering photons, as shown by MANDEL and PFLEEGOR ([18]) and persists even in a highly attenuated laser light where only one photon at a time is involved in the experiment.

([17]) C. DEWDNEY: Ph. D. Thesis, London (1983).
([18]) L. MANDEL and R. L. PFLEEGOR: *Phys. Rev.*, **159**, 1084 (1967).

Furthermore, since the quantum potential is no longer zero ($Q \neq 0$), we know that only the sum $E_{ph} + Q$ is conserved, the photon energy is no more a constant of the motion and depends on the value of the quantum potential which oscillates violently in the interference region (see fig. 1*b*) and can accordingly enhance or reduce, respectively, the photon energy. Conclusively, although originally $h\nu < W$ the enhanced photon at the minima of the quantum potential Q can still produce a photoelectric effect if the enhanced photon energy now exceeds W.

From this reasoning we can readily deduce that the nonlinear photo-electric effect is definitely not an intensity effect, as already suggested by ALLEN [9], but merely an effect depending on the interference pattern, which is related to the focusing parameters of the lens. Hence, it would survive even at low intensities of the laser, provided the interference pattern persists and the quantum potential could supply the photon with the needed energy to produce the photoelectric effect. These conclusions lead us to the following experimental proposal which is thought of as a test of our approach and has the merit of decoupling this experiment from any association with over-proportional intensity increase.

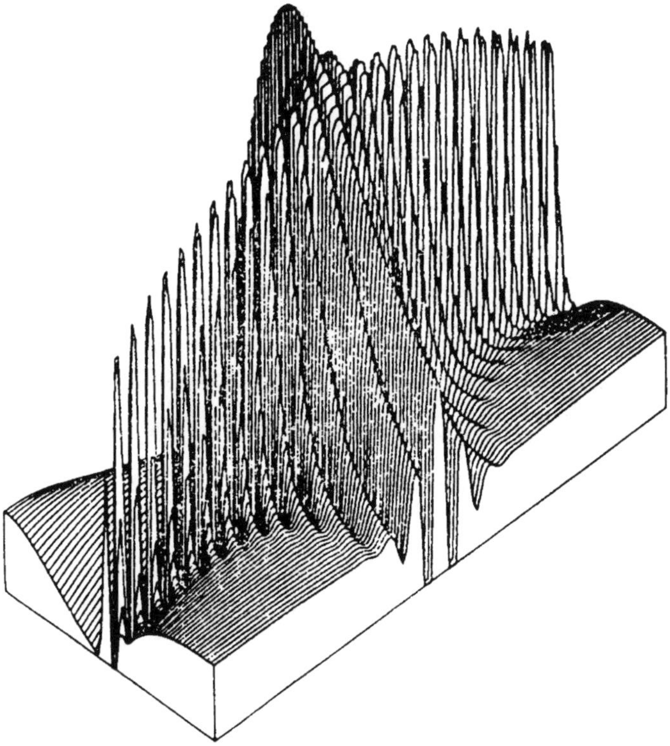

Fig. 1*b*. – Quantum potential in the interference region of the beam.

We propose a Michelson interferometer set-up of fig. 2. An incident laser beam, which is linearly polarized in order to exclude any spin effects, passes through a system of filters F which enables us to vary the intensity of the beam at will. The photon energy of the beam $h\nu$ should be well below the work function W of the target material. The beam is now split by a semi-transparent mirror A with equal reflection/

transmission probability into two equal parts and these partial beams after passing through the indicated system of mirrors are brought to interference and then diverge. The mirror D is, furthermore, allowed to turn around an axis perpendicular to the plane of the figure so that the angle of incidence of the two beams θ and the fringe pattern becomes variable. A target material is now successively put in the incident or partial beams and interference region, at the indicated positions a to h, and should be completely insensitive to the beams with respect to photoelectric emission in all locations except for the interference area f. Here the foregoing analysis predicts that an electron current should be detected as a consequence of the enhanced photon energy due to the oscillations of the quantum potential (see fig. 1b). This set-up is favourable to eliminate noise problems: if the signal obtained with two detectors blocked is S and with only one blocked $S + E$, then we expect to observe a signal $S + \eta$ with $\eta \gg 2E$ when neither of the two lasers is blocked. In the very low-intensity limit, this advantage is very important.

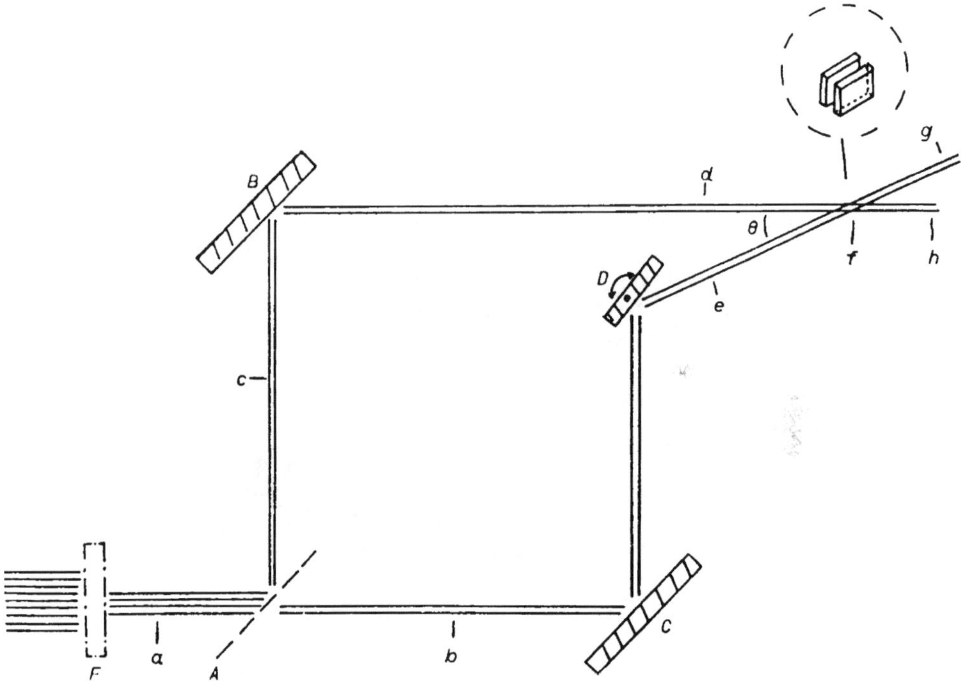

Fig. 2. – Proposed experimental set-up for the test of laser-induced nonlinear photoelectric effect.

Furthermore, the effect should survive if we attenuate the laser light by means of the filter system and there is even a chance of observing the effect in the highly reduced laser intensity in which only one photon at a time is present in the apparatus. The latter should happen when the photon reaches the target material in a position of the interference region where the quantum potential has a sufficiently low value to provide the photon with sufficient energy needed for the overcoming of the threshold W.

This last remark enables us to make a definite prediction as to where the increased energy photons arrive in the interference pattern and consequently from which areas of the target material the photo-electrons should emerge. Since the photons gain energy in the regions between the quantum potential maxima, it is this area of lower luminosity in the fringe system that can produce the nonlinear photoelectric effect and a refined detection technique could help to test this definite prediction.

We conclude with a brief comment on the theoretical implications of the preceding analysis which can be summarized in three points. The first point is that the authors do not believe that the formalism of quantum theory is erroneous, but only that the wave packet collapse introduced by BOHR in the Copenhaguen interpretation of quantum mechanics is not correct and conflicts with experiment in specific situations such as certain experiments discussed on light ([19]) and on neutron interference ([20]). This view does not conflict with facts, since CINI has shown that one could construct a realistic quantum measurement model without wave packet collapse ([21]), the latter being a concept which evidently conflicts with relativity theory.

The second point is that the prediction that the focusing of laser beams (or their convergence on an intersection region) pushed some photons above detection threshold is not in contradiction with the CIQM, since this possibility arises if one considers the introduction of a target (photocathode) as a measuring device. Indeed, as shown by ALLEN ([9]), the Heisenberg uncertainty relations imply that some photons go above threshold.

The third and last remark is that with respect to this prediction the SIQM (which also predicts the rise above threshold) predicts something more, *i.e.* that the photons which present frequency shifts appear on the edges of the bright fringes ... a fact which can be tested with the set-up of superposed $\lambda/2$ plates with which MANDEL and PFLEEGOR have shown that interference occurs even when only one photon at a time is present in the experiment. Alternatively a photographic plate can be used.

Photographic emulsions, being only blue sensitive, should be affected by the high-energy photons produced in the interference of a split beam from a laser which has a frequency lower than that of the cut-off sensitivity of the emulsion. Since only the enhanced energy photons are recorded, an immediate record of their position in the fringes is made in the emulsion which can be compared with the predictions of the approach advocated here.

Additionally we note what follows:

The photon frequency may be raised or lowered in the area of overlap, since the total average energy remains constant and so intersecting laser beams may be used to produce higher- or lower-frequency laser beams by simple triggering higher- or lower-frequency lasers.

The substitution of an incoherent light source for the laser in this experiment should make no difference to the effect, all other things being equal, according to the interpretation given by ALLEN based on the Heisenberg uncertainty relation ([9]). However, the prediction of the causal stochastic theory is that such a substitution should lead to a reduction in the observed effect, since the broadening of the fringes leads to fewer enhanced energy photons. Clearly this can be tested by experiment.

Thus what we, in fact, claim is that the presented model is perfectly compatible with quantum theory, but includes additional information with respect to it. In the usual

---

([19]) K. A. POPPER, A. GARUCCIO, J. P. VIGIER: *Phys. Lett. A*, **86**, 326 (1982); A. GARUCCIO, V. A. RAPISARDA, J. P. VIGIER: *Lett. Nuovo Cimento*, **32**, 451 (1981); A. GARUCCIO, A. KYPRIANIDIS, D. SARDELIS and J. P. VIGIER: *A possible experimental test of the wave packet collapse*, in *Lett. Nuovo Cimento*, in press.

([20]) G. BADUREK, H. RAUCH, J. SUMMHAMMER: *Phys. Rev. Lett.*, **51**, 1015 (1983); D. DEWDNEY, PH. GUERET, A. KYORIANIDIS and J. P. VIGIER: *Time dependent neutron interferometry: evidence against wave packet collapse* (to be published). *Testing wave particle dualism with time-dependent neutron interferometry*: *Phys. Lett. A*, in press.

([21]) M. CINI: *Nuovo Cimento B*, **73**, 27 (1983).

quantum theory the effect is predicted as a consequence of the uncertainty relation, but the presented model enables a detailed calculation of the path-by-path photon energy distribution and the number and maximum kinetic energy of the emitted electrons as well as the areas on the target material that contribute to the photoelectric current. This is a fact that goes beyond ordinary quantum predictions and turns to be in favour of Einstein's idea that the quantum-mechanical description of physical reality is not complete (*).

* * *

One of us (CD) wants to thank the Royal Society for the European Exchange Fellowship which enabled him to carry out this work and another (AK) wants to thank the French government for a grant which made this research possible.

(*) During the completion of this work we have learned that experiments along the same lines performed in Limoges by Prof. FROEHLY have preliminary results that confirm our predictions.

This page is deliberately left blank.

Reprinted from *Apeiron*, Vol. 2, No. 4, pp. 114-115 (1995).

Fundamental Problems of Quantum Physics

Jean-Pierre Vigier
Laboratory for Relativistic Gravitation and Cosmology
University of Paris-Pierre and Marie Curie
75252 Paris Cedex 05

1. Quantum theory vs. hidden variables

It must be realized that quantum mechanics in its present state, as it is taught in universities and utilized in laboratories, is essentially a mathematical formalism which makes statistical predictions. In all known experiments, those statistical properties have been confirmed. The interpretation of such a statistical formalism is a different matter. The dominant, so-called Copenhagen interpretation of quantum mechanics states that there are no hidden variables behind these statistical predictions. In other words, the particle aspect of matter that appears in all experiments does not correspond to motions in space and time. There is nothing behind quantum mechanics and the statistical information provided by quantum mechanics represents an ultimate limit of all scientific knowledge in the microworld.

Since 1927, with ups and downs, alternative interpretations of the formalism have been developed in terms of real physical hidden parameters. It must be realized that these alternative hidden variable models are of two different, conflicting natures. In one version, the initial de Broglie-Bohm model, individual micro-objects are waves and particles simultaneously, the individual particles being piloted by the waves. This realistic model must of course be extended to many-body entangled particle systems. In this situation, the so-called hidden variable models split into two. In the first, local model, there is no such thing as superluminal correlations between the particles. This has led to Bell's research and the discovery of the so-called Bell inequality, which should be violated by non-local hidden variable models. This position has been defended to his last days by de Broglie himself, and some of his followers (Lochak, Selleri, Andrade da Silva, *etc.*).

In the second version the hidden variables engender non-local interactions between entangled particle states. This is the view defended by Bohm, myself, *etc.* These non-local correlations (which, by the way also appear in the quantum mechanical formalism) correspond to the superluminal propagation of the real phase wave packets which were introduced by de Broglie as the basis of his discovery of wave mechanics.

The present situation is very exciting because for the first time one can make experiments that detect photons and other particles one by one, and therefore, we are going to be able to test in an unambiguous way, the existence or not of superluminal correlations. In my opinion, the existence of these correlations has already been established not only by Aspect's experiments (I believe the improved version now underway will confirm his initial results), but they have also been established by down-converted photon pair experiments (Maryland experiments).

2. Most important unresolved issue in quantum physics today

There are two crucial questions in quantum physics today:

1. Do particles always travel in space and time along timelike trajectories? This of course implies the existence of quantum potentials and variation of particle energy along the path which results from the particle-wave gearing.
2. Do superluminal interactions conflict or not with relativity theory, in other words, are the observed non-local correlations compatible with Einstein's conception of causality?

Both points can now be answered by experiment. On the first point, experiments can now be made with neutrons, one by one, to test Einstein's *einweg* assumption in the double-slit experiment. It is also now possible to perform photo-electric experiments to show the existence of the quantum potential. On the second point, calculations started by Sudarshan and other people have shown that non-local correlations preserve Einstein causality provided the Hamiltonians of entangled particles commute, and it has been shown that the quantum potential in the many-body system built by Bohm, myself, *etc.*, satisfies this causality condition. In other words, quantum non-locality can now be considered as an experimental fact which satisfies Einstein's causality in the non-local realistic interpretation of hidden variables.[1]

3. The earlier debate (Solvay 1927)

On the third question, the present debate is an extension of the Solvay controversy. At that time, there was no possibility to realize in the laboratory, Einstein's or Bohr's gedanken experiments. The situation is now different, so that the Bohr-Einstein controversy and the discussion between proponents of local or non-local realistic quantum mechanical models are going to be settled by experiments.

4. Future developments of foundations

In my opinion the most important development to be expected in the near future concerning the foundations of quantum physics is a revival, in modern covariant form, of the ether concept of the founding fathers of the theory of light (Maxwell, Lorentz, Einstein, *etc.*). This is a crucial question, and it now appears that the vacuum is a real physical medium which presents surprising properties (superfluid, *i.e.* negligible resistance to inertial motions) so that the observed material manifestations correspond to the propagation of different types of phase waves and different types of internal motions within the extended particles themselves. The transformation of particles into each other would correspond to reciprocal transformations of such motions. The propagation of phase waves on the top of such a complex medium first suggested by Dirac in his famous

---

[1] This non-locality rests on the idea that the particles and the wave constitutive elements are not delta functions, but correspond to extended hypertubes (which contain real clock-like motions) which can thus carry superluminal phase waves.

If the existence of a gravitational field which determines the metric is confirmed, gravitational interactions could also correspond to spin-two phase waves moving faster than light.

1951 paper in *Nature* yields the possibility to bring together relativity theory and quantum mechanics as different aspects of motions at different scales. This ether, itself being built from spin one-half ground-state extended elements undergoing covariant stochastic motions, is reminiscent of old ideas at the origin of classical physics proposed by Descartes and in ancient times by Heraclitus himself. The statistics of quantum mechanics thus reflects the basic chaotic nature of ground state motions in the Universe.

Of course, such a model also implies the existence of non-zero mass photons as proposed by Einstein, Schrödinger, and de Broglie. If confirmed by experiment, it would necessitate a complete revision of present cosmological views. The associated tired-light models could possibly replace the so-called expanding Universe models. Non-velocity redshifts could explain anomalous quasar-galaxy associations, *etc.*, and the Universe would possibly be infinite in time. It could be described in an absolute spacetime frame corresponding to the observed 2.7 K microwave background Planck distribution. Absolute 4-momentum and angular momentum conservation would be valid at all times and at every point in the Universe.

This page is deliberately left blank.

Bibliography of Works by Jean-Pierre Vigier

1. Photon Mass and Heaviside Force, J-P Vigier, accepted for publication by *Physics Letters A*, 2000.
2. New "hidden" parameters describing internal motions within extended particle elements associated with a Feynman-Gell-Mann type causal electron model, J-P Vigier, in preparation.
3. Quantum dynamics from the Brownian recoil principle, Garbaczewski, P.; Vigier, J.P. Journal: *Physical Review A* (Statistical Physics, Plasmas, Fluids, and Related Interdisciplinary Topics), vol.46, no.8, p.4634-8, 1998.
4. *Causality and locality in modern physics*: proceedings of a symposium in honour of Jean-Piere Vigier / edited by Geoffrey Hunter, Stanley Jeffers, and Jean-Pierre Vigier, 1998, Kluwer Academic, Dordrecht; Boston
5. Possible consequences of an extended charged particle model in electromagnetic theory, J.P. Vigier, *Physics Letters A*, vol.235, no.5 p.419-31, 1997.
6. *The Enigmatic Photon*: *New Directions* (Fundamental Theories of Physics, Vol 90) by Myron W. Evans (Editor), Jean-Pierre Vigier (Editor), Sisir Roy, Kluwer Academic Publishers
7. New non-zero photon mass interpretation of the Sagnac effect as direct experimental justification of the Langevin paradox, J.P.Vigier, *Physics Letters A*, vol.234, no.2, p.75-85, 1997.
8. On cathodically polarized Pd/D systems, J.P.Vigier, *Physics Letters A*, vol.221, no.1-2, p.138-40, 1996.
9. *The Enigmatic Photon: Theory and Practice of the B3 Field* (Fundamental Theories of Physics, Vol 77) by Myron W. Evans, Jean-Pierre Vigier, Sisir Roy, Stanley Jeffers, Kluwer Academic Publishers
10. Variation of local heat energy and local temperatures under Lorenz transformations, Fenech, C.; Vigier, J.P., *Physics Letters A*, vol.215, no.5-6 p.247-53, 1996.
11. Possible test of the reality of superluminal phase waves and particle phase space motions in the Einstein-de Broglie-Bohm causal stochastic interpretation of quantum mechanics Vigier, J.P., *Foundations of Physics*, vol.24, no.1, p.61-83, 1994.
12. *The Enigmatic Photon*: *The Field B(3)* by Myron W. Evans, Jean-Pierre Vigier, Kluwer Academic Publishers
13. Thermodynamical properties/description of the de Broglie-Bohm pilot-waves, Fenech, C.; Vigier, J.P., *Physics Letters A*, vol.182, no.1, p.37-43,1993.
14. David Joseph Bohm: 1917-1992, P.R. Holland and J-P Vigier, *Foundations of Physics*, Vol. 23, No 1, 1993.
15. From Descartes and Newton to Einstein and De Broglie, J-P Vigier, *Foundations of Physics*, Vol. 23, No. 1, 1993.

16. Preliminary observations on possible implications of new Bohr orbits (resulting from electromagnetic spin-spin and spin-orbit coupling) in 'cold' quantum mechanical fusion processes appearing in strong 'plasma focus' and 'capillary fusion' experiments, Antanasijevic, R.; Lakicevic, I.; Maric, Z.; Zevic, D.; Zaric, A.; Vigier, J.P., *Physics Letters A* vol.180, no.1-2 p.25-32,1993.

17. De Broglian probabilities in the double-slit experiment, Bozic, M.; Maric, Z.; Vigier, J.P., *Foundations of Physics*, vol.22, no.11, p.1325-44, 1992.

18. Brownian Motion and its Descendants according to Schrodinger, P.Garbaczewski and J-P Vigier, *Physics Letters A*. 167, pp. 445-451, 1992.

19. Reply to the Comment on 'Proposed neutron interferometry test of Einstein's 'Einweg' assumption in the Bohr-Einstein controversy', Rauch, H.; Vigier, J.P., *Physics Letters A*, vol.157, no.4-5, p.311-13, 1991.

20. Do Quantum Particles travel in real space time? Experimental Evidence and Theoretical Implications, J-P Vigier, *Information Dynamics*, eds H. Atmanspacher and H. Scheinraber, Plenum Press, New York, 1991.

21. Explicit Mathematical Construction of Relativistic Nonlinear de Broglie Waves Described by Three-Dimensional (Wave and Electromagnetic) Solitons 'Piloted' (Controlled) by Corresponding Solutions of Associated Linear Klein Gordon and Schrodiner Equations, J-P Vigier, *Foundations of Physics*, Vol. 21, No 2, 1991.

22. Comment on 'Experimental test of the de Broglie guided-wave theory for photons' (with reply), Holland, P.R.; Vigier, J.P., *Physical Review Letters*, vol.67, no.3 , p.402-3, 1991.

23. Does a possible laboratory observation of a frequency anisotropy of light result from a non-zero photon mass? Narlikar, J.V.; Pecker, J.C.; Vigier, J.P., *Physics Letters A*, vol.154, no.5-6, p.203-9, 1991.

24. Ampere forces considered as collective non-relativistic limit of the sum of all Lorentz interactions acting on individual current elements: possible consequences for electromagnetic discharge stability and tokamak behaviour, Rambaut, M.; Vigier, J.P. Journal: *Physics Letters A*, vol.148, no.5, p.229-38, and 1990.

25. The simultaneous existence of EM Grassmann-Lorentz forces (acting on charged particles) and Ampere forces (acting on charged conducting elements) does not contradict relativity theory, Rambaut, M.; Vigier, J.P., *Physics Letters A*, vol.142, no.8-9, p.447-52, 1989.

26. Evolution time Klein-Gordon equation and derivation of its nonlinear counterpart, Kaloyerou, P.N.; Vigier, J.P., *Journal of Physics A (Mathematical and General)*, vol.22, no.6, p.663-73, 1989.

27. Real Physical Paths in Quantum Mechanics. Equivalence of the Einstein-de Broglie and Feynman Points of View on Quantum Particle Behaviour, J-P Vigier, Proc. 3rd Symp. *Foundations of Quantum Mechanics*, 1989.

28. Comments on the üncontrollable' character of non-locality, J.P.Vigier, in the *Concept of Probability*, eds E. I. Bitsakis and C. A. Nicolaides, Kluwer Academic Publishers, 1989.

29. Comments on 'Violation of Heisenberg's Uncertainty Relations by Z. Maric, K. Popper and J.P. Vigier , *Found. Phys. Lett.* 2 (1989) 403 (with J. Hilgevoord)

30. Particular solutions of a non-linear Schrodinger equation carrying particle-like singularities represent possible models of de Broglie's double solution theory, Vigier J.P., *Physics Letters A*, vol.135, no.2, p.99-105, 1989.

31. Quantum Physics: Gdansk '87. Recent and Future Experiments and Quantum particle motions in real physical space-time, Vigier, J.P., Conference Title: *Problems in Interpretations*, p.317-49, Editor(s): Kostro, L.; Posiewnik, A.; Pykacz, J.; Zukowski, M. 1988.

32. Second-Order Wave Equation for Spin-1/2 Fields: 8-Spinors and Canonical Formulation, N. Cufaro-Petroni, P. Gueret and J.P. Vigier, *Foundations of Physics*, Vole 18, No 11, 1988.

33. Spin and non-locality in quantum mechanics, Dewdney, C.; Holland, P.R.; Kyprianidis, A.; Vigier, J.P, *Nature*, vol.336, no.6199, p.536-44, 1988.

34. The quantum potential and signalling in the Einstein-Podolsky-Rosen experiment, Holland, P.R.; Vigier, J.P., *Foundations of Physics*, vol.18, no.7, p.741-50, 1988.

35. New theoretical implications of neutron interferometric double resonance experiments, Vigier, J.P., *Physica B & C*, vol.151, no.1-2, p. 386-92, Conference Title: *International Workshop on Matter Wave Interferometry in the Light of Schroedinger's Wave Mechanics* Conference Sponsor: Hitachi; Erwin Schroedinger Gesellschaft; Siemens; et al, 1988.

36. Kyprianidis and J. P. Vigier in "Quantum Mechanics Versus Local Realism" ed. Franco Selleri, New York, Plenum, 1988, p. 273.

37. Derivation of a non-linear Schrodinger equation describing possible vacuum dissipative effects, Kaloyerou, P.N.; Vigier, J.P., *Physics Letters A*, vol.130, no.4-5, p.260-6, 1988.

38. Einstein's materialism and modern tests of quantum mechanics, Vigier, J.P. *Annalen der Physik*, vol.45, no.1, p.61-80, 1988.

39. EPR Version of Wheeler's Delayed Choice Experiment, J-P Vigier, in *Microphysical Reality and Quantum Formalism*, Kluwer Academic Publishers, 1988.

40. Violation of Heisenberg's Uncertainty Relations on Individual Particles within subset of gamma photons in e^+e^-2 gamma pair creation, Z. Maric, K. Popper and J.P. Vigier, *Foundations of Physics*, Letters, Vole 1, no 4, 1988.

41. A Possible Tired-Light Mechanism, Jean-Claude Pecker/Jean-Pierre Vigier, *Apeiron* Number 2 (February 1988)

42. Theoretical implications of time-dependent double resonance neutron interferometry, Vigier, J.P., Conference Title: *Quantum Uncertainties, Recent and Future Experiments and Interpretations.* Proceedings of a NATO Advanced Research Workshop on Quantum Violations: Recent and Future Experiments and Interpretations, p.1-18 Editor(s): Honig, W.M.; Kraft, D.W.; Panarella, E. Publisher: Plenum, New York, NY, USA, Publication Date: 1987.

43. Einstein-Podolsky-Rosen constraints on quantum action at a distance: the Sutherland paradox, Cufaro-Petroni, N.; Dewdney, C.; Holland, P.R.; Kyprianidis, A.; Vigier, J.P., *Foundations of Physics*, vol.17, no.8, p.759-73, 1987.

44. Trajectories and causal phase-space approach to relativistic quantum mechanics, Holland, P.R.; Kyprianidis, A.; Vigier, J.P., *Foundations of Physics*, vol.17, no.5 p.531-47, 1987.

45. Quantum properties of chaotic light in first-order interference experiments, Kyprianidis, A.; Vigier, J.P., *Europhysics Letters*, vol.3, no.7 p.771-5, 1987.

46. Distinguishability or indistinguishability in classical and quantum statistics, Kyprianidis, A.; Roy, S.; Vigier, J.P., *Physics Letters A*, vol.119, no.7, p.333-36, 1987.

47. *Quantum Implications*, Vigier, J.P., Dewdney, C, Holland, P.R., Kyprianidis, A, ed. B.J. Hiley and F.D. Peat, pp. 169-204, London: Routledge & Kegan Paul, 1987.

48. A non-negative distribution function in relativistic quantum mechanics, Holland, P.R.; Kyprianidis, A.; Vigier, J.P., *Physica A*, vol.139A, no.2-3, p. 619-28, 1986.

49. Theoretical implications of neutron interferometry derived from the causal interpretation of the Pauli equation, Kyprianidis, A.; Vigier, J.P. *Hadronic Journal Supplement*, vol.2, no.3, p.534-56, and 1986.

50. Causal phase-space approach to fermion theories understood through Clifford algebras, Holland, P.R.; Kyprianidis, A.; Vigier, J.P., *Letters in Mathematical Physics*, vol.12, no.2, p.101-10, 1986.

51. Relativistic generalization of the Wigner function and its interpretation in the causal stochastic formulation of quantum mechanics, Holland, P.R.; Kyprianidis, A.; Maric, Z.; Vigier, J.P. *Physical Review A (General Physics)*, vol.33, no.6, p.4380-3, 1986.

52. Relativistic Wigner function as the expectation value of the PT operator, Dewdney, C.; Holland, P.R.; Kyprianidis, A.; Vigier, J.P., *Physics Letters A*, vol.114A, no.8-9, p.440-4, 1986.

53. Second-order wave equation for spin-/sup 1///sub 2/ fields. II. The Hilbert space of the states, Petroni, N.C.; Gueret, P.; Vigier, J.P.; Kyprianidis, *Physical Review D (Particles and Fields)*, vol.33, no.6, p.1674-80, 1986.

54. Cosmology and the causal interpretation of quantum mechanics, Dewdney, C.; Holland, P.R.; Kyprianidis, A.; Vigier, J.P., *Physics Letters A*, vol.114A, no.7, p.365-70, 1986.

55. Superfluid vacuum carrying real Einstein-de Broglie waves, Sinha, K.P.; Sudarshan, E.C.G.; Vigier, J.P., *Physics Letters A*, vol.114A, no.6, p.298-300, 1986.

56. A review of extended probabilities Muckenheim, W.; Ludwig, G.; Dewdney, C.; Holland, P.R.; Kyprianidis, A.; Vigier, J.P.; Cufaro Petroni, N.; Bartlett, M.S.; Jaynes, E.T., *Physics Reports*, vol.133, no.6, p. 337-401, 1986.

57. Stochastic physical origin of the quantum operator algebra and phase space interpretation of the Hilbert space formalism: the relativistic spin zero case, Dewdney, C.; Holland, P.R.; Kyprianidis, A.; Maric, Z.; Vigier, J.P., *Physics Letters A*, vol.113A, no.7, p.359-64, 1986

58. Vigier: L'onde et la particule *Science et Vie* Feb. 86

59. Testing for non-locally correlated particle motions in the hydrogen atom, Dewdney, C.; Dubois, M.A.; Holland, P.R.; Kyprianidis, A.; Laurent, L.; Pain, M.; Vigier, J.P., *Physics Letters A*, vol.113A, no.3, p.135-8, 1985.

60. The Redshift Distribution Law of Quasars Revisited, S. Depaquit, J-C Pecker and J-P Vigier, *Astron. Nachr.*, 306, 1985.

61. Non-locality and space time in quantum N-body systems, J.P Vigier, in *Determinism in Physics*, eds E.I.Bitsakis and N.Tambakis, Gutenberg Publishing Company, 1985.

62. Time-dependent neutron interferometry: evidence against wave packet collapse?, Dewdney, C.; Garuccio, A.; Gueret, Ph.; Kyprianidis, A.; Vigier, J.P., *Foundations of Physics*, vol.15, no.10, p.1031-42, 1985.

63. Depaquit, J.C. Pecker, J.P. Vigier, *Astr. Nach.*, t.306, p.1, 1985.

64. Rauch's experiment and the causal stochastic interpretation of quantum statistics, Vigier, J.P.; Roy, S., Indian Stat. Inst., Calcutta, India Journal: *Hadronic Journal Supplement*, vol.1, no.3 p.475-501, 1985.

65. Causal stochastic interpretation of quantum statistics Vigier, J.P., *Pramana* vol.25, no.4, p.397-418, 1985.

66. Realistic physical origin of the quantum observable operator algebra in the frame of the casual stochastic interpretation of quantum mechanics: the relativistic spin-zero case, Cufaro-Petroni, N.; Dewdney, C.; Holland, P.; Kyprianidis, T.; Vigier, J.P., *Physical Review D (Particles and Fields)*, vol.32, no.6, p.1375-83, 1985.

67. Nonlocal quantum potential interpretation of relativistic actions at a distance in many-body problems, Vigier, J.P. Conference Title: *Open Questions in Quantum Physics.* Invited Papers on the Foundations of Microphysics p.297-32 Editor(s): Tarozzi, G.; van der Merwe, A. Publisher: Reidel, Dordrecht, Netherlands, 1985.

68. Positive probabilities and the principle of equivalence for spin-zero particles in the causal stochastic interpretation of quantum mechanics, Holland, P.R.; Vigier, J.P., *Nuovo Cimento B* vol.88B, ser.2, no.1, p.20-8, 1985.

69. Second-order wave equation for spin-1/2 fields, Petroni, N.C.; Gueret, P.; Vigier, J.P.; Kyprianidis, A., *Physical Review D (Particles and Fields)*, vol.31, no.12, p.3157-61, 1985.

70. Causal action at a distance in a relativistic system of two bound charged spinless particles: hydrogenlike models Dewdney, C.; Holland, P.R.; Kyprianidis, A.; Vigier, J.P., *Physical Review D (Particles and Fields)*, vol.31, no.10, p.2533-8, 1985.

71. An alternative derivation of the spin-dependent quantum potential, Petroni, N.C.; Gueret, P.; Kyprianidis, A.; Vigier, J.P., *Lettere al Nuovo Cimento*, vol.42, ser.2, no.7, p.362-4, 1985.

72. Causal space-time paths of individual distinguishable particle motions in N-body quantum systems: elimination of negative probabilities, Petroni, N.C.; Dewdney, C.; Holland, P.; Kyprianidis, A.; Vigier, J.P., *Lettere al Nuovo Cimento*, vol.42, ser.2, no.6, p.285-94, 1985.

73. Positive energy-positive probability density association in second-order fermion theories, Gueret, P.; Holland, P.R.; Kyprianidis, A.; Vigier, J.P., *Physics Letters A*, vol.107A, no.8, p.379-82, 1985.

74. On the association of positive probability densities with positive energies in the causal theory of spin 1 particles, Holland, P.R.; Kyprianidis, A.; Vigier, J.P., *Physics Letters A*, vol.107A, no.8, p.376-8, 1985.

75. Elimination of negative probabilities within the causal stochastic interpretation of quantum mechanics, Cufaro-Petroni, N.; Dewdney, C.; Holland, P.; Kypriandis, T.; Vigier, J.P., *Physics Letters A*, vol.106A, no.8, p.368-70, 1984.

76. *Quantum, space and time - the quest continues: studies and essayes in honour of Louis de Broglie, Paul Dirac, and Eugene Wigner* / edited by Asim O. Barut, Alwyn van der Merwe, Jean-Pierre Vigier, 1984, Cambridge University Press, Cambridge (Cambridgeshire); New York

77. Relativistic predictive quantum potential: the N-body case, Garuccio, A.; Kyprianidis, A.; Vigier, J.P., *Nuovo Cimento B*, vol.83B, ser.2, no.2, p.135-44, 1984.

78. The anomalous photoelectric effect: quantum potential theory versus effective photon hypothesis, Dewdney, C.; Garuccio, A.; Kyprianidis, A.; Vigier, J.P. *Physics Letters A*, vol.105A, no.1-2, p.15-18, 1984.

79. Causal stochastic prediction of the nonlinear photoelectric effects in coherent intersecting laser beams, Dewdney, C.; Kyprianidis, A.; Vigier, J.P.; Dubois, M.A., *Lettere al Nuovo Cimento*, vol.41, ser.2, no.6, p.177-85, 1984.

80. Illustration of the causal model of quantum statistics, Dewdeny, C.; Kyprianidis, A.; Vigier, J.P., *Journal of Physics A (Mathematical and General)*, vol.7, no.14, p. L741-4, 1984.

81. Energy conservation and complementarity in neutron single crystal interferometry, Dewdney, C.; Garuccio, A.; Kyprianidis, A.; Vigier, J.P., *Physics Letters A*, vol.104A, no.6-7, p.325-8, 1984.

82. Time-dependent neutron interferometry: evidence in favour of de Broglie waves, Dewdney, C.; Kyprianidis, A.; Vigier, J.P.; Garuccio, A.; Gueret, P., *Lettere al Nuovo Cimento*, vol.40, ser.2, no.16, p.481-7, 1984.

83. Form of a spin-dependent quantum potential, Petroni, N.C.; Gueret, P.; Vigier, J.P., *Physical Review D (Particles and Fields)*, vol.30, no.2, p.495-7, 1984.

84. Testing wave-particle dualism with time-dependent neutron interferometry, Dewdney, C.; Gueret, P.; Kyprianidis, A.; Vigier, J.P., *Physics Letters A*, vol.102A, no.7, p.291-4, 1984.

85. Possible experimental test of the wave packet collapse, Garuccio, A.; Kypriandis, A.; Sardelis, D.; Vigier, J.P. *Lettere al Nuovo Cimento*, vol.39, ser.2, no.11, p.225-33, 1984.

86. Causal stochastic interpretation of Fermi-Dirac statistics in terms of distinguishable non-locally correlated particles, Cufaro-Petroni, N.; Kyprianidis, A.; Maric, Z.; Sardelis, D.; Vigier, J.P., *Physics Letters A*, vol.101A, no.1, p.4-6, 1984.

87. An interpretation of the Dirac rigid electron model by the theory of general relativity, Vigier, J.P.; Dutheil, R., *Bulletin de la Societe Royale des Sciences* vol.52, no.5p.331-5, 1983.

88. Relativistic Wave Equations with Quantum Potential Nonlinearity, Ph.Gueret and J-P Vigier, *Lettere Al Nuovo Cimento,* Vol.38, no4, 1983.

89. Random motions at the velocity of light and relativistic quantum mechanics, Petroni, N.C.; Vigier, J.P., *Journal of Physics A (Mathematical and General)*, vol.17, no.3, p.599-608, 1984.

90. Relativistic wave equations with quantum potential nonlinearity, Gueret, Ph.; Vigier, J.P., *Lettere al Nuovo Cimento,* vol.38, ser.2, no.4, p.125-8, 1983.

91. Some possible experiments on quantum waves, Andrade e Silva, J.; Selleri, F.; Vigier, J.P. *Lettere al Nuovo Cimento* vol.36, ser. 2. p.503-8, 1983.

92. Sur l'anomalie de décalage spectral des paires mixtes de galaxies, E.Giraud, J-P Vigier, *C.R. Acad.Sc.Paris,* t. 296, 193, 1983.

93. Sur Une Interpretation par la theorie de la relativite generale du modele délectron rigide de Dirac, J-P Vigier and R.Dutheil, *Bull. Soc. Roy.des Sces de Liege,* 1983.

94. Pairs of spiral galaxies with magnitude differences greater than one, Arp, H.; Giraud, E.; Sulentic, J.W.; Vigier, J.P., *Astronomy and Astrophysics,* vol.121, no.1, pt.1, p.26-8, 1983.

95. Dirac's aether in relativistic quantum mechanics, Petroni, N.C.; Vigier, J.P., *Foundations of Physics,* vol.13, no.2, p.253-86, 1983.

96. New Experimental Set-Up for The Detection of De Broglie Waves, A.Garrucio, V.Rapisarda, J-P Vigier, *Physics Letters,* Vol 90 A, number 1, 1982.

97. Louis de Broglie-physicist and thinker, Vigier, J.P. *Foundations of Physics,* vol.12, no.10, p.923-30, 1982.

98. J.-P. Vigier, *Astr.Nach.,* t.303, p.55., 1982.

99. *Quantum Space and Time* (in collaboration with N. Cufaro-Petroni). Cambridge University Press, p.505., 1982.

100. Nonlinear Klein-Gordon equation carrying a nondispersive solitonlike singularity, Gueret, P.; Vigier, J.P., *Lettere al Nuovo Cimento,* vol.35, ser.2, no.8, p.256-9, 1982.

101. Soliton model of Einstein's 'nadelstrahlung' in real physical Maxwell waves, Gueret, P.; Vigier, J.P., *Lettere al Nuovo Cimento,* vol.35, ser.2, no.8, p.260-4, 1982

102. Une corrélation entre le décalage spectral et le type morphologique pour les galaxies binaires, E. Giraud, M.Moles and J-P Vigier, *C.R.Acad.Sc.Paris,* t. 294, 1982.

103. Stochastic model for the motion of correlated photon pairs, Petroni, N.C.; Vigier, J.P., *Physics Letters A,* vol.88A, no.6, p.272-4, 1982.

104. Relativistic Hamiltonian Description of the Classical Photon Behaviour: A Basis to Interpret Aspect's Experiments, F.Halbwachs, F.Piperno and J-P Vigier, *Lettere al Nuovo Cimento,* vol 33, no 11, 1982.

105. De Broglie's Wave Particle Duality in the Stochastic Interpretation of Quantum Mechanics: A Testable Physical Assumption, Ph.Gueret and J-P Vigier, *Foundations of Physics*, vol 12, no 11, 1982.

106. Action-at-a-distance and causality in the stochastic interpretation of quantum mechanics, Petroni, N.C.; Droz-Vincent, P.; Vigier, J.P., *Lettere al Nuovo Cimento*, vol.31, ser.2, no.12 p.415-20, 1981.

107. Stable states of a relativistic bilocal stochastic oscillator: a new quark-lepton model, Petroni, N.C.; Maric, Z.; Zivanovic, Dj.; Vigier, J.P. *Journal of Physics A (Mathematical and General)* vol.14, no.2 p.501-8, 1981

108. Analyse du potentiel quantique de De Broglie dans le cadre de l'intérpretation stochastique de la théorie des quanta, C.Fenech and J-P Vigier, *C.R.Acad.Sc.Paris*, t. 293, 1981.

109. Description of spin in the causal stochastic interpretation of Proca-Maxwell waves: theory of Einstein's 'ghost waves', Garuccio, A.; Vigier, J.P., *Lettere al Nuovo Cimento*, vol.30, ser.2, no.2, p.57-63, 1981.

110. Possible Direct Physical Detection of De Broglie Waves, Augusto Garuccio and Jean-Pierre Vigier, *Physics Letters*, 1981 December 7.

111. An Experiment to Interpret E.P.R. Action-at-a-Distance: The Possible Detection of Real De Broglie Waves (with Augusto Garuccio and Jean-Pierre Vigier), *Epistemological Letters*, 1981 July. Includes reply by O. Costa de Beauregard

112. Stochastic derivation of the Dirac equation in terms of a fluid of spinning tops endowed with random fluctuations at the velocity of light, Cufaro Petroni, N.; Vigier, J.P., *Physics Letters A*, vol.81A, no.1, p.12-14,1981.

113. Baryon octet magnetic moments in an integer-charged-quark oscillator model, Cufaro Petroni, N.; Maric, Z.; Zivanovic, Dj.; Vigier, J.P., *Lettere al Nuovo Cimento* vol.29, ser.2, no.17 p.565-71, 1980.

114. Garuccio, K. Popper and J.-P. Vigier, *Phys.Lett.,* t.86A, 1981, p.397 1980.

115. De Broglie waves on Dirac aether: a testable experimental assumption, Vigier, J.P., *Lettere al Nuovo Cimento*, vol.29, ser.2, no.14 p.467-75, 1980.

116. On the physical nonexistence of signals going backwards in time, and quantum mechanics, Garuccio, A.; Maccarrone, G.D.; Recami, E.; Vigier, J.P. *Lettere al Nuovo Cimento* vol.27, ser.2, no.2 p.60-4, 1980.

117. Unacceptability of the Pauli-Jordan propagator in physical applications of quantum mechanics, Selleri, F.; Vigier, J.P., *Lettere al Nuovo Cimento*, vol.29, ser.2, no.1, p.7-9, 1980.

118. Après le colloque de Cordoue, un accusé nommé Einstein, *Raison Présente*, Vigier J.P. (1980), N°56 ("La parapsychologie, oui ou non?"), 4ème trimestre, p. 77

119. and M. Andrade e Silva, *C. R. Acad. Sci. Paris*, 290, 501 (1980): A. Garuccio, V. Rapisarda and J. P. Vigier, Phys.

120. On a contradiction between the classical (idealised) quantum theory of measurement and the conservation of the square of the total angular momentum in Einstein-Podolsky-Rosen paradox, Cufaro-Petroni, N.; Garuccio, A.; Selleri, F.; Vigier, J.P., *Comptes Rendus Hebdomadaires des Seances de l'Academie des Sciences, Serie B (Sciences Physiques)*, vol.290, no.6, p.111-14, 1980.

121. Empirical Status in Cosmology and the Problem of the Nature of Redshifts, T.Jaakkola, M.Moles and J-P Vigier, *Astron.Nachr*, 1979.

122. *On the physical non-existence of signals going backwards in time and quantum mechanics*, Garuccio, A.; Maccarrone, G.D.; Recami, E.; Vigier, J.P., Istituto Nazionale Fisica Nucleare, Catania, Italy, 1979.

123. Markov process at the velocity of light: The Klein-Gordon statistic, Curfaro Petroni, N.; Vigier, J.P., *International Journal of Theoretical Physics* vol.18, no.11 p.807-18, 1979.

124. Causal superluminal interpretation of the Einstein-Podolsky-Rosen paradox, Cufaro Petroni, N.; Vigier, J.P *Lettere al Nuovo Cimento*, vol.26, ser.2, no.5, p.149-54, 1979.

125. Stochastic derivation of Proca's equation in terms of a fluid of Weyssenhoff tops endowed with random fluctuations at the velocity of light, Cufaro Petroni, N.; Vigier, J.P., *Physics Letters A*, vol.73A, no.4, p.289-9, 1979.

126. On two conflicting physical interpretations of the breaking of restricted relativistic Einsteinian causality by quantum mechanics, Petroni, N.C.; Vigier, J.P., *Lettere al Nuovo Cimento* vol.25, ser.2, no.5 p.151-6, 1979.

127. Stable states of a relativistic harmonic oscillator imbedded in a random stochastic thermostat, Gueret, P.; Merat, P.; Moles, M.; Vigier, J.P. *Letters in Mathematical Physics*, vol.3, no.1, p.47-56, 1979.

128. Model of quantum statistics in terms of a fluid with irregular stochastic fluctuations propagating at the velocity of light: a derivation of Nelson's equations, Vigier, J.P., *Lettere al Nuovo Cimento*, vol.24, ser.2, no.8, p.265-72, 1979.

129. Superluminal propagation of the quantum potential in the causal interpretation of quantum mechanics, Vigier, J.P. *Lettere al Nuovo Cimento*, vol.24, ser.2, no.8, p.258-64, 1979.

130. Internal rotations of spinning particles, Fenech, C.; Moles, M.; Vigier, J.P. *Lettere al Nuovo Cimento*, vol.24, ser.2, no.2, p.56-62, 1979.

131. Interpretation of the apparent north-south asymmetry and fluctuations of galactic rotation, Jaakkola, T.; Moles, M.; Vigier, J.P., *Astrophysics and Space Science*, vol.58, no.1, p.99-102, 1978.

132. Remarks on the impact of photon-scalar boson scattering on Planck's radiation law and Hubble effect (and reply), Moles, M.; Vigier, J.P. *Astronomische Nachrichten*, vol.298, no.6, p.289-91, 1977.

133. Liste de supernovae de type I dont la détermination du maximum de luminosité permet l'établissement d'établissement d'échantillons homogénes(in collaboration with S. Depaquit, G. Le Denmat). *C.R. Acad. Sci.*, t.285 B, p.161, 1977.

134. Continuous increase of Hubble modulus behind clusters of galaxies, Nottale, L.; Vigier, J.P., *Nature* vol.268, no.5621 p.608-10, 1977.

135. Red-shifting of light passing through clusters of galaxies: a new photon property?, Maric, Z.; Moles, M.; Vigier, J.P., *Lettere al Nuovo Cimento*, vol.18, ser.2, no.9, p.269-76, 1977.

136. A Peculier Disribution of Radial Velocities of Faint Radio-Galaxies with $13.0 < m_{corr} < 15.5$, H.Karoji, L.Nottale and J-P Vigier, *Astrop. Sp.Sc*, 44, 1976.

137. Possible measurable consequences of the existence of a new anomalous redshift cause on the shape of symmetrical spectral lines, Maric, Z.; Moles, M.; Vigier, J.P., *Astronomy and Astrophysics*, vol.53, no.2, pt.1, p.191-6, 1976.

138. A set of working hypotheses towards a unified view of the Universe, Pecker, J.-C.; Vigier, J.P., *Astrofizika*, vol.12, no.2 , p.315-30, 1976.

139. Charmed quark discovery in antineutrino-nucleon scattering? Vigier, J.P., *Lettere al Nuovo Cimento*, vol.15, ser.2, no.2, p.41-8, 1976.

140. Observation of excess redshifts when light travels through clusters of galaxies, Karoji, H.; Nottale, L.; Vigier, J.P., *Comptes Rendus Hebdomadaires des Seances de l'Academie des Sciences, Serie B (Sciences Physiques)*, vol.281, no.1 p.409-12, 1975.

141. Les supernovae de type I et l'anisotropie de la <<constant>> de Hubble, G.Le Denmat, and J-P Vigier, *C.R.Acad.Sc.Paris*, t. 280, 1975.

142. Cosmological implications of anomalous redshifts-A possible working hypothes, Jaakkola, T.; Moles, M.; Vigier, J.P.; Pecker, J.C. Yourgrau, W., *Foundations of Physics*, vol.5, no.2, p.257-69, 1975.

143. Sur les dangers d'une approximation classique dans l'analyse des décalages spectraux vers le rouge (in collaboration with J.C. Pecker). *C.R. Acad.Sci.*, t.281 B, p.369, 1975.

144. Déplacements anormaux vers le rouge liés à la traversée des amas de galaxies par la lumière (in collaboration with H. Karoji and L. Nottale). *C.R. Acad. Sci.*, t.281 B, p. 409, 1975.

145. Possible local variable of the Hubble constrant in Van Den Bergh's calibration of Sc- type galaxies (in collaboration with G. Le Denmat, M. Moles and J.L. Nieto). *Nature*, t.257, p.773, 1975.

146. Calcul et tables d'une intégrale utile dans 'intérprétation de phénomènes voisins du bord solaire (ou stellaire) (in collaboration with J. Borsenberger and J.C. Pecker). C.R. *Acad. Royale de Liège.*, 1975.

147. Are Bell's inequalities concerning hidden variables really conclusive?, Flato, M.; Piron, C.; Grea, J.; Sternheimer, D.; Vigier, J.P., Helvetica *Physica A*cta, vol.48, no.2, p.219-25, 1975.

148. Anisotropic redshift distribution for compact galaxies with absorption spectra, Jaakkola, T.; Karoji, H.; Moles, M.; Vigier, J.P. *Nature* vol.256, no.5512 p.24-5, 1975.

149. Type I supernovae and angular anisotropy of the Hubble 'constant', Le Denmat, G.; Vigier, J.P., *Comptes Rendus Hebdomadaires des Seances de l'Academie des Sciences, Serie B (Sciences Physiques)*, vol.280, no.14 p.459-61, 1975.

150. Three recent experiments to test experimentally realizable predictions of the hidden variables theory, Vigier, J.P., *Comptes Rendus Hebdomadaires des Seances de l 'Academie des Sciences, Serie B (Sciences Physiques)*, vol.279, no.1 p.1-4, 1974.

151. On The Geometrical Quantization of the electric charge in five dimensions and its numerical determinationas a consequence of asymptotic SO (5,2) group invariance, J-P Vigier, *Colloques Int. C.N.R.S.*, no 237, 1974.

152. Sur trois vérifications experimentales recentes de consequences mesurables possibles de la theorie des parametres caches, J-P Vigier, *C.R. Acad. Sc. Paris*, t. 279,1974.

153. Observed deflection of light by the Sun as a function of solar distance, Merat, P.; Pecker, J.C.; Vigier, J.P.; Yourgrau, W., *Astronomy and Astrophysics*, vol.32, no.4, pt.1, p.471-5, 1974.

154. Anomalous redshifts in binary stars, Kuhi, L.V.; Pecker, J.C.; Vigier, J.P. *Astronomy and Astrophysics* vol.32, no.1, pt.2, p.111-14, 1974.

155. Possible interpretation of solar neutrino and Mont Blanc muon experiments in terms of neutrino-boson collisions, Moles, M.; Vigier, J.P. *Lettere al Nuovo Cimento* vol.9, ser.2, no.16, p.673-6, 1974.

156. Comparaison de deux observations de déplacements anormaux vers le rouge observés au voisinage du disque solaire, S. Depaquit, J-P Vigier and Jean Claude Pecker, *C.R.Acad.Sc.Paris*, t. 279, 1974.

157. A symmetry scheme for hadrons, leptons and intermediate vector bosons, Gueret, P.; Vigier, J.P.; Tait, *Nuovo Cimento A*, vol.17A, no.4, p. 663-80, 1973.

158. Photon mass. quasar redshifts and other abnormal redshifts (in collaboration with J.C. Pecker, W.Tait). *Nature,* t.241, p.338., 1973.

159. Conséquences physiques possibles de l 'existence d'une masse non nulle du photon sur les interactions de la lumière avec la matière et la théorie du corps noir (in collaboration with M.Moles). *C.R. Acad. Sci.,* t.276 B, p.697., 1973.

160. Theoretical determination of alpha =e/sup 2//h (cross) c deduced from asymptotic group invariance properties of high-energy charged Dirac particles in a constant external vector potential, Vigier, J.P *Lettere al Nuovo Cimento* , vol.7, ser.2, no.12 p.501-6, 1973.

161. Calcul théorique de la valeur de alpha=$e^2/\hbar c$ á partir du groupe d'invariance asymptotique de particules de Dirac en mouvement dans un champ extérieur constant, J-P Vigier, *C.R.Acad.Sc.Paris*, t. 277, 1973.

162. Are vector potentials measurable quantities in electromagnetic theory? Vigier, J.P.; Marcilhacy, G. *Lettere al Nuovo Cimento*, vol.4, no.13, ser.2, p.616-18, 1972.

163. Sur une interprétation possible du déplacement vers le rouge des raies spectrales dans le spectre des objets astonomiques. (in collaboration with J.-C. Pecker, A.-P. Roberts). *C.R.Acad. Sci.,* t.274 B, p.765., 1972.

164. Sur une interprétation possible du déplacement vers le rouge des raies spectrales dans le spectre des objets astronomiques. Suggestions en vue d,expériences directes (in collaboration with J.C. Pecker and A.P. Roberts). *C.R. Acad. Sci.,* t.274 B, p.1159.1972.

165. Non-velocity redshifts and photon-photon interactions, Pecker, J.C.; Roberts, A.P.; Vigier, J.P., Nature, vol.237, no.5352, p.227-9, 1972.

166. Photon mass and new experimental results on longitudinal displacements of laser beams near total reflection, de Broglie, L.; Vigier, J.P. *Physical Review Letters*, vol.28, no.15, p.1001-4, 1972.

167. Classical spin variables and classical counterpart of the Dirac- Feynman-Gell-Mann equation, Depaquit, S.; Gueret, Ph.; Vigier, J.P. *International Journal of Theoretical Physics*, vol.4, no.1, p.19-32, 1972.

168. Photon mass. Imbert effect and Goos-Hanchen effect in polarized light, De Broglie, L.; Vigier, J.P., *Comptes Rendus Hebdomadaires des Seances de l'Academie des Sciences*, Serie B (Sciences Physiques), vol.273, no.25 p.1069-73, 1971.

169. Sur la réciprocité de la réponse de la caméra électronique aux très faibles flux de lumière cohérente (in collaboration with M. Duchesne). *C. R. Acad. Sci.,* t.273 B, p.911.,1971.

170. Masse du photon et expériences de Kunz sur l'effet photoélectrique du plomb supraconducteur, J-P Vigier, *C.R.Acad.Sc.Paris*, t. 273, 1971.

171. Remarks on a possible dynamical and geometrical unification of external and internal groups of motions of elementary particles and their application to SO/sub 6,1/ global dynamical symmetry, Gueret, P.; Vigier, J.P. *Nuovo Cimento A*, vol.67, no.1, p. 23-8, 1970.

172. Nouvelles expériences d,interférence en lumière faible (in collaboration with M.M.P. Bozec, M. Cagnet, M. Duchesne and J.-M. Leconte). *C.R. Acad. Sci.,* t.270, p.324., 1970.

173. Masse du photon et expériences de Kunz sur l,effet photoélectrique du plomb supraconducteur. *C.R. Acad. Sci.,* t. 273, p.993, 1970.

174. Masse du photon. Effet Imbert et effet Goos-Hänchen en lumière incidente polarisée (in collaboration with L. De Broglie). *C.R. Acad. Sci.,* t.273 B, p.1069, 1970.

175. Unification des quarks et des leptons dans la représentation de la base de SO (6,1) (in collaboration with Ph. Guéret). *C.R. Acad. Sci.,* t.270, p.653, 1970.

176. Remarks on a possible dynamical and geometrical unification of external groups of motions of elementary particles and their application to SO (6, 1) global dynamical symmetry (in collaboration with Ph. Guéret). *Nuovo Cimento,* t.67 A, p.23, 1970.

177. Phenological spectroscopy of baryons and bosons considered as discrete quantized states of an internal structure of elementary particles, Depaguit, S.; Vigier, J.P., *Comptes Rendus Hebdomadaires des Seances de l'Academie des Sciences*, Serie B (Sciences Physiques), vol.268, no.9 p.657-9, 1969.

178. Formule de masse associée aux multiplets de SU (3) et équations d'ondes des baryons, obtenues à partir du groupe dynamique global d,unification G = SEAU (6,1) (in collaboration with Ph. Guéret). CR *Acad. Ceci.,* t.268, p.1153, 1969.

179. Unification on External and Internal motions within SO (6,1) an possible mass splitting on SU (3) baryon multiplets without symmetry breaking. *Lettre al Nuov. Cim., Série* 1, t.1, p.445, 1969.

180. Spectroscopie phénoménologique des baryons et des bosons considérés comme états quantifiés discrets d,une structure interne des particules élémentaires (in collaboration with S. Depaquit). *CR Acad. Ceci., t.268, p.657, 1969.*

181. Formule de masse associée aux multiplets de SU (3) et équations d'ondes des baryons, obtenues á partir du groupe dnamique global d'unification G=SO (6,1) XU (1), P.Gueret and J-P Vigier, *C.R.Acad.Sc.Paris.*,t. 268, 1969.

182. Interprétation géométrique et physique de la formule du guidage en relativité générale. *C.R. Acad. Ceci.,* t.266, p.598, 1968.

183. On a possible extension to the motion of Pauli spin to the groups SO (p, 1) and In (SO (p, 1), Guenet, P.; Vigier, J.P. *Comptes Rendus Hebdomadaires des Seances de l'Academie des Sciences,* Serie B (Sciences Physiques), vol.267, no.19p.997-9, 1968.

184. Sur une extension possible de la notion de spin de Pauli aux groupes SO (p, 1) et In (SO (p, 1)) (in collaboration with Ph. Guéret). CR *Acad. Ceci.,* t.267, p.997, 1968.

185. Diagonalisation de l'opérateur de masse au carré dans les modéles étendus de particules élémentaires, P.Gueret and J-P Vigier, *C.R. Acad. Sc. Paris,* t. 265, 1967.

186. Fonctions d'ondes définies sur une variété de Riemann V5 admettant localement le groupe conforme SU (2,2) comme groupe d'isométrie, P.Gueret and J-P Vigier, *C.R.Acad.Sc.Paris,* t.264, 1967.

187. Hidden Parameters Associated with Possible Internal Motions of Elementary Particles, J-P Vigier, *Studies in Founds. Methodology, and Philosophy of Sciences,* vol 2, ed M.Bunge, Springer-Verlag, 1967.

188. On the masses on non-strange pseudo-scalar mesons and the generalized Klein- Gordon equation (in collaboration with M.M.M. Flato, D. and J. Sternheimer, G. Wathaghin), 1966.

189. Unification possible des mouvements internes et externes des particules élémentaires déscrits comme groupes de mouvements isométriques de déplacement sur des variétiés de Riemann, J-P Vigier, *C.R.Acad. Sc.Paris,* t. 262, 1966.

190. Unification possible des mouvements internes et externes des particules élémentaires décrits comme groupes de mouvements isométriques de déplacements sur des variétes de Riemann. *C.R. Acad. Ceci.,* t.262, p.1239-1241, 1966.

191. Possible external an internal motions on elementary particles on Riemannian manifolds. J-P Vigier, *Physical Review Letters,* t.17. n° 1, 1966.

192. Possibilité d'unification des comportements internes et externes des particules élémentaires au moyen de groupes de mouvements isométriques sur des variétés de Riemann. (Communication au *Colloque International du C.N.R.S. sur l'extension du groupe de Poincaré aux symétries internes des particules élémentaires*), 1966.

193. Une représentation paramétrique de l'algébre de Lie du groupe SU (2,2) considéré comme groupe d'isométrie sur une variete de Riemann á cinq dimensions, P.Guret and J-P Vigier, *C.R.Acad.Sc.Paris*, t. 263, 1966.

194. Sur une généralisation conforme possible de l'équation de Dirac, H-F Gautrin, R.Prasad and J-P Vigier, *C.R.Acad.Sc.Paris*, t. 262, 1966.

195. L'extension du groupe de Poincaré aux symétries internes des particules élémentaires, J-P Vigier, *Colloques Internationaux de Centre National de la Recherche Scientifique*, No 159, 1966.

196. Définition on PCT operators for the relativistic-rotator model an the Bronzon-Low symmetry (in collaboration with M. Flato and G. Rideau). *Nuclear Phys.,* t. 61, p.250-256, 1965.

197. High energy electron-positron scattering (in collaboration with V. Wataghin). *Nuovo Cimento,* t.36, p.672, 1965.

198. On The 'Space-time character' of Internal Symmetries of Elementary Particles, D.Bohm,M.Flato,F.Halbwachs,P.Hillion and J-P Vigier, *Il Nuovo Cimento*, Serie X, vol 36, 1965.

199. Conformal group symmetry on elementary particles (in collaboration with D. Bohm, M. Flato, D. Sternheimer). *Nuovo Cimento,* t.38, 1965, p.1941, 1965.

200. Le groupe conforme comme possibilité de symétrie unifée en physique des interactions fortes (in collaboration with M. Flato and D. Sternheimer). CR *Acad. Ceci.,* t.260, p.3869-3872, 1965.

201. High Energy Electron-Positron Scattering, J-P Vigier and V.Wataghin, *Il Nuovo Cimento*, Serie X, Vol 36, 1965.

202. Un test possible de la symmetrié d'hypercharge, P.Hillion and J-P Vigier, *C.R.Acad.Sc.Paris*, t. 258, 1964.

203. Signification physique des potentiels vecteurs et abandon de la notion d'invariance de jauge en théorie des mésons vectoriels intérmediares, J-P Vigier, *C.R.Acad. Sc.Paris*, t.259, 1964.

204. An approach to the unified theory of elementary particles (en coil. avec MM. Yukawa and Katayama). *Progr. of Theor. Phys.,* ix 29, p. 468., 1963.

205. Theory of weak interactions based on a rotator model (en coil. avec MM. Yukawa and Katayama). *Progr. of Theor. Phys.,* t. 29. , p. 470, 1963.

206. Rotator model of elementary particles considered as relativistic extended structures in Minkowski space (in collaboration with L. De Broglie, D. Bohm. P. Hillion. Halbwachs and Takabayasi). *Physical Review,* t. 129, p. 438, 1963.

207. Space time model of relativistic extended particles in Minkowski space II (in collaboration with De Broglie. Bohm, Hillion, Halbwachs and Takabayasi). *Physical Review.* t. 129. p. 451, 1963.

208. Application de la théorie de la fusion au nouveau modèle étendu de particules élémentaires (in collaboration with L. De Broglie). *C. R. Acad. Sci.,* t. 256, p. 3390, 1963.

209. Table des particules élémentaires associées au nouveau modèle étendu des particules élémentaires (in collaboration with L. De Broglie). *C. R. Acad. Sci*, t. 256, p. 3351, 1963.

210. Spectroscopy of Baryons and Resonances Considered as Quantized Levels of a Relativistic Rotator, G.Barbieri, S.Depaquit and J-P Vigier, *Il Nuovo Cimento*, Serie X, vol 30, 1963.

211. Space time Model of Relativistic Extended Particles in Minkowski Space.II Free Particle and Interaction Theory, L.De Broglie, F.Halbwachs, P.Hillion, T.Takabayasi and J-P Vigier, *Physical Review*, Vol 129, Jan 1963.

212. *Introduction to the Vigier theory of elementary particles* (translated by Arthur J. Knodel), Louis de Broglie, Elsevier Pub. Co., 1963, Amsterdam; New York.

213. Théorie des résonances isobariques considérées comme états excités interns du modélé du rotateur relativiste dés particules élémentaires, P.Hillion and J-P Vigier, *C.R.Acad.Sc.Paris*, t. 255, 1962.

214. Mass of the Yang-Mills Vector Field, J-P Vigier, *Il Nuovo Cimento*, Serie X, Vol 23, 1962.

215. Les Ondes associées à une structure inteme des particules. Théorie relativiste. *Anna/es Institut Henri Poincaré*, fasc. 111, p. 149, 1962.

216. Une Seconde Extension du Formalisme de Utiyama, P.Hillion and J-P Vigier, excerpt from *Cahier de Physique* no 137, Jan 1962.

217. Sur Les Equations D'Ondes Associees a la structure des fermions, P.Hillion and J-P Vigier, *Annales de l'Institut Henri Poincare*, pp 229-254, 1962.

218. Sur le champ de Yang et Mills. *C. R. Acad. Sci.*, t. 252, p. 1113., 1961.

219. Sur un groupe de transformations isomorphe en tant que groupe du groupe de Lorentz, excerpt from *Cahier de Physique*, no 127, 1961.

220. On the physical meaning of negative probabilities (in collaboration with Y. P. Terletski). *J. F. T. P.*, t. 13, p. 356.,1961.

221. Sur l'interprétation causale de l'équation non linéaire de Heisenberg (in collaboration with M. Hillion). *Cahiers de Phiys.*, t. 130-131, p. 315, 1961.

222. New Isotopic Spin Space and Classification of Elementary Particles, P.Hillion and J-P Vigier, *Il Nuovo Cimento*, Serie X, Vol 18, 1960.

223. Forme possible des fonctions d'ondes relativistes associées au mouvement et a la struture des particules elementaires, au niveau nucleaire, P.Hillion and J-P Vigier, *C.R.Acad.Sc.Paris*,t. 250, 1960.

224. Application des groupes d'invariance relativistes aux modeles de particules etendues en Relativite restreinte, P.Hillion and J-P Vigier, P.Hillion and J-P Vigier, *C.R.Acad.Sc.Paris*, t. 250, 1960.

225. Formalisme hamiltonien associé au rotateur de Nakano, F.Halbwachs, P.Hillion and J-P Vigier, *C.R.Acad.Sc.Paris*, t. 250, 1960.

226. Relativistic hydrodynamics of rotating fluid masses moving with the velocity of light, P.Hillion, T.Takabayasi and J-P Vigier, *Acta Physica Polonica*, Vol XIX, 1960.

227. Relativistic Rotators and Bilocal theory: in collaboration with D. Bohm, Hillion and Takabayasi). *Progr. of Theor. Phys.*. t. 23, p. 496.1960.

228. Les fonctions d'onde non relativistes associées au corpuscle étendu, P.Hillion and J-P Vigier, *C.R.Acad.Sc.Paris*, t.250 1960.

229. Internal quantum states of hyperspherical (Nakuano) relativistic rotators (in collaboration with Bohm and Hillion). *Progr. of Theor. Phys.*, t. 16, p. 361.,1960.

230. Elementary particle waves and irreductible representations of the Lorentz group (in collaboration with M. Hillion). *Nuclear Physics*, t. 16, p. 361., 1960.

231. Internal Motions of Relativistic Fluid Masses, F.Halbwachs, P.Hillion and J-P Vigier, *Il Nuovo Cimento*, Serie X, vol 15, 1960.

232. *Représentation hydrodynamique des fonctions d'ondes.* Travaux résumés dans la thèse de Halbwachs. Théorie relativiste des fluides à spin. Paris, Gauthier-Villars, 1960.

233. Interprétation de la valeur moyenne des opérateurs internes dans la théorie des masses fluides relativistes, P.Hillion, B.Stepanov and J-P Vigier, excerpt from *Cahier de Physique*, pp 283-289, 1959.

234. Quadratic Lagrangians in relativistic hydrodynamics (in collaboration with Halbwachs and Hillion). *Nuovo Cimento*. t. 11, p. 882., 1959.

235. Formalisme lagrangien pour une particule relativiste isolée étendue (in collaboration with M. Halbwachs). *C. R.*, t. 248. p. 490., 1959.

236. Relativistic hydrodynamics of rotating fluid masses moving with the velocity of light (in collaboration with P. Hillion and Takabayasi). *Acta Physica Polonica.*, 1959.

237. Propriétés classiques et représentation bilocale du rotateur de Nakano (in collaboration with Louis De Broglie and P. Hillion). *C. R. Acad. Sci.* t. 249, p. 2255., 1959.

238. Fonctions propres des opérateurs quantiques de rotation associés aux angles d'Euler dans l'espace-temps (in collaboration with M. Hillion). *Annales Institut Henri Poincaré,* t. XVI. fasc. III, p. 161., 1959.

239. Quelques précisions sur la nature et les propriétés des retombées radioactives résultant des explosions atomiques depuis 1945 (in collaboration with Pauling, Sakato. Tomonaga and Yukawa). *C.R.*, t. 245, p. 982., 1959.

240. Étude mathématique des fonctions propres des moments cinétiques internes des masses fluides relativistes, P.Hillion and J-P Vigier, excerpt from *Cahier de Physique*, pp 257-282, 1959.

241. Lagrangian formalism in relativistic hydrodynamics of rotating fluid masses (in collaboration with Halbwachs and Hillion). *Nuovo Cimento,* t. 90, 58, p. 818., 1958.

242. Description of Pauli Matter as a continuous Assembly of small rotating bodies (in collaboration with M. Takabayasi). *Progress of Theor. Phys.,* t. 18, n^0 6, p. *573., 1958.*

243. Introduction des paramètres relativistes d'Einstein, Kramers et de Cayley-Klein dans La théorie relativiste des fluides dotés de moment cinétique interne *(spin). C. R.,* t. 245, p. 1787 En coll. avec M. Unal., 1957.

244. Introduction des paramètres relativistes d'Einstein-Klein dans l'hydrodynamique relativiste du fluide à spin de Weyssenhoff. *C.R.* 245, p. 1891., 1957.

245. Tendance vers un état d' Équilibre stable de phénomènes soumis à evolution markovienne (in collaboration with M. Fuchs). *Note aux C. R.,* t. 9. 27 février 1956, p. 1120., 1956.

246. Interprétation de l'équation de Dirac comme approximation linéaire de l'équation d'une onde se propageant dans un fluide tourbillonnaire en agitation chaotique du type éther de Dirac (in collaboration with Bohm and Lochack). Séminaire de M.-L. De Broglie, *Institut Henri Poincaré.,* 1956.

247. Structure des micro-objets dans l'interprétation causale de la théorie des quanta. *Thèse no 3600,* série A, n^0 2727; Gauthier-Villars, Paris.

248. Décomposition en fonction de variables dynamiques du tenseur d'énergie impulsion des fluides relativistes dotés de moment cinétique interne (en coil. avec MM. Halbwachs and Lochack). *Note aux C. R.* t. 241. 19 septembre . p. 692., 1955.

249. Model of the causal interpretation of quantum theory in terms of a fluid with irregular fluctuations (in collaboration with D. Bohm). *Physical Review,* t. 2, 96, p. 208-216., 1955.

250. La Physique Quantique restera-t-elle indéterministe (in collaboration with L. De Broglie). Paris, Gauthier-Villars. 1953.

251. Sur la relation entre l'onde à singularité et l'onde statistique en théorie unitaire relativiste. *Note aux C. R.,* t. *235,* 20 octobre, p. 869., 1952.

252. Forces s'exercant sur les lignes de courant des particules de spin 0. ½ et 1 en théorie de l'onde pilote. *Note aux C. R.,* t. 235, 10 novembre. p. 1107., 1952.

253. Remarque sur la théorie de l'onde pilote. *Note aux C. R. Acad. Sci.,* t. 233, 17 Septembre, p. 641., 1951.

254. Introduction géométrique de l'onde pilote en théorie unitaire affine. *Note aux C. R.,* t. 233, 29 Octobre, p. 1010., 1951.

255. Rapport sur la théorie générale de la diffusion des neutrons thermiques (*Rapport interne du C. E. A.,* 1949).

256. *Étude sur les suites infinies d'opérateurs hermitiens (*Thèse *n^0 1089,* presented in Geneva to obtain the Docteur ès-Sciences Mathématiques, Geneva, 1946).

43133620R00168

Made in the USA
Lexington, KY
18 July 2015